Titles in This Series

Volume

1 **Markov random fields and their applications,** Ross Kindermann and J. Laurie Snell

2 **Proceedings of the conference on integration, topology, and geometry in linear spaces,** William H. Graves, Editor

3 **The closed graph and P-closed graph properties in general topology,** T. R. Hamlett and L. L. Herrington

4 **Problems of elastic stability and vibrations,** Vadim Komkov, Editor

5 **Rational constructions of modules for simple Lie algebras,** George B. Seligman

6 **Umbral calculus and Hopf algebras,** Robert Morris, Editor

7 **Complex contour integral representation of cardinal spline functions,** Walter Schempp

8 **Ordered fields and real algebraic geometry,** D. W. Dubois and T. Recio, Editors

9 **Papers in algebra, analysis and statistics,** R. Lidl, Editor

10 **Operator algebras and K-theory,** Ronald G. Douglas and Claude Schochet, Editors

11 **Plane ellipticity and related problems,** Robert P. Gilbert, Editor

12 **Symposium on algebraic topology in honor of José Adem,** Samuel Gitler, Editor

13 **Algebraists' homage: Papers in ring theory and related topics,** S. A. Amitsur, D. J. Saltman, and G. B. Seligman, Editors

14 **Lectures on Nielsen fixed point theory,** Boju Jiang

15 **Advanced analytic number theory. Part I: Ramification theoretic methods,** Carlos J. Moreno

16 **Complex representations of GL(2, K) for finite fields K,** Ilya Piatetski-Shapiro

17 **Nonlinear partial differential equations,** Joel A. Smoller, Editor

18 **Fixed points and nonexpansive mappings,** Robert C. Sine, Editor

19 **Proceedings of the Northwestern homotopy theory conference,** Haynes R. Miller and Stewart B. Priddy, Editors

20 **Low dimensional topology,** Samuel J. Lomonaco, Jr., Editor

21 **Topological methods in nonlinear functional analysis,** S. P. Singh, S. Thomeier, and B. Watson, Editors

22 **Factorizations of $b^n \pm 1$, $b = 2, 3, 5, 6, 7, 10, 11, 12$ up to high powers,** John Brillhart, D. H. Lehmer, J. L. Selfridge, Bryant Tuckerman, and S. S. Wagstaff, Jr.

23 **Chapter 9 of Ramanujan's second notebook—Infinite series identities, transformations, and evaluations,** Bruce C. Berndt and Padmini T. Joshi

24 **Central extensions, Galois groups, and ideal class groups of number fields,** A. Fröhlich

25 **Value distribution theory and its applications,** Chung-Chun Yang, Editor

26 **Conference in modern analysis and probability,** Richard Beals, Anatole Beck, Alexandra Bellow, and Arshag Hajian, Editors

27 **Microlocal analysis,** M. Salah Baouendi, Richard Beals, and Linda Preiss Rothschild, Editors

28 **Fluids and plasmas: geometry and dynamics,** Jerrold E. Marsden, Editor

29 **Automated theorem proving,** W. W. Bledsoe and Donald Loveland, Editors

30 **Mathematical applications of category theory,** J. W. Gray, Editor

31 **Axiomatic set theory,** James E. Baumgartner, Donald A. Martin, and Saharon Shelah, Editors

32 **Proceedings of the conference on Banach algebras and several complex variables,** F. Greenleaf and D. Gulick, Editors

Titles in This Series

Volume

Titles in This Series

Volume

Geometry of Group representations

CONTEMPORARY MATHEMATICS

Volume 74

Geometry of Group Representations

Proceedings of the AMS-IMS-SIAM Joint Summer Research Conference held July 5–11, 1987 with support from the National Science Foundation

William M. Goldman and
Andy R. Magid, Editors

AMERICAN MATHEMATICAL SOCIETY
Providence · Rhode Island

The AMS-IMS-SIAM Joint Summer Research Conference in the Mathematical Sciences on Geometry of Group Representations was held at the University of Colorado, Boulder, Colorado, on July 5–11, 1987, with support from the National Science Foundation, Grant DMS-8613199.

1980 *Mathematics Subject Classification* (1985 *Revision*). Primary 20C15, 22E40,30F40, 57M05, 57R15.

Library of Congress Cataloging-in-Publication Data

Geometry of group representations: proceedings of a summer research conference held July 5–11, 1987 / William M. Goldman, Andy R. Magid, editors.
p. cm. -- (Contemporary mathematics, ISSN 0271-4132; v. 74)
ISBN 0-8218-5082-2
1. Representations of groups -- Congresses. 2. Geometry, Differential -- Congresses. I. Goldman, William Mark. II. Magid, Andy R. III. Series: Contemporary mathematics (American Mathematical Society); v. 74.
QA171.G435 1988
512′.22 -- dc19 88-14490

TABLE OF CONTENTS

PREFACE

The representations of a finitely generated group in a topological group form a topological space which is an analytic or algebraic variety if the topological group is a Lie group or an algebraic group. This study draws from and contributes to a wide range of mathematical subjects: Algebra, Analysis, Topology, Differential Geometry, Representation Theory, even Mathematical Physics. In some cases the space of representations is the object of the study, in others it is a tool in a program, in many cases it's both.

The AMS-IMS-SIAM Summer Research Conference on the Geometry of Group Representations, held at the University of Colorado in Boulder, Colorado July 5 to 11, 1987, was designed to bring together researchers from the diverse areas of mathematics whose work involved spaces of group representations in order to share methods and results. The fifty registered participants, from thirty-eight colleges, universities, and research institutes in the United States, Europe, and the Near and Far East heard twenty lectures and actively engaged in informal discussion. Perhaps because of the diversity of the primary mathematical interests of the audience, the speakers made a special effort to keep their lectures accessible to non-specialists, an effort which the organizers and audience greatly appreciated.

Most of the papers in this volume are based on talks delivered at the conference. In keeping with the spirit of the conference, they survey their chosen topic with the non-specialist in mind. In addition, they contain technical developments to bring the subject to the current research frontier. Some of the papers here also contain the first publication of new results. The organizing committee is grateful for those authors' decision to choose this outlet for those results.

While no work of this kind can truly delimit the current state of knowledge in the Geometry of Group Representations in a form accessible to novice and expert alike, we hope that this volume, like the conference it records, does provide an entry point into some of these ideas.

The conference organizing committee was Andy Magid, William Goldman, and Hyman Bass. The administrative burden for the conference was born by Carole Kohanski of the AMS and her assistants. Carole's efficient management was an important factor in the success of the conference and the organizing committee is pleased to thank her here for it.

Andy Magid
William Goldman

LIST OF PARTICIPANTS

Silvana Abeasis
William Abikoff
Hyman Bass
Max Bauer
Gregory Brumfiel
Beifang Chen
Kevin Corlette
Randall Dahlberg
Richard Dipper
Daniel Farkas
Benjamin Fine
Robert Fisher
Charles Frohman
Murray Gerstenhaber
William Goldman
Andrew Haas
William Harvey
Luis Hernandez
Craig Hodgson
Renfang Jiang
Yoshinobu Kamishima
Eric Klassen
Hanspeter Kraft
Walter Lok
Alex Lubotzky
Andy Magid
John McCarthy
Geoffrey Mess
John Millson
Kent Morrison
David Mullins
Joyce O'Halloran
Robert Penner
Claudio Procesi
Robert Riley
Gerhard Rosenberger
Ze'ev Rudnick
John Ryan
Peter Shalen
Patricia Sipe
Lance Small
Martha Smith
Dennis Snow
Joanne Snow
Christopher Stark
Perry Susskind
Ser Tan
Marvin Tretkoff
Karen Vogtmann
Andrew Winkler

LIST OF TALKS

Embedding rings into matrix rings, L. Small
The Teichmuller theory of punctured surfaces, R. Penner
Geometric structures and varieties of representations, W. Goldman
Connected components of representation varieties, K. Morrison
A characterization of orbit closure, J. O'Halloran
Invariant theory and matrices, C. Procesi
The role of representation varieties in a 3-manifold theory, P. Shalen
Real algebra and Teichmuller space, G. Brumfiel
Affine homogeneous spaces, D. Snow
Spaces of discrete subgroups of Lie groups, J. Millson
Local structure of representation varieties, A. Magid
Complex representations and one-relator groups, B. Fine
Automorphism groups of free groups, K. Vogtmann
Non-arithmetic lattices over local fields, A. Lubotzky
Kleinian groups, W. Abikoff
Closed curves on a genus 2 surface, P. Susskind
Geometry of representations of quivers, H.-P. Kraft
Modular groups and representation spaces, W. Harvey
Rigidity of flat bundles over Kahler manifolds, K. Corlette
Casson's invariant for homology 3 spheres (a tutorial), J. McCarthy

Contemporary Mathematics
Volume **74**, 1988

Kleinian groups—
geometrically finite and geometrically perverse

by

William Abikoff

This article is an expanded version of a talk which introduced and surveyed some of the basic geometry of Kleinian groups and then proceeded to some speculations on the nature of geometrically infinite, finitely generated Kleinian groups.

The emphasis of the talk was on finitely generated groups, that is, the class of groups which can be finitely specified. Here one can hope that the algebraic parametrization techniques which, as we shall see, are quite appropriate to the study of families of geometrically finite groups, may have geometric analogues.

The examples I chose to examine are directed toward the formally Kleinian model for geometrically infinite, finitely generated, groups. This class of groups lies in a gray area between the geometrically finite groups, that is those with a finite volume convex core, and the infinitely generated ones. For the latter class, we demonstrate that there are possibilities for the geometry which are inconceivable for finitely generated groups.

Of the work discussed here, both successful and speculative, much is due to people other than this lecturer. Hopefully, credit is given where due. I would particularly like to thank Andy Haas for suggesting many improvements to any early version of this manuscript.

I should like to take the opportunity to dedicate this paper to Lars V. Ahlfors whose eightieth birthday we share in celebrating this year. It was the clarity of his vision of the geometry and function theory of finitely generated Kleinian groups that has lead to much of our current interest and understanding.

1 Some examples and a few definitions

The study of Kleinian groups has spawned many mathematical disciplines. Much of the interesting geometry arose in the classical period which we can take as starting in the early nineteenth century and running up to about 1930.

1.1 The triangle groups

The first examples of Kleinian groups were the triangle groups which are usually attributed to Schwarz. One can easily make a case for attributing earlier studies of at least some of these groups to either Riemann or Gauss. The latter two were intensely interested in the hypergeometric equation. This equation appears in the study of holomorphic functions invariant under, that is automorphic with respect to, triangle groups. These functions arise from the Schwarz-Christoffel construction of conformal mappings of disks to curvilinear triangles. I will not quibble over the origins of these groups, however I note with pleasure that they are of great contemporary interest as is shown by their appearance in several of the lectures given at this meeting.

To define a triangle group one starts with three integers l, m and n, with $1 < l \leq m \leq n$, and tries to find a 3-sided planar figure whose sides are circular arcs and whose vertex angles are π/l, π/m and π/n.[1] We denote that figure by $T(l,m,n)$ if it exists. $T(2,2,n)$, $T(2,3,3)$, $T(2,3,4)$, $T(2,3,5)$ all exist as triangles in spherical geometry (see e.g. Figure 1 where the construction is direct; in other cases a continuity argument is appropriate). $T(2,3,6)$ and $T(2,4,4)$ can be constructed using Euclidean segments as sides. The remaining triangles $T(l,m,n)$ can be realized as triangles in the hyperbolic plane.

Suppose that $T(l,m,n)$ exists as a triangle in some geometry $\mathcal{G}$. We can consider the group of motions of $\mathcal{G}$ generated by reflecting $T(l,m,n)$ in each of its sides. We call the resulting group $G(l,m,n)$. In the study of discontinuous groups, we are interested in the situation where the translates of $T(l,m,n)$ tile some set without overlap—it is for this reason that we have restricted our attention to triangles whose vertex angles are submultiples of

[1] The restriction to integral submultiples of π is not a necessary condition for finding the figures but is necessary for the remainder of our construction.

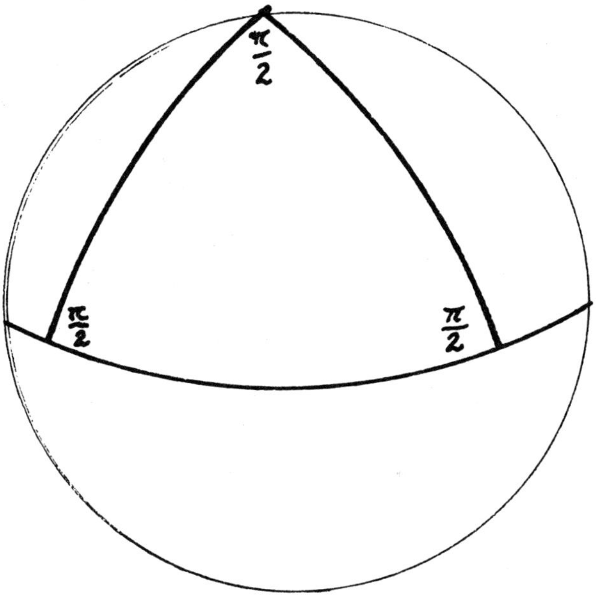

Figure 1: $T(2, 2, n)$ in spherical geometry

π. In the three spherical cases, we observe as did our ancestors that the groups of motions are extensions of the motions of the the Platonic solids.[2]

As one can discover in the seminal book by Ford [For51], the finite triangle groups are the spherical ones. The Euclidean triangle groups consist of those that are finite extensions of a free abelian group A on two generators. A consists of translations of the Euclidean plane which are not all linearly dependent (see the example in Figure 2).

Schwarz first came upon the hyperbolic triangle groups when he noted that the Schwarz-Christoffel maps can be analytically continued to maps from a disk or half-plane into the Riemann sphere. The Schwarz-Christoffel maps onto curvilinear triangles can be written as quotients of linearly independent solutions to the hypergeometric equation. The monodromy groups of the differential equations are often Fuchsian groups.[3]

---

[2]Here we are permitting motions which may be orientation reversing. These are not usually considered as lying in the groups of the Platonic solids.

[3]The description in Jeremy Gray's article [Gra84] is particularly enlightening. It shows the connection between Fuchs' studies of differential equations and Poincaré's work on Fuchsian and Kleinian groups (see below).

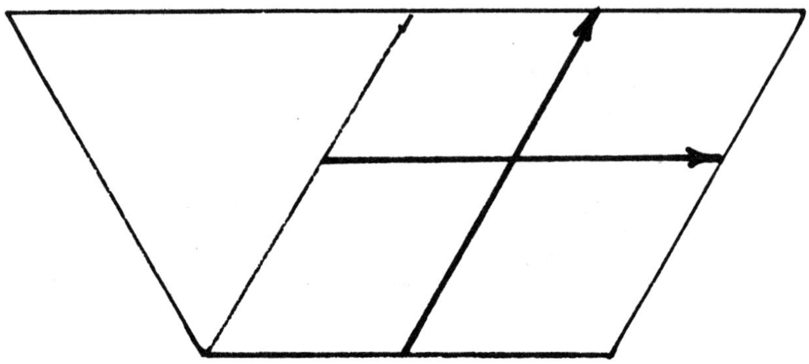

Figure 2: the translations in the Euclidean triangle group T(3,3,3).

The triangle groups are not examples of Kleinian groups. Kleinian groups are groups of möbius, hence orientation–preserving, transformations while reflections are orientation–reversing motions. However every group of motions, which is generated by reflections, has a subgroup of index two. This subgroup consists of directly conformal motions, that is, of möbius transformations. The möbius subgroups of the triangle groups are our first examples of Kleinian groups. In fact one of these möbius subgroups was probably the first Kleinian group to be studied. The group is the elliptic modular group and it was certainly known in the first half of the eighteenth century. It is the group of möbius transformations $z \mapsto (az+b)/(cz+d)$ with $a, b, c, d \in \mathbf{Z}$ and $ad - bc = 1$.[4]

It is not difficult to discover the geometry of these index two subgroups. First consider two circular arcs A_1 and A_2 which meet at angle π/n. Reflection in the same arc twice yields the identity map. Let us denote by R_i the

[4]The elliptic modular group parametrizes the different choices of bases for a rank two abelian subgroup L of $\mathbf{C}$. It is assumed here that L cannot be conjugated (in $\mathbf{C}$) into a subgroup of $\mathbf{R}$. The groups L are those which support doubly periodic meromorphic functions. $\mathbf{C}/\mathbf{L}$ is a compact Riemann surface of genus one. In other contexts, these surfaces are called *elliptic curves*.

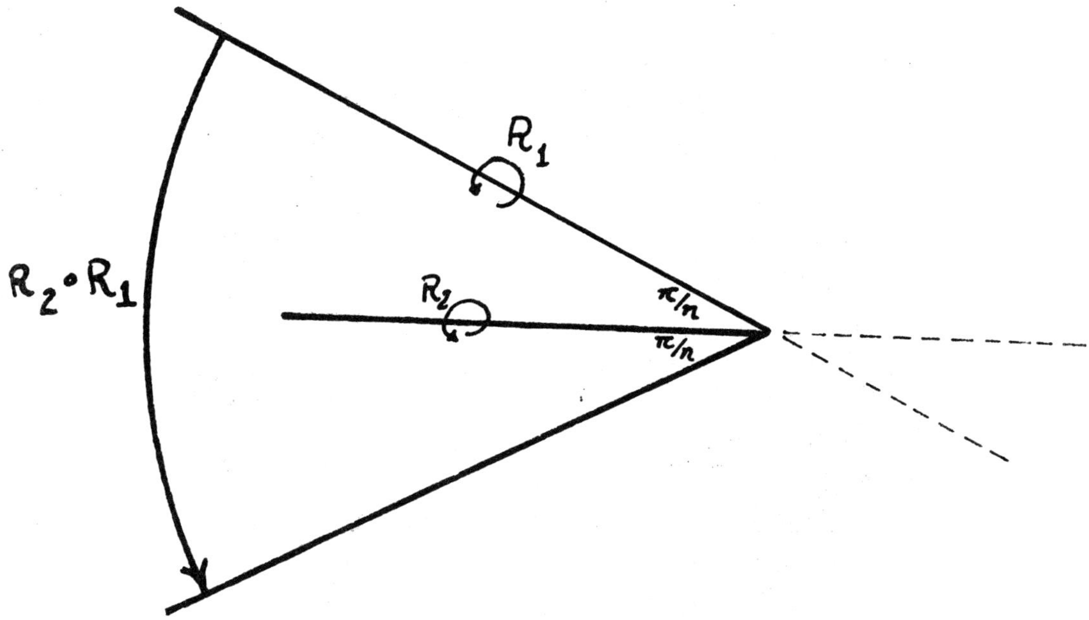

Figure 3: Two reflections make a rotation

reflection in the arc A_i. The only non-trivial products of the reflections R_i may be written as alternating the symbols R_1 and R_2. The product $R_2 \circ R_1$ is a rotation by angle $2\pi/n$. (See Figure 3.) The product is therefore a directly conformal map. We also note that the reflection in two tangent arcs yields a parabolic möbius transformation whose fixed point is the point of tangency. This last remark gives us

Truism 1 *The geometry of a parabolic transformation γ is the limit of elliptic transformations of period n as $n \to \infty$. γ has infinite order and one fixed point.*

The triangle group $G(l, m, n)$ may be thought of as generated by the reflections $R_1, \ldots, R_3$ in the sides of $T(l, m, n)$. The following relations between the reflections hold. The square of each reflection is the identity. If A_i and A_j meet at angle π/α_{ij} then $(R_i \circ R_j)^{\alpha_{ij}} = \text{id}$. These relations define the presentation of the triangle group.

If we think of the triangle group $G(l, m, n)$ as tiling a geometric space by the translates of $T(l, m, n)$ under the group then a subgroup of index two must have a tile of twice the size. The tile can be taken as $T(l, m, n) \cup R_1(T(l, m, n))$. The index two subgroup $H(l, m, n)$ is generated by $\gamma_1 = R_1 R_2$

Reprinted with the courtesy of B. G. Teubner, Stuttgart, Germany.

Figure 4: A tesselation from Fricke-Klein (1897)

and $\gamma_2 = R_2R_3$. The only relations then are $\gamma_1^l = \gamma_2^m = (\gamma_1\gamma_2)^n =$ identity. I find it amazing that the triangle group tesselations of the hyperbolic plane, appearing in some of the nineteenth century books, are accurate and complete. It must have required monumental effort to produce these illustrations (see Figure 4) by hand.

So far the groups I have discussed have nothing to do with the subject of this talk. The hyperbolic triangle groups are not examples of Kleinian groups. Their index two subgroups are Kleinian but they only admit deformations which are trivial. Here triviality means obtained by conjugation in the group $\mathcal{M}$öb of möbius transformations. It does provide a first example of the rigidity phenomenon—the group theory defines the geometric action up to conjugation. In this case, deformation varieties, that is spaces of deformations, reduce to finite sets of points and, more remarkably, there is only one point in each variety.

Our next example is one of an infinitely generated group. Again it is obtained as an index two subgroup of a reflection group, but this time we get lots of nontrivial deformations. First we give some of the basic definitions used in the theory of Kleinian groups.

1.2 Some definitions

Kleinian groups are groups of möbius transformations with complex entries. The groups must act properly discontinuously somewhere, that is on some open set, on the Riemann sphere $\hat{\mathbf{C}}$ (see below). This definition was Poincaré's. My preference is to be somewhat more encompassing in the definition and also to avoid trivialities. To wit, I will adopt the definitions given below—they are not the classical ones, although they are quite common in the current literature. Every möbius transformation acts on the unit sphere in $\mathbf{R}^3$—when the sphere is viewed as the Riemann sphere. We note that Poincaré had already demonstrated that each möbius transformation acting on the unit sphere can have its action extended into the unit ball $\mathbf{B} = \mathbf{B}^3$.[5]

Probably the most elementary examples of these concepts are in the section on Schottky groups which we will get to shortly.

Let G be a group of möbius transformations with complex entries. We assume that G is not a finite extension of an abelian group and that G is discrete in the uniform topology on homeomorphisms of $\hat{\mathbf{C}}$. Then G is called a *Kleinian group*.

A group G acting on any topological space X acts *properly discontinuously at a point* $x \in X$ if there is some neighborhood N of x so that the equation $\gamma(N) \cap N \neq \emptyset$ has at most finitely many solutions $\gamma \in G$. The (necessarily open) set on which G acts properly discontinuously is called the *set of proper discontinuity* of G.

If G is a Kleinian group then the set of points, if any, in $\hat{\mathbf{C}}$ on which G acts properly discontinuously is called the *set of discontinuity* of G and is denoted $\Omega(G)$. $\hat{\mathbf{C}} \setminus \Omega(G)$ is called the *(topological) limit set* of G and is denoted $\Lambda(G)$. Whenever possible we simplify the notation by dropping the reference to G. Ω/G is a union of connected Riemann surfaces called the *quotient surface(s)*. The celebrated finiteness theorem of Lars Ahlfors (see e.g. Kra [Kra72]) states that when G is finitely generated, Ω/G is a (possibly

[5]The extension is currently proved by noting that the 3-ball is the symmetric space for the group $Gl_2(\mathbf{C})$ hence $SL_2(\mathbf{C})$ and $PSL_2(\mathbf{C})$ act naturally on that ball. $PSL_2(\mathbf{C})$ is naturally isomorphic to $\mathcal{M}$öb. Poincaré's idea is more direct. He notes that every möbius transformation can be written as a product of reflections in circles C. We could just as easily reflect in the sphere S which intersects the unit sphere orthogonally precisely in the circle C. This extends the action of the product by extending the action of the factors. A more computationally tractable method is given in Abikoff [Abi84].

void) finite union of connected Riemann surfaces S of finite conformal type each of which admits a hyperbolic metric. This last condition is equivalent to S having the unit disk as holomorphic universal cover.

If G has no torsion, the orbit space of the action of G on $\mathbf{B}^+ = \mathbf{B}_G^+ = \mathbf{B} \cup \Omega$ is a 3-dimensional manifold $\mathcal{M} = \mathcal{M}_G$ called the *quotient manifold* of G. If $\Omega \neq \emptyset$, $\mathcal{M}$ is a bordered 3-manifold. $\mathcal{M}$ is called a *Kleinian manifold*.

When G has torsion, the orbit space $\mathcal{M}_G$ is a 3-dimensional orbifold, i.e. a space locally modelled on $\mathbf{R}^3$ modulo a finite group. $\mathcal{M}$ is then called a *Kleinian orbifold*. Even when G has torsion, $\mathcal{M}$ is topologically a manifold. This is a peculiarity of dimension three—higher dimensional orbifolds need not be manifolds. The main difference between the torsion and torsion-free cases is that when G is torsion-free, it is isomorphic to the fundamental group of the manifold $\mathcal{M}$ and, in the other case, it is not. When G has torsion, we say that it is, or is isomorphic to, the *fundamental group of the orbifold* $\mathcal{M}$. The fundamental group of the manifold is then a quotient of G. (See Section 7.1).

A *fundamental set* for the action of G on X is a subset F of X with the properties that $\gamma(F) \cap F = \emptyset$ for all $\gamma \in G \setminus \{id\}$ and the translates of $\bar{F}$ form a cover of X. Fundamental sets which are used for purposes of analysis are usually chosen to be measurable with boundaries of zero measure. It is often convenient to think of a fundamental set as the interior of a polygon or polyhedron together with some of its sides. The sides are identified to form the orbit space of the action of G on X. We then use the terms *fundamental polygon* or *fundamental polyhedron*. In the context of hyperbolic spaces, one generally assumes that F is convex and locally finite sided.

The translates of a fundamental set form the tiling or *tesselation* of a space. The fundamental set acts as the tile.

A *Fuchsian group* G is a Kleinian group which leaves invariant a disk or half-plane in $\hat{\mathbf{C}}$. G is said to be of the *first kind* if its limit set is the boundary of the disk; otherwise we say it is of the *second kind*.

2 The deformations of Kleinian groups

A Kleinian group is about as simple an object as one could conceivably deform. After all, a Kleinian group G is a group of möbius transformations and a transformation $\gamma : z \mapsto (az+b)/(cz+d)$ can be deformed or, less precisely, wiggled by wiggling the numbers a, b, c and d. So we can embed the space of deformations of G into $\mathbf{C}^{4n}$ in the following way. Let $\gamma_i : z \mapsto (a_i z + b_i)/(c_i z + d_i)$ for $i = 1, \ldots, n$ be generators of G. G is then mapped to the vector $(a_1, b_1, c_1, d_1, \ldots, a_n, b_n, c_n, d_n)$. Of course, we should maintain the normalization $ad - bc = 1$ but that presents no difficulty. We can then wiggle a whole group by wiggling its generators. In terms of the parametrization, this means wiggling the numbers a_i, b_i, c_i and d_i. Usually, we want the perturbed group G' to have a close relationship with G—in particular we would like $\mathcal{M}_G$ to be homeomorphic to $\mathcal{M}_{G'}$ as an orbifold—this requires a definition.

Two Kleinian orbifolds $\mathcal{M}_G$ and $\mathcal{M}_{G'}$ are *homeomorphic* if there is a homeomorphism of $\mathcal{M}_G$ to $\mathcal{M}_{G'}$ which lifts to a homeomorphism $f : \mathbf{B} \cup \Omega(G) \to \mathbf{B} \cup \Omega(G')$ with the property that $fGf^{-1} = G'$. Since G is isomorphic to the fundamental group of $\mathcal{M}_G$, we want the wiggled group G' to be isomorphic to G. This places the following additional requirement on the allowable wiggles. Each relation in G is a polynomial relation $P = 0$ in the a, b, c, d entries of the generators. Thus when we perturb G we only permit wiggles where $P = 0$ for the wiggled entries. Technically, this may be restated as: *all perturbations must lie along the affine algebraic variety defined by* $P = 0$. In particular, this forces any perturbation of an elliptic transformation to be elliptic of the same order. Here we are using the fact that elliptic transformations in Kleinian groups have finite order. It is also common to require that a perturbation of a parabolic transformation be parabolic and that all parabolic transformations in the wiggled group be wiggles of parabolics. This last requirement is geometric; generally it is not forced by the group theory.

The deformations that we have described so far are obtained by topological conjugacy of groups. In general we don't know how to parametrize these deformations. In particular, the topological deformation variety is only known when it is the same as the space of deformations obtained from a restricted class of homeomorphisms—the quasiconformal ones. The equivalence is known for the class of geometrically finite groups (see Section 6). As we shall see in the next section, there is a natural way to parametrize quasiconformal homeomorphisms of a sphere which conjugate one group of möbius transformations into another.[6] Two Kleinian orbifolds are *quasiconformally* (respectively, *quasi-isometrically*) *equivalent* if they are homeomorphic via a map which is quasiconformal (respectively, quasi-isometric). A deformation is *allowable* if $\mathcal{M}_{G'}$ is quasiconformally equivalent to (or, equivalently, quasi-isometric to) $\mathcal{M}_G$.

There is one type of deformation which we regard as the identity—we will then identify all deformations which can be obtained from a fixed one by these trivial deformations. The geometry of a group G' obtained from G by conjugation by a möbius transformation is the same as that for G; we have simply changed a choice of base frame in the tangent bundle to $\mathcal{M}$. (To see this note that a möbius automorphicsm of $\mathbf{B}$ will generally move 0 and rotate a frame relative to the usual frame on $\mathbf{R^3}$.) We therefore take the class of all allowable deformations of G and mod out the action of $\mathcal{M}$öb by conjugation to obtain the *deformation space* $V(G)$ of G. It should be noted that our deformations always allow us to keep track of a system of generators of the groups.

In the above discussion we have delineated a collection of conditions for deformations to be allowable. Now we must describe the methods used to obtain deformations.

A Kleinian group can be described either by its action on $\hat{\mathbf{C}}$ or on $\mathbf{B}^3$. A border component of $\mathcal{M}$ carries the natural structure of a Riemann surface S because it is the quotient of a plane domain D by a group of möbius (hence conformal) automorphisms of D. If we can wiggle the conformal geometry of S we must have changed the Kleinian group. This is the traditional method for deforming Kleinian groups.

More formally, let G be a Kleinian group with $\Omega(G) \neq \emptyset$. Further, let $f\colon \hat{\mathbf{C}} \to \hat{\mathbf{C}}$ be a homeomorphism so that $G_f = fGf^{-1}$ is again a group of möbius transformations. G_f must then be a Kleinian group and $\Sigma = f(\Omega(G))/G_f$ is a union of Riemann surfaces. If Σ is not conformally equivalent to Ω/G via a map covered by f, then we have constructed a non-trivial deformation of G. We have thus reduced the problem of finding (at least some) deformations of G to constructing homeomorphisms of $\hat{\mathbf{C}}$ which realize deformations of conformal structure of surfaces.

[6]Informally, a *quasiconformal* or *qc mapping* of the sphere is a homeomorphism with the property that, almost everywhere, it maps infinitesimal circles to infinitesimal ellipses (see e.g. Ahlfors [Ahl66b] or Lehto and Virtanen [LV73]).

We next give an outline of the "μ-trick" which is the standard method for changing the conformal structure of a surface. Each Riemann surface structure S' on the surface S can be obtained from a Riemannian metric which we write in local coordinates in the form

$$ds^2 = g_{11}dx^2 + 2g_{12}dxdy + g_{22}dy^2.$$

If we use the complex differentials

$$(1) \qquad \begin{aligned} dz &= dx + idy \\ d\bar{z} &= dx - idy \end{aligned}$$

then a complicated juggle allows us to write

$$ds^2 = \lambda(z)|dz + \mu(z)d\bar{z}|^2$$

with $|\mu| < 1$. Under change of local coordinate, λ and μ transform as differential forms. Precisely, the expressions $\lambda(z)|dz|^2$ and $\mu(z)d\bar{z}/dz$ are invariant. In particular, $|\mu|$ is a function.

Notice that λ rescales the metric pointwise but is independent of direction. It therefore does not affect the conformal geometry of S. The changes of pointwise conformal structure are therefore parametrized by the forms $\mu(z)$ which lie in the open unit ball in L^∞. μ is called a *Beltrami differential.* If S is a border surface of $\mathcal{M}(G)$, then we may lift μ to $\Omega(G)$. It is traditional to also label the lift μ; it must satisfy the invariance relation

$$(\mu \circ \gamma)\frac{\overline{(\gamma')}}{\gamma'} = \mu.$$

It is then a fact (see e.g. Abikoff [Abi80]) that any solution to the Beltrami equation

$$f_{\bar{z}} = \mu f_z$$

in $\hat{\mathbb{C}}$ has all the desired properties. It is a quasiconformal mapping which conjugates G into a group G'. Further, if S is a border surface of $\mathcal{M}_G$ covered by a component D of $\Omega(G)$ then $S'' = f(D)/H$, where H is the isotropy subgroup of D in G', is conformally equivalent to the surface S'.

The μ-trick consists of finding an L^∞ Beltrami coefficient μ, with norm less than one, on $\hat{\mathbb{C}}$. It must satisfy the above invariance condition with

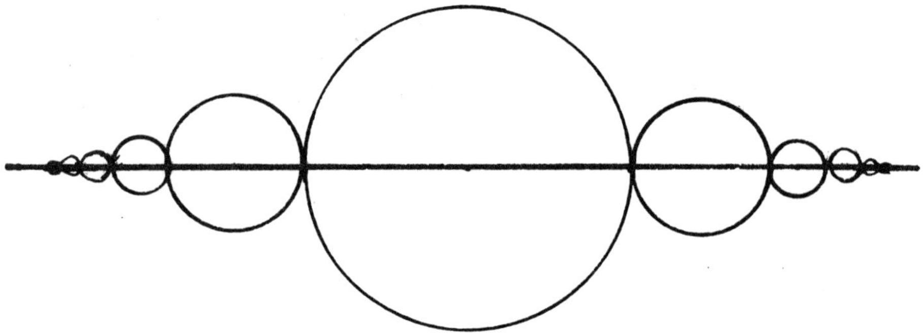

Figure 5: Packing a segment with circles

respect to G. We then solve the Beltrami equation using that coefficient. We are guaranteed that the resulting solution is a quasiconformal homeomorphism which conjugates G into an isomorphic Kleinian group G'. This process depends holomorphically on μ and is therefore stable. We shall something to say about finding more bizarre μ's in the next section.

2.1 a digression on some infinitely generated groups

There exist very few general results on infinitely generated Kleinian groups. Some of the most interesting concern the dynamics of the action of G on its limit set. Many of these results are due to Sullivan (see e.g. [Sul81]). The details of the example we give here may be found in Abikoff [Abi71].

Take a Jordan curve C in $\hat{\mathbf{C}} = \mathbf{C} \cup \{\infty\}$ with the following property: it is the closure of a countable number of Euclidean segments L_n and has positive area. C is called an *Osgood curve*. Clearly each point in the interior of L_n is at positive distance from the remaining segments hence also from the remainder of C. Therefore each L_n can be covered by circles as in Figure 5.

Define a reflection group G generated by the reflections in all of these circles. It is, of necessity, infinitely generated. Every accumulation point of the segments is a limit point of the group and it can be easily shown that the limit set $\Lambda(G)$ is a very jagged (dare I say *fractal*) approximation to C. By using sufficiently small circles, the approximations can be made arbitrarily close to C in the uniform topology on the defining maps from S^1.

The metrical, potential-theoretic and measure-theoretic properties of both this group and its quotient manifolds are bizarre. The existence of the group leads, using the method described below, to sobering examples of infinite-dimensional complex manifolds of deformations of G.

The exterior of the circles (respectively, spheres) is a fundamental set for the action of the reflection group on the 2-sphere (respectively, 3-ball). The group H consisting of the orientation-preserving maps in G is a group of möbius transformations and there is a natural fundamental set which contains the complement of the original collection of circles. It follows that there are points of areal density for the Osgood curve, hence also for the limit set $\Lambda(H)$ of H, which lie in a fundamental set for H.

There are two distinct types of deformations of H within the category of Kleinian groups. Some arise from Beltrami coefficients that live on the border surfaces of $\mathcal{M}$. We have described this construction in the previous section. There are other deformations of the complex structure of $\hat{\mathbf{C}}$ which arise from Beltrami coefficients with support in the limit set of H. First take any L_∞ function μ on the intersection B of the limit set and the fundamental set. Assume that the norm of μ is less than one. μ can be extended to almost every point in the limit set as a $(-1,1)$-form for H because B is disjoint from its translates. Solving the Beltrami equation with data μ gives a mapping f with the property that fHf^{-1} is a deformation of H and $f|\Omega(H)$ is conformal. Further, distinct μ's give distinct deformations.

The group H also has the following bizarre dynamics: H does not act ergodically on $\Lambda(H)$. The topological limit set $\Lambda(H)$, and the dynamical limit set (the recurrent or conservative set) $REC(H)$ are not almost everywhere identical.[7] Both of these statements are impossible for finitely generated groups (Ahlfors, Sullivan) because finitely generated groups cannot have infinite dimensional deformation spaces (see Sullivan [Sul81]).

One can do other bizarre things with H. For example, take its free product with a finitely generated Fuchsian group F of the first kind. Assume C lies in the exterior component of F. Then we may densely pack the exterior

[7] The recurrent set for the action of a group on a space is defined up to sets of measure zero. A group G *acts recurrently* on a measure space X if any set $A \subset X$, having positive measure, has positive measure intersection with infinitely many of its translates. The *recurrent set* for the action of G on X is the subset $REC(G)$ where G acts recurrently.

component of F with circles. A Kleinian group G' can be constructed as the index two subgroup of the free product of H, F and the group generated by reflection in the packing circles. The quotient, $\Omega(G')/G'$ consists of a single conformally finite surface. In addition, G' has spaces of differentials which appear, algebraically, to come from a very nice group, i.e. a finitely generated one. It follows that the complex analysis of G' arising from the quotient surfaces looks like it comes from a finitely generated group. But it doesn't.[8]

The analytic deformation varieties of infinitely generated groups have not been extensively studied for lack of results of any generality. Infinite dimensional *deformation spaces* for some suitable notion of Riemann surface seem to be arising naturally in string theory, but definitive results are yet to appear (at least to my knowledge).

3 Schottky groups

In the 1870s, Schottky wrote his doctoral dissertation. There were three versions of this work. They differed mainly in the problems they claimed to solve. The main idea was the construction of a class of Kleinian groups which are of great contemporary interest. We will give an example of the construction of these groups and then show the relationship to the deformation theory of Riemann surfaces.

3.1 Constructing a Schottky group

Start with $2g$ disjoint circles a_i and a_i' in $\mathbf{C}$. In our drawings, g will always be two but it can be arbitrary. Assume that the circles have disjoint interiors as in Figure 6. Since $\mathcal{M}$öb acts transitively on disks in $\hat{\mathbf{C}}$, for each $i \leq g$ we may find a möbius transformation γ_i so that $\gamma_i(\operatorname{Ext} a_i) = \operatorname{Int} a_i'$. Let G be the group generated by the γ_i and let F be the common exterior of the circles. Notice that $\gamma_1(C) \subset \operatorname{Int} a_1'$ whenever $C = a_1'$, a_2, or a_2'. Similarly, $(\gamma_1)^{-1}(C) \subset \operatorname{Int} a_1$ for $C = a_1$, a_2, or a_2'.

[8]The main interest of this example is historic in that it shot down some of the approaches to Ahlfors' measure problem which used differentials on quotient surfaces. Ahlfors' measure problem asks whether the limit set of any finitely generated Kleinian

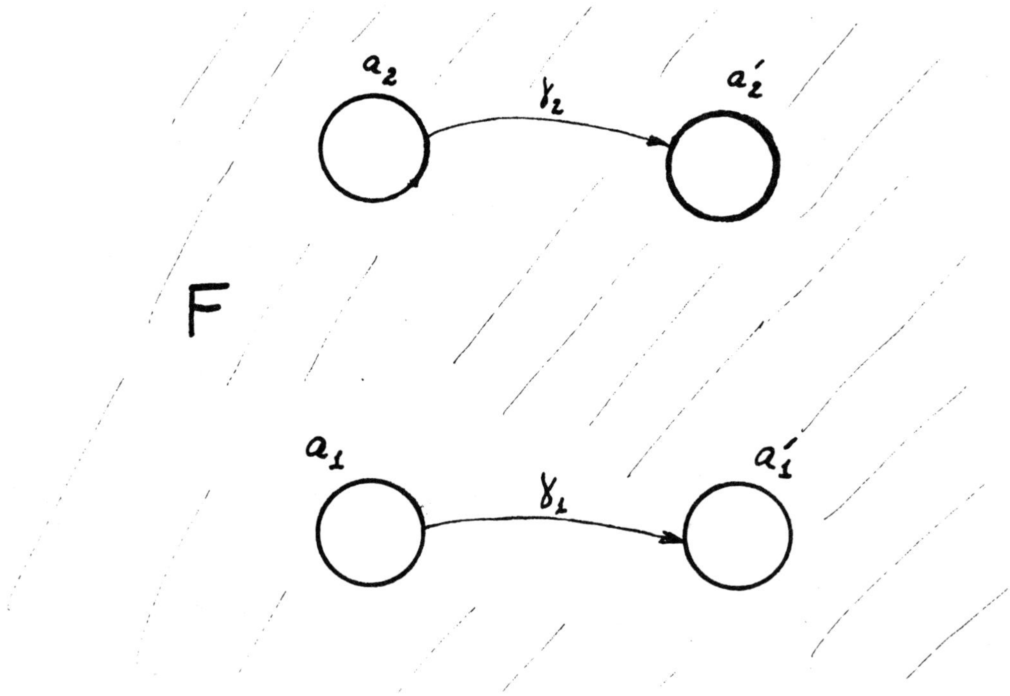

Figure 6: A fundamental set F for a Schottky group

Let W be any finite product (or, word in) the γ_i's and their inverses. If we choose any point z in F, we can chase its image under W. For example, assume $W = \gamma_1 \circ \gamma_2^2$. Then $W(z)$ lies inside a_1'. More generally, $\gamma(z) \in F$ if and only if γ is the identity. This proves that G is a free group on g generators and F is a fundamental set for the action of G on

$$\Omega(G) = \bigcup \left\{ \overline{\gamma(F)} | \gamma \in G \right\}.$$

The limit set $\Lambda(G)$ consists of the points not covered by the translates of $\overline{F}$. They are the points which are nested inside infinitely many translates of the original circles a_i and a_i'.

The quotient $\Omega(G)/G$ can be obtained by identifying the boundary circles of F hence it is a compact surface of genus g. $\mathcal{M}(G)$, the quotient 3-manifold, can also be simply described. γ_i identifies the hemisphere in $\mathbf{B}^3$ bounded by a_i to the hemisphere bounded by a_i'. The common exterior in $\mathbf{B}^3$ of these hemispheres is then a fundamental set for the action of G on $\mathbf{B}^3$. The identification space is homeomorphic to the region bounded, in 3-space, by a compact surface of genus g. This object is called a *handlebody of genus g*.

group must be 0 or 4π, that is, must the measure of the limit set be everything or nothing.

3.2 Wiggling a Schottky group

In our construction of a Schottky group we observe that there is little sacred about our initial choice of circles. Additionally, we have a wide choice of maps which take a given oriented circle to another oriented circle. We therefore have a large choice of the generating möbius transformations γ_i. In fact, if we make a sufficiently small perturbation of each γ_i we again obtain möbius transformations which pairwise identify circles and are equally good candidates for the above construction. Notice that each möbius transformation may be specified by 3 complex numbers so we need $3g$ numbers to specify G.

While the $3g$ numbers specify G uniquely, G doesn't uniquely determine the numbers. Recall that our choice of numbers used a particular set of generators for G. There are, however, infinitely many sets of generators of G. We are therefore not parametrizing groups but rather systems of generators for the groups. The system of generators is called a *marking* and G together with the distinguished system of generators is called a *marked group.*

We also note that conjugate groups have the same geometry. For example, if $G' = \gamma G(\gamma)^{-1}$ then $\Omega(G') = \gamma(\Omega(G))$, $F(G') = \gamma(F(G))$. The precise statement is the general fact that conjugation of groups is equivalent to translation of geometry by the conjugating transformation or motion.

As we have already noted, to obtain truly distinct geometries, we must regard conjugate groups, or systems of generators, as equivalent. To choose a single representative of a conjugacy class we can resort to the following normalization: conjugate G so that γ_1 has attracting (respectively, repelling) fixed point at -1 (respectively, $+1$) and γ_2 has attracting fixed point at i. Before, we had $3g$ parameters that could be wiggled to vary the choice of marked Schottky group. The normalization removes 3 free parameters and we are left with a space of $3g-3$ parameters which, at least locally parametrizes the Schottky groups. This is intimately tied to Riemann's problem of the moduli of Riemann surfaces to which we next turn.

3.3 Riemann's problem of moduli

Riemann's method of branch cuts allows us to construct compact Riemann surfaces of any genus. For many purposes it is reasonable to consider two

Riemann surfaces as identical if there is a biholomorphic homeomorphism, that is, a holomorphic or conformal equivalence, between them. In this section, we restrict our attention to compact Riemann surfaces. When viewed as algebraic curves, two Riemann surfaces are holomorpically equivalent precisely when they are birationally equivalent or when they have isomorphic fields of meromorphic functions. Riemann knew this and, based on a parameter count, he observed that the space of holomorphic equivalence classes of compact Riemann surfaces of genus $g > 1$ has complex dimension $3g - 3$. This observation became known as *Riemann's problem of moduli* because it was difficult to make it mathematically meaningful.

In fact about a century passed before the concepts and tools were available for the problem to be solved. The space is called the *moduli space of Riemann surfaces of genus g* and is denoted M_g. It is a normal complex of complex dimension $3g - 3$. Because it is known that M_g is not a manifold, although it is an orbifold, its properties are usually studied in some branched covering space. One of these covering spaces is the Teichmüller space whose analytic geometric structure was discovered, in various stages, by Teichmüller, Rauch, Ahlfors, Bers, Kravetz, Royden *et al.* As a real analytic manifold, the Teichmüller space was discovered by Klein (see below) and was probably known to Riemann (see Klein's lectures [Kle63]).

Our reason for mentioning the moduli problem here is that Schottky essentially solved it using Schottky groups. More important, he thought that he had solved it but Weierstrass convinced him not to publish the result.[9]

[9] Our source of this information is Felix Klein [Kle23]. It seems that when Klein thought he had proved the uniformization theorem he noted the relationship to Schottky's earlier work. In response to Klein's questions, Schottky stated that physical considerations led him to study the Schottky groups. Klein then asserts that the strength of Weierstrass' personality caused Schottky to weaken some of the claims in the original manuscript. I've been told by highly reliable sources that the original version of his thesis essentially contained the approach to moduli theory outlined in the next section but I've never seen the original manuscript.

4 The role of Schottky groups in uniformization theory

The original uniformization problem may be simply stated. Given an algebraic curve or equivalently a compact Riemann surface, we seek to holomorphically parametrize it by a function π whose domain of definition is some subset D of the Riemann sphere $\hat{\mathbf{C}}$. I have discussed the uniformization problem at length elsewhere (see Abikoff [Abi81]), here I only examine Schottky's idea.

Let S be the compact Riemann surface of genus g and choose a family of simple closed analytic curves C_i on S which, when cut, open all the handles. We further assume that the removal of the curves leaves the surface S connected (see Figure 7). It then follows that there are exactly g curves C_i and $S' = S \setminus \cup_i C_i$ is topologically $\hat{\mathbf{C}}$ with $2g$ holes removed. Equivalently, S' may be topologically embedded in $\hat{\mathbf{C}}$. It is much deeper, but also true that S' may be holomorphically embedded in $\hat{\mathbf{C}}$ as a domain D bounded again by analytic curves (see Figure 7). With luck D was chosen so that analytic continuation of π across the boundary curves is possible.[10]

S' is not really the object we want since it has edges. The parametrization using $\pi|S'$ then is not continuous along the dissecting curves C_i. If we walk on S' at bounded speed starting at some point z_0, we can reach an edge or dissecting curve of S in finite time. To avoid the discontinuity, we can sew another copy of S' along each border curve. If we start walking at the same place we will have to cross a dissecting curve at least once before we come to a new edge. (See Figure 8).

To avoid ever having a discontinuity we must make the paths to the edges infinitely long. This is possible on the infinite sheeted cover of S which is constructed by infinitely repeating the gluing process described above. The resulting surface we call R. Using a normal family argument we can prove that R embeds conformally in the Riemann sphere as a domain D. Its complement is a planar Cantor set Λ (see Figure 9). One can show that every holomorphic function f on D, which is locally bounded near Λ, extends to a meromorphic function on $\hat{\mathbf{C}}$. If f is a deck transformation for the covering,

[10] Luck really isn't completely necessary—this choice is always possible.

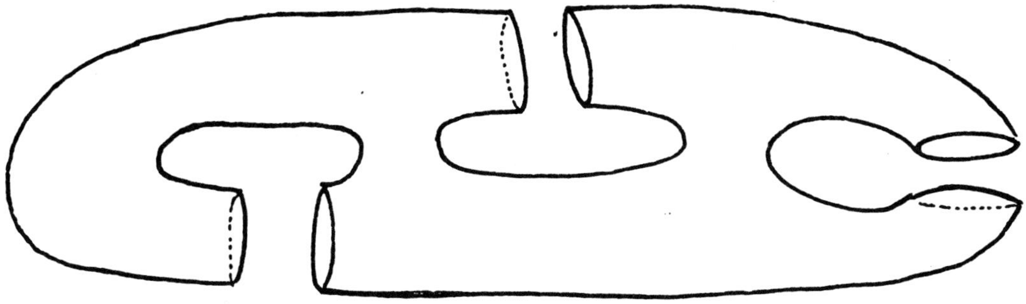

Figure 7: A Schottky dissection of a compact Riemann surface

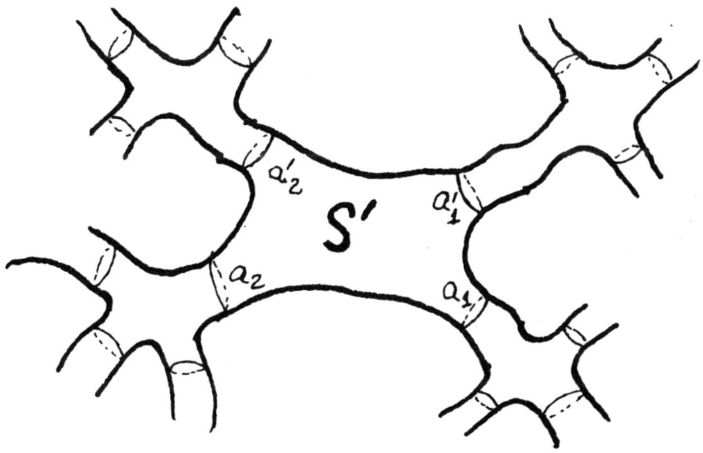

Figure 8: Moving the border curves away from S'

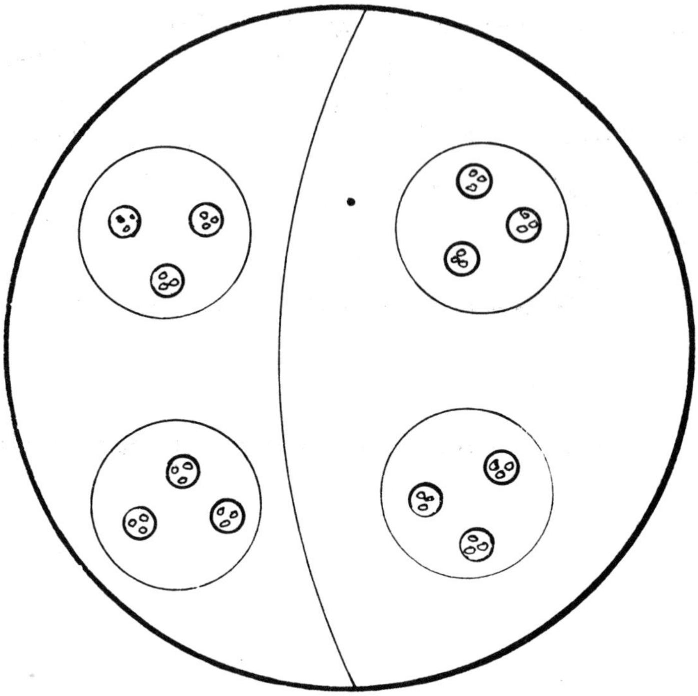

Figure 9: Moving the border curves to ∞

then it must extend to an injective meromorphic function on $\hat{\mathbf{C}}$, that is, it extends to a möbius transformation.

We have shown that D is the set of discontinuity of a group of möbius transformations and the fundamental set S' is topologically the same as that for the Schottky group constructed previously with one minor difference. D is bounded by analytic Jordan curves which aren't necessarily circles— we have no way to insure that D is bounded by Euclidean circles. In the general definition, the circles a_i are permitted to be analytic Jordan curves, otherwise the definition is exactly the same as in our first example. Marden [Mar74b] has proved that some Schottky groups cannot be realized using geometric circles. The Schottky groups whose fundamental sets are bounded by disjoint circles are now called *classical Schottky groups*.

5 Fuchsian groups and uniformization

In a remarkable confluence of disparate ideas, Klein and Poincaré both discovered general Fuchsian groups and uniformizations in the year 1882 (see Abikoff [Abi81]). Here I will briefly mention some of their ideas which are relevant to the present discussion. Both Klein and Poincaré were trying to uniformize Riemann surfaces. We must however recall that the notion of Riemann surface was not then formulated as clearly as it is today. So we must be more specific and say that Klein was trying to uniformize algebraic curves while Poincaré was after the uniformization of the Riemann surfaces of solutions of Fuchsian differential equations.[11]

In the process of studying this problem, Klein developed the combinatorial approach to Kleinian groups—he studied the question of when the group formed from the free product of two groups of möbius transformations is again a Kleinian group.[12] His approach to the general uniformization problem was to uniformize a single surface of a given genus and then to wiggle the uniformizing group. This is a valid approach to moduli theory of more general deformable objects and has, in later years, been called the *continuity method*.

Poincaré's approach emphasized the geometry of the space of solutions of differential equations and, in particular, the ratios of linearly independent solutions. He also mostly proved the polygon theorem which states conditions under which a hyperbolic polygon can be a tile or fundamental set for some Fuchsian group. The first complete proof of this theorem was given by Maskit [Mas71].

In general, Fuchsian groups are studied as groups of möbius transformations with real entries. Both Klein and Poincaré studied the Fuchsian or real wiggles of a Fuchsian group G. In the case where $\Omega(G)/G$ is compact and G is torsion-free, there are $2g$ generators $\gamma_1, \dots, \gamma_{2g}$ which are subject to the relation

$$\prod_1^g \gamma_i \gamma_{g+i} \gamma_i^{-1} \gamma_{g+i}^{-1} = \text{id}.$$

Poincaré noted that it is possible to perturb G so that the entries become complex and the relation remains valid. Here we are interested in the case where the perturbed group is no longer conjugate to a group of möbius transformations with real entries. For sufficiently small perturbations, the group remains Kleinian but has a limit set which is a non-circular Jordan curve.

[11]Recall that a second order linear differential equation, whose coefficients are rational functions, generally has multivalued solutions in $\mathbf{C}$.

[12]His analysis is the direct precursor of the algorithmic method for studying Kleinian groups which has been developed into a powerful technique by Maskit (See e.g. [Mas87]).

It is self-similar, has rather jagged spirals in a neighborhood of every point and has Hausdorff measure greater than one. These groups are called *quasi-Fuchsian*. Poincaré underestimated their importance—he felt that they were too complicated to be interesting. They are central to much of the modern research in moduli theory and the geometrization of 3-manifolds. Our first examples of geometrically infinite, finitely generated groups are limits of quasi-Fuchsian groups.

6 a quick glance at geometrically finite groups

Geometrically finite Kleinian groups are groups which admit a convex finite-sided fundamental polyhedron for the action of the group on hyperbolic 3-space. We will first need some more definitions.

The unit ball $\mathbf{B}^3$ is a model of hyperbolic 3-space with the hyperbolic or Poincaré metric

$$ds_P^2(x) = \frac{1}{4}(1 - \|x\|^2)^{-2}dx^2.$$

Here $\|x\|$ is the Euclidean norm of x and dx^2 is the flat or Euclidean metric on $\mathbf{R}^3$. As I've already mentioned, the action of a Kleinian group G on $\mathbf{B}^3$ is an extension of the action of G on $\hat{\mathbf{C}}$ when $\hat{\mathbf{C}}$ is viewed as the unit sphere in $\mathbf{R}^3$. That sphere is at infinite hyperbolic distance from any point in the unit ball. Recall that $\mathbf{B}_G^+$ is the union of the unit ball with the set of discontinuity $\Omega(G)$ for the action of G on $\hat{\mathbf{C}}$. The *3-manifold or 3-orbifold* of G is the quotient space $\mathcal{M} = \mathcal{M}(G) = \mathbf{B}_G^+/G$. The hyperbolic metric projects to $\mathcal{M}(G)$. The border surfaces of $\mathcal{M}(G)$ lie at infinite distance from any interior point. $\mathcal{M}(G)$ contains closed geodesics in most free homotopy classes–the missing closed geodesics may be assigned zero length. The latter are thought of as lying at infinity and are covered in G by parabolic transformations.

The *convex core* $\mathcal{C}(G)$ of $\mathcal{M}(G)$ is the smallest convex subset of $\mathcal{M}(G)$ which contains all closed geodesics.[13] It is easiest to visualize $\mathcal{C}(G)$ as the image under the quotient map $\pi : \mathbf{B}^3 \to \mathcal{M}(G)$ of the hyperbolic convex hull in $\mathbf{B}^3$ of $\Lambda(G)$.

[13]There are a number of equivalent definitions.

Thurston was first to observe that a group G is geometrically finite precisely when its convex core has finite volume.[14]

When G is geometrically finite and torsion-free, the topology of $\mathcal{M}(G)$ is particularly easy to describe. It is obtained from a compact (bordered) 3-manifold X by removing a finite number of closed solid tori and a finite number of closed solid cylinders whose ends (disks) lie in the border of the initial 3-manifold. The tori are called *cusped tori* and the cylinders are called *doubly cusped cylinders.* In case G has torsion, the quotient orbifold still has an underlying manifold structure, but the suborbifolds which must be removed can be finite quotients of the above omitted manifolds.

Our picture of $\mathcal{M}(G)$ has not yet mentioned its geometry. A neigborhood of the border surfaces looks like a horn whose volume is growing exponentially as a function of distance in the transverse direction. The tubes bounding the omitted cylinders connect two, not necessarily distinct, border surfaces. As the radius of the cylinder shrinks, the whole cylinder moves to infinity. In addition, the circles on the ends of the cylinder shrink to punctures on the border surfaces of $\mathcal{M}(G)$. After we remove a torus from the compact manifold, we have a neighborhood which is topologically a product $[0,\infty[\times T$ where T is a torus. Geometrically, the diameter of $\{x\}\times T$ decreases to zero. The picture is given in Figure 10.

There is yet another characterization of geometrically finite groups in terms of the G-orbits of points in $\mathbf{B}^3$. It is a standard result[15] that a point $\lambda\in\hat{\mathbf{C}}$ is a limit point of G if and only if, for every $x\in\mathbf{B}^3$, λ is an accumulation point of the G-orbit of x. Call λ a *conical limit point* or *point of approximation* of G if: the intersection of the G-orbit of any x and some small cone with vertex at λ contains a sequence converging to λ. A group G is geometrically finite if each limit point is a conical limit point or a parabolic fixed point.

7 Isomorphisms of Kleinian groups

Since the subject matter of this conference is group representations I would be remiss were I not to mention the subject of isomorphism classes of Kleinian

[14]To him it was an observation. Many of us had spent years trying to prove it.

[15]A *standard result* is one known by, and only by, experts. We abuse terminology here since this result is in every standard reference, e.g. Beardon [Bea83].

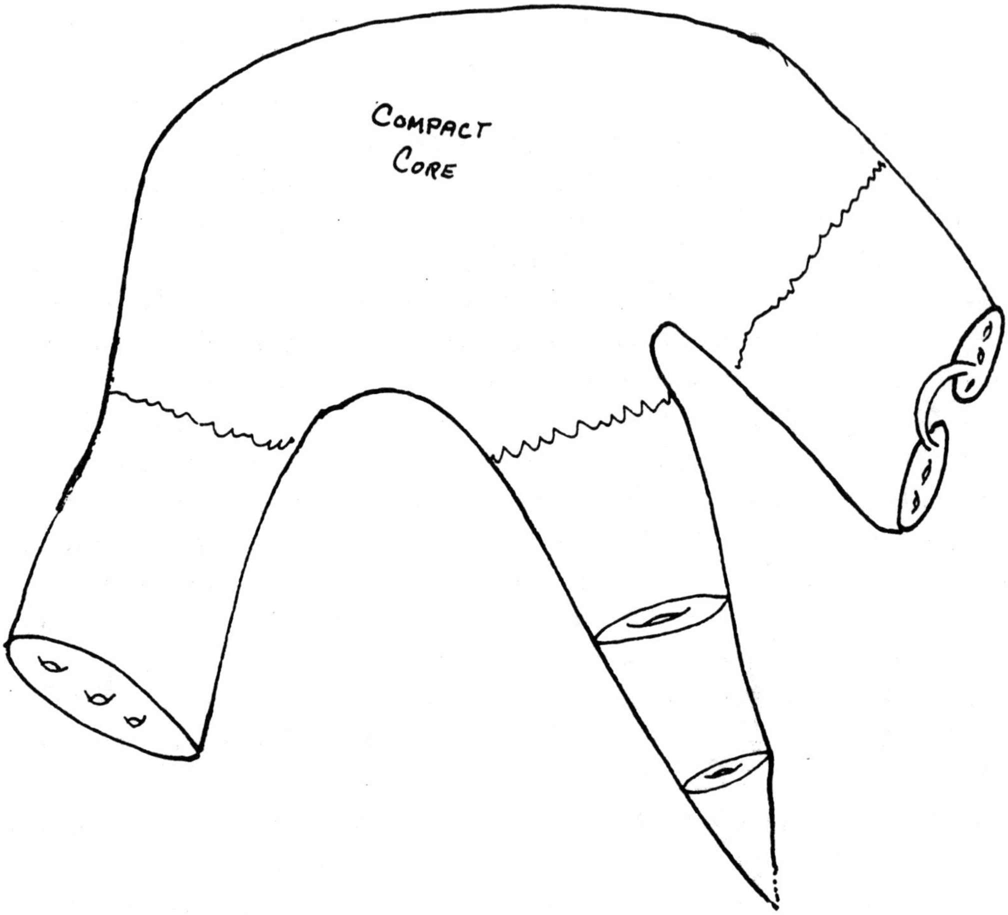

Figure 10: The ends of a geometrically finite manifold

groups. There are several classes of results which have been proved. They generally fall into two types: first, which groups can be faithfully represented as Kleinian groups and, second, what relationship does the group-theoretic structure impose on the geometry of the group and its quotient manifold.

7.1 Group subpresentations

With regard to the first question, it is convenient to introduce the notion of a subpresentation of a group. This allows us to easily relate Kleinian groups with torsion to the topological manifold which underlies the quotient orbifold.

Suppose $\mathcal{S}$ is some collection of symbols and, in the usual way, we form the free group $F_{\mathcal{S}}$ whose elements are the words in the symbols and their formal inverses. Let $\mathcal{W}$ be some collection of words in $F_{\mathcal{S}}$. A *group subpresentation* is a pair $(\mathcal{S}, \mathcal{W})$. A *group presentation subordinate to* $(\mathcal{S}, \mathcal{W})$ is a assignment of positive integers $n(W)$, one to each $W \in \mathcal{W}$, so that the pair

$$\left\{\mathcal{S} : \{W^{n(W)} | W \in \mathcal{W}\}\right\}$$

is a group presentation. Equivalently, a group presentation is a triple $(\mathcal{S}, \mathcal{W}, f)$ where $f : \mathcal{W} \to \mathbf{Z}^+$.

It is central to the discussion here that the underlying space of a Kleinian orbifold is a manifold. Suppose we have two Kleinian orbifolds $\mathcal{M}_G$ and $\mathcal{M}_H$ which have homeomorphic quotients. It then follows that G and H possess presentations which are subordinate to the same subpresentation—although it might not be the given presentation.[16]

Thurston has announced that every closed orientable 3-manifold can be realized as the topological space underlying some Kleinian orbifold.[17] By a

[16]The statment is a consequence of the fact that the fundamental group of the manifold, $\pi_1 M$ can be presented as $(S, \mathcal{W}, f)$ where $f(W) = 1$ for all $W \in \mathcal{W}$. You triangulate the manifolds M_i and get a refinement for which the branch locus $L_i \subset M_i$ is in the edge set. A further refinement will satisfy the property that $f(L_1)$ and $f^{-1}(L_2)$ also lie in the edge set. Then the branch locus for both of the coverings lies in the edge set and the relation obtained by circling the edge is a word to a positive power.

[17]The proof turns out to be a relatively easy consequence of the the existence of hyperbolic structures on mapping tori of irreducible surface diffeomorphisms. Brooks has proved—and even published (see [Bro85])—that any closed orientable 3-manifold admits a 2-sheeted branched cover by such a mapping torus. The branch locus is a link.

change of generators, we may assume that the manifold fundamental group of that space and the orbifold fundamental group share the same subpresentation. It is not necessarily true that the existence of a faithful Kleinian representation of a subpresentation using one function f means that any other choice of function will also yield a faithful representation.

7.2 Rigid groups

Suppose G is a Kleinian group with $\mathcal{M}_G$ having finite volume in the hyperbolic metric. The extensions to Mostow's rigidity theorem due to Margulis and Prasad (see Gromov's proof in Thurston [Thu79]) imply that any Kleinian group isomorphic to G in the sense of group theory is conjugate to G in the group of isometries of hyperbolic space. Here, in addition to möbius isometries, we must permit orientation-reversing conjugating mappings.

Sullivan has shown that a wider class of groups is rigid. His conditions can be stated in many ways (see Sullivan [Sul81]).

8 geometrically infinite groups

After Ahlfors proved his finiteness theorem, most people believed–for lack of evidence to the contrary–that all finitely generated Kleinian groups must be geometrically finite. Bers [Ber70] discovered (totally or singly) degenerate Kleinian groups in 1965 and Leon Greenberg [Gre66] soon proved these must be geometrically infinite.[18] Greenberg's proof is quite complicated; a more direct approach is in Marden's pioneering article [Mar74a]. This latter article realizes a geometric description of quotient manifolds of many torsion-free Kleinian groups—Poincaré and, much later, Ahlfors had speculated on the possibility of such a description.

8.1 Bers' construction of a geometrically infinite group

The idea underlying Bers' argument is quite simple and is summarized below. Suppose G is a finitely generated Kleinian group and $\chi_n : G \to \mathcal{M}\text{öb}$

[18]There is a list in Fricke-Klein in which it is asserted that degenerate groups exist! There is no proof and no explanatory text. The list is in [FK97, p.169ff].

is a sequence of isomorphisms of G onto discrete subgroups of the group $\mathcal{M}$öb of möbius transformations with complex entries. Jørgensen, following Chuckrow, has shown that if χ_n has a limiting homomorphism χ, then χ is an isomorphism onto a discrete group. $\chi(G)$ is called the *algebraic limit* of the sequence $G_n = \chi_n(G)$.

Jørgensen, as reported by and expanded upon by Thurston, showed that the limiting geometry need not be the geometry of the algebraic limit. There are two different but equivalent notions of the convergence of geometry. One is obtained from limits of groups rather than, as above, limits of isomorphisms. The other is convergence of quotient manifolds. Complementary discussions are given in Thurston [Thu79] and Abikoff [Abi].

The question of existence of algebraic limits is quite elusive. In Bers' case, the group G has a 3-manifold which is topologicially the product of a surface S and an interval. He holds fixed the conformal geometry of one end surface $\bar{S}$ of $\mathcal{M}_G$ and wiggles the other. (This is a simplified description of the Bers embedding of the Teichmüller space or, equivalently, of the Bers' slices of the space of quasi-Fuchsian groups.) When we lift these deformations to $\hat{\mathbf{C}}$, we find that the isomorphisms of the Kleinian groups G_n are induced by a conformal mapping of the unit disk onto the simply connected domains Δ_n which cover $\bar{S}$. Now one need only recall that the space $\mathcal{S}$, of properly normalized conformal maps into $\mathbf{C}$ of the unit disk, is compact in the topology of uniform convergence on compacta. Consequently, sequences in the Bers embedding always have convergent subsequences.

To obtain the desired geometrically infinite limits, we still need do some work. First we must force the geometry of the sequence to diverge. Bers does this by making the wiggled surface diverge in the Teichmüller space. The limit G therefore cannot be quasi-Fuchsian (the product of a surface and a closed interval) because small perturbations of quasi-Fuchsian groups, which preserve parabolicity, are again quasi-Fuchsian. He shows that there are only two other possibilities. First some sequence $\gamma_n \in G_n$ of non-parabolic transformations must have a parabolic limit, called an *accidental parabolic transformation*, or the set of discontinuity of G must be connected and simply connected. In the first case, the limit group lies on a positive codimension subset of the boundary of a parametrization space.[19] In the latter case, the

[19]The Bers parametrization space is an open subset of a complex vector space of rep-

limit manifold has the homotopy type of the product of a (possibly bordered) surface with a closed interval but the border has the wrong Euler characteristic for the 3-manifold to be the orbit space of a geometrically finite group (see the discussion of compact cores below).

8.2 the topology of the orbit space

We have already noted that geometrically infinite, finitely generated groups span a region between the geometrically finite and the infinitely generated groups. That the structure of these groups more mimicks the geometrically finite than the infinitely generated ones is one of the great lessons taught us by Thurston. For simplicity, I will assume unless otherwise stated that *all groups considered henceforth are torsion-free.* Their quotient spaces are then 3-manifolds rather than orbifolds.

There is one general compelling result that leads us to hope that, at least topologically, the quotient manifolds of geometrically infinite groups are fairly simple. It is the slightest generalization of a theorem of G. P. Scott [Sco73] (This statement appears in Abikoff [Abi86], but is a trivial consequence of Scott's result and the fact that Kleinian manifolds are $K(G,1)$'s.)

Theorem 1 *Suppose M is the 3-manifold of a finitely generated Kleinian group. Then there exists a compact submanifold $i: N \hookrightarrow M$ so that i is a homotopy equivalence.*

Any submanifold N which satisfies the conclusion of the theorem is called a *compact core* of M. It is tempting, but generally false in the topological setting, to believe that the ends, i.e. the components of $M \setminus N$, of M are products of border surfaces of N and half-open intervals. Such manifolds are called *missing boundary manifolds.* Bonahon [Bon86] has shown that a wide class of Kleinian manifolds are, in fact, missing boundary manifolds. His condition is slightly complicated in the presence of parabolic transformations; in their absence it is precisely that the Kleinian group be indecomposable with respect to non-trivial free products.

---

resentations of $\pi_1 S$. In that representation space, the *accidental parabolic* set lies on a countable union of analytic varieties.

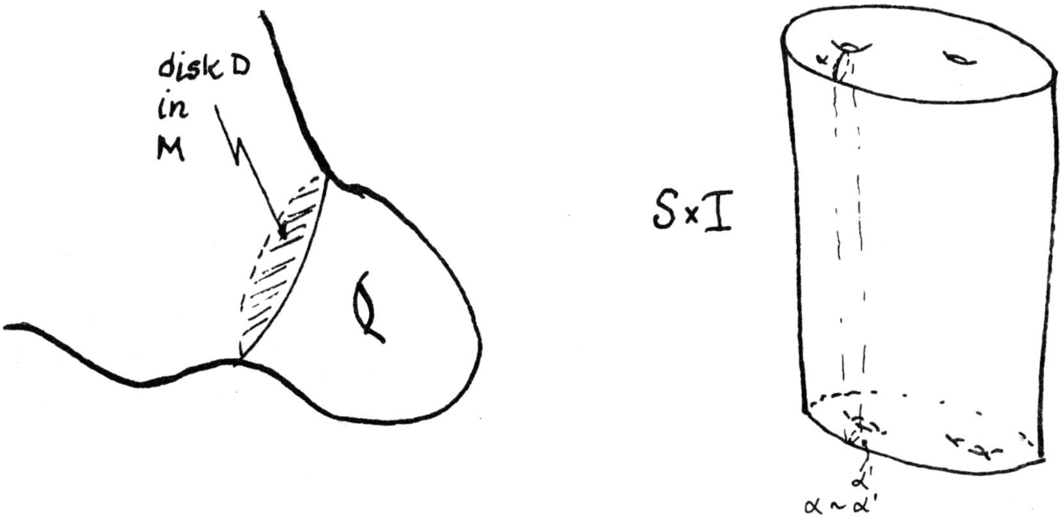

Figure 11: compressible and incompressible boundaries of 3-manifolds

Conjecture 1 *Suppose G is a finitely generated Kleinian group without torsion, then $\mathcal{B}^3/G$ is a missing boundary manifold.*

As far as I know, the earliest reference for this conjecture is Marden [Mar74a].

Free product decompositions of compact 3-manifolds arise from gluing two 3-manifolds together along a simply connected surface. For hyperbolic manifolds, the gluing surface cannot be a sphere unless the sphere is the boundary of a ball. Figure 11 shows an example of one 3-manifold with incompressible boundary and another whose boundary is compressible.

If we have a compact 3-manifold $\mathcal{C}$ with incompressible boundary, the fundamental group holds the manifold together. When the boundary is compressible the manifold may be split along a disk whose boundary lies in the boundary of the manifold.[20] These notions are discussed at length in Abikoff [Abi].

In the preceding discussion we have restricted our attention to the compact core of a geometrically infinite Kleinian manifold. In general, to decom-

[20]The parabolic transformations can sometimes hold them together even when the boundary of $\mathcal{C}$ is compressible.

pose $\mathcal{M} = \mathcal{M}(G)$, we need to extend the disk which splits $\mathcal{C}$ to a disk which is properly embedded in $\mathcal{M}$. Otherwise we don't split the whole Kleinian manifold and there is little geometric information we can get from the group theoretic splitting.

8.3 The topology of the group action

We suppose G is a geometrically infinite, but finitely generated group. We can view its action on either $\hat{\mathbf{C}}$ or on (some subset of) the closure $\bar{\mathbf{B}}^3$ of $\mathbf{B}^3$. Each of these has some interesting features and we will discuss some of them next.

8.3.1 the shape of fundamental polyhedra

For geometrically infinite groups, the work of Beardon-Maskit on points of approximation shows that there exist limit points which are not in the conical limit set and are not parabolic fixed points.

For the moment let us ignore groups with parabolic elements. Then we know that any convex fundamental polyhedron meets the limits set at points which are not conical limit points. Note that this is impossible for a geometrically finite group. In his seminal paper [Ahl66a] on the relationship between the shape of fundamental polyhedra and potential theory, Ahlfors showed that the measure of the limit set of a Kleinian group must be zero unless the fundamental set admits an incompressible flow of bounded flux (speed) whose source is the limit points in the boundary and whose sink is the set of discontinuity on $\hat{\mathbf{C}}$. There is a clear intuition[21] that such a flow cannot be of bounded speed if the asymptotic cross-sectional area near the limit set remains bounded.[22]. Here we see that there is a strong correlation between the asymptotic geometry of a fundamental set and the metrical properties of the limit set.

Thurston has shown that there is a class of geometrically infinite groups, which he calls groups with tame ends, whose quotients are missing boundary manifolds and satisfy the abovementioned criterion of Ahlfors. The easiest

[21] At least once you've thought about it for a few years.

[22] This is a higher dimensional analog of Riemann's theorem on removable singularities in the context of Riemannian geometry

way to describe these groups is to utilize their compact cores. Let $C(G)$ be the (or any) compact core of $M(G)$ and S be a border surface of $C(G)$. The end of $M(G)$ bounded by S is tame if either of the following two conditions holds:

(i) the injection $i : S \hookrightarrow M(G)$ induces an isomorphism on π_1.

(ii) The condition, due to Bonahon, on the location of parabolic elements holds.

It still remains to discover what is happening in the case of decomposable groups. Again later we will see that our inability to control the structure of the decomposable manifolds lies in the fact that the fundamental group of $M(G)$ does not hold together the geometry of $M(G)$.

Because it has now been shown by various authors that there are only finitely many conjugacy classes of parabolic transformations in finitely generated Kleinian groups, the following two conjectures make sense. Let G be an arbitrary finitely generated Kleinian group or more generally a finitely generated discrete group of isometries of hyperbolic 3-space.

Conjecture 2 *Every locally finite sided convex polyhedron for G intersects the limit set in at most finitely many points.*

Conjecture 3 *The number of G-orbits of limit points which are not points of approximation is finite.*

In the first case it is known (Ahlfors [Ahl66c]) that the intersection set has area zero. Otherwise the μ-trick would give an infinite dimensional space of deformations of G into $\mathcal{M}$öb—we have already seen that this is impossible for finitely generated groups. As far as I know, nothing is known about the second conjecture.

9 Geodesic Laminations—a quick tour

Most of this section is a quick introduction to concepts which were developed by Thurston [Thu79]. His original development, while terse, is perhaps the most elegant treatment currently available.

A conformally finite Riemann surface S is the complement in a compact surface S', having genus g, of a finite number n of points (punctures). S is said to have *type* (g, n). We further assume that S admits a Poincaré metric or, equivalently, that S has negative Euler characteristic.

A *geodesic lamination* $\mathcal{L}$ or, more simply, a *lamination* of a hyperbolic Riemann surface S is a collection of disjoint simple geodesics on S whose union is a closed subset of S. A maximal connected subset of a lamination is called a *leaf* of the lamination and a complementary component of $\mathcal{L}$ is called a *face* of $\mathcal{L}$. We include among the simple geodesics the simple closed geodesics.

There are two elementary examples of laminations which immediately come to mind. Both consist of a single geodesic ℓ. In the first case, ℓ is a simple closed geodesic. In the second case, it is a geodesic which starts and ends at punctures. These two types of laminations are illustrated in Figure 12. We note that the latter type of lamination is not useful for our purposes.[23]

Definition 1 A lamination of a surface S is said to have *compact support* if it forms a compact subset of S.

Unless otherwise stated, all of our laminations of conformally finite surfaces will have compact support.

There is a natural topology on the laminations of a surface. Here's the definition. If we lift a lamination $\mathcal{L}$ of S to the the unit disk Δ, the universal cover of S, each leaf lifts to a countable number of unoriented hyperbolic lines in Δ. Any oriented line can be represented by an ordered pair of distinct points on S^1. Thus any family of oriented lines defines a subset of $Cylinder = (S^1 \times S^1) \setminus Diagonal$. To parametrize the unoriented lines we must mod out by the interchange of endpoints. Therefore each lamination corresponds to a closed subset of the möbius band $Cylinder/\mathbf{Z}_2$. It is elementary to show that the laminations form a compact subset of the subsets of the möbius band.

In the remainder of this section, we will observe the convention that S is a conformally finite hyperbolic surface.

We note that any finite union of disjoint simple closed geodesics forms a lamination and these laminations are dense in the set of all laminations of S

[23]Following Thurston, we are only interested in those laminations which can be approximated uniformly, as an embedded submanifold, by a simple closed curve.

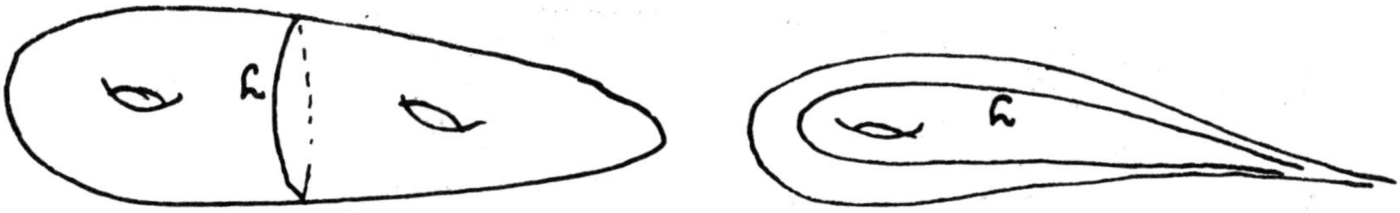

(1) (2)

Figure 12: The simplest laminations—the second has non-compact support

having compact support. Even the laminations consisting of a single simple closed curve are dense. This reduces the problem of finding all laminations with compact support to finding all limits of simple closed curves. To classify laminations, it is necessary to consider the possible simple closed geodesics, the geodesics asymptotic to simple closed geodesics and the nature of the faces of the lamination. Here's a list of the possible laminations $\mathcal{L}$ with compact support. The first possibility is generic.

(Type–I) $\mathcal{L}$ contains no simple closed curves and each face is a finite sided asymptotic polygon.

(Type–II) $\mathcal{L}$ consists solely of a finite number of simple closed geodesics.

(Type–III) $\mathcal{L}$ has simple closed curves and only faces which are finite sided polygons.

(Type–IV) $\mathcal{L}$ has some faces which are not simply connected. If S has punctures, they lie inside the faces, some polygons may be monogons and any of the above conditions may hold

In Figures 13–16, we show examples of these laminations on a compact surface.

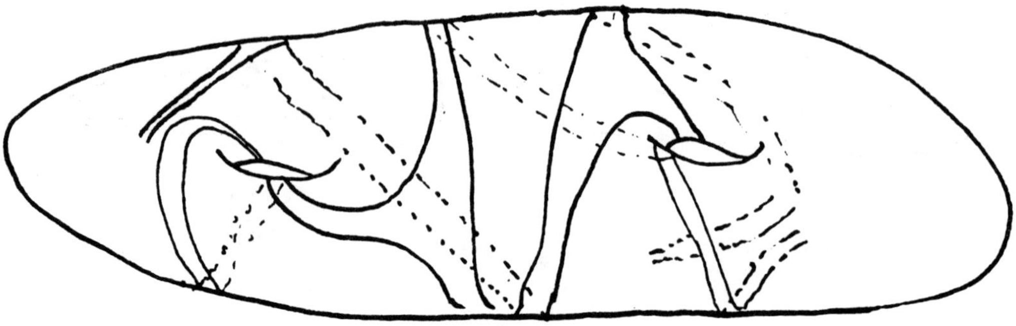

Figure 13: A Type I lamination

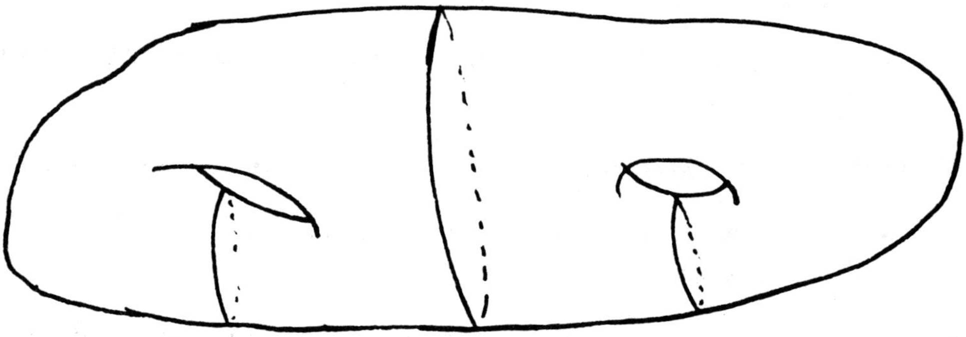

Figure 14: A Type II lamination

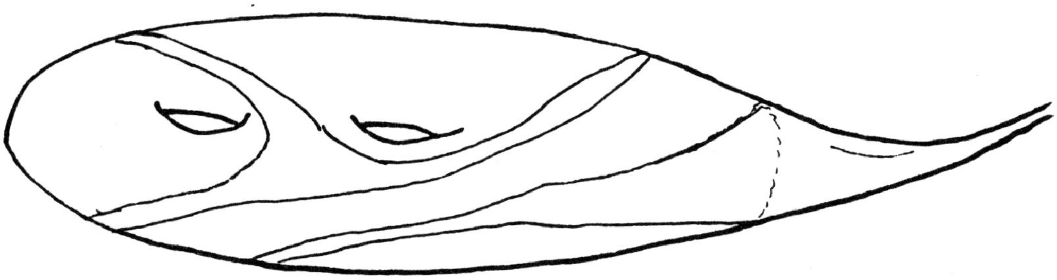

Figure 15: A Type III lamination

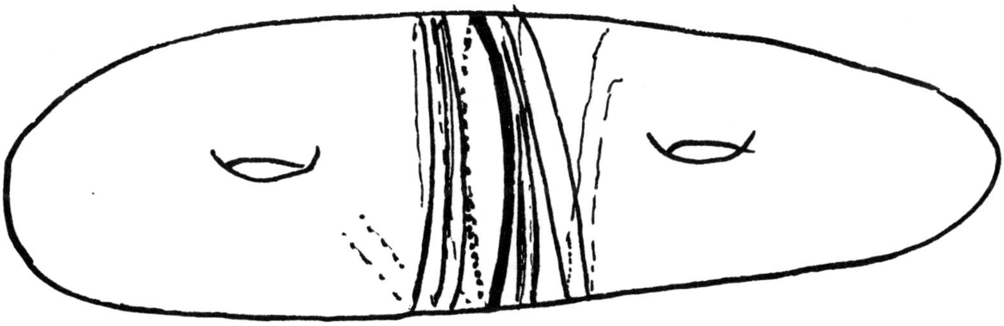

Figure 16: A Type IV lamination

Notice that there are two kinds of faces of a lamination. The first, and most common, is a finite sided asymptotic n-gon Π. Π is a cell whose vertex angles are zero hence the area is $(n-2)\pi$. Thus there is an absolute bound on the number of polygonal faces. Further, on lifting the surface to some covering, each component of the the lift of the polygon has the same area upstairs as downstairs. This is decidedly not true in the second and fourth cases. Here, a face is not a cell and its lifts to some coverings have infinite area. It is known (cf. Thurston [Thu79]) that any lamination occupies no area on a finite hyperbolic surface. It follows that we can determine the number of faces from combinatorial data and there are absolute bounds depending only on the Euler characteristic.

It is a basic fact that each lamination $\mathcal{L}$, with compact support, admits a non-trivial transverse measure μ. The pair $(\mathcal{L}, \mu)$ is called a *measured lamination* and μ is said to be *supported on* $\mathcal{L}$. It is easiest to describe this measure by duality when $\mathcal{L}$ is a single simple closed curve. Let ℓ be any simple smooth curve intersecting $\mathcal{L}$ transversally. $\mu(\ell)$ is the number of intersection points of ℓ and $\mathcal{L}$. If we multiply μ by an arbitrary real number, we define a one dimensional vector space of measured laminations. By taking weak* limits of measured laminations supported on a single curve, we obtain measured laminations supported on arbitrary laminations with compact support. Generically, the topology, that is the lamination, uniquely determines the measured lamination up to scale factor. To make this construction better defined, we consider only the projective classes of measured laminations. As Thurston has shown, these form a natural boundary for the Teichmüller space and the resulting compactified space is a topological cell.

10 Limit sets of singly degenerate groups

I have already described Bers' proof of the existence of singly degenerate groups. When I first learned of them, I tried to understand them using techniques that had already proved useful in studying some geometrically finite limits of quasi-Fuchsian groups. These simpler groups are called *regular b-groups* and the analysis relied heavily on the fact that they possess limit sets which are locally connected. For regular *b*-groups, the group action is topologically determined by the set of cut points of the limit set. A more explicit description is given below. It is important to note that local connectedness is central to this form of analysis.

In [Abi76], I gave an incomplete case analysis which led to the conclusion that degenerate groups have non-locally connnected limit sets.

The proof is not correct.[24]

I, along with most other practitioners of Kleinian group theory, believe that the limit sets are locally connected. Any other conclusion means that the limit set is complicated to an extent not matched by any other property of these groups. Proofs, however, are not yet available.[25] In this section, we will examine the implications of the locally connected limit set hypothesis and give the one serious proof in this paper.

Recall that a singly degenerate group G has a set of discontinuity $\Omega(G)$ which is a dense, open, connected and simply connected subset of $\hat{\mathbf{C}}$. Its complement in $\hat{\mathbf{C}}$, $\Lambda(G)$, is thus a continuum and $\Lambda(G)$ contains more than two points. The Riemann mapping theorem then provides us with a conformal map $f : \Delta \to \Omega(G)$. It is elementary that $\Lambda(G) = \partial\Omega(G)$. Classical conformal mapping theory, see e.g. Tsuji, Golusin, Pommerenke or Duren, includes the following standard result.

Theorem 2 *Suppose D is a simply connected plane domain conformally equivalent to the unit disk Δ. Further assume that ∂D is locally connected. Then the Riemann map $f : \Delta \to D$ has a continuous extension to $\bar{\Delta}$.*

The converse to this theorem is trivially true since the continuous image of an interval or of the unit circle is locally connected.

For the remainder of this section, assume G is a singly degenerate group with a locally connected limit set. Since the complement of Λ is open and dense, Λ itself is nowhere dense. It is known (see Whyburn [Why42, p.46]) that a locally connected continuum in a complete metric space is arcwise connected. We therefore know that, given any two points $x, y \in \Lambda(G)$, there is an arc $\overline{xy}$ connecting them. This arc must be unique—otherwise the union of two would contain a Jordan curve C. A Jordan curve in the limit set

[24] I am grateful to Bill Floyd, Steve Kerckhoff and Bill Thurston for pointing out the error to me.

[25] Thurston has announced that some singly degenerate groups have locally connected limit sets.

of a singly degenerate group is impossible because the limit set of a singly degenerate group does not separate the sphere. We have therefore shown that locally connected limit sets must have the structure of what is termed a dendrite but is more commonly called a tree. We shall later give a formal definition to the type of structure we have encountered here.

We need the following lemma.

Lemma 1 *Suppose x and y are distinct fixed points of a loxodromic transformation γ, then the arc $\overline{xy} \subset \Lambda(G)$ is not a proper subset of any other arc in $\Lambda(G)$.*

Proof Assume x is the attracting fixed point and y is the repelling fixed point of a loxodromic transformation γ and $\overline{xyu}$ is a arc in $\Lambda(G)$ with y an interior point of the arc $\overline{xu}$. If $u = x$ then we have found a Jordan curve in $\Lambda(G)$ which is impossible by the previous argument. Therefore the arcs $A_n = \gamma^n(\overline{yu})$ connect y to points u_n which converge to x.

Since x has a connected neighborhood base in Λ, for n suffiently large, u_n must lie in a small connected neighborhood N of x. Hence there must exist an arc $A' \subset N$ connecting x to u_n. Since $A' \subset N$, $y \notin A'$ and we have found two distinct paths connecting x and y. This is impossible and proves the lemma. ∎

We recall some graph theoretic terminology.

Definition 2 Let X be a locally connected space with the property that there is a unique arc connecting any two points. The *valence* or *cut order* of a point x is the number of path components of $X \setminus \{x\}$. A point of valence one is called an *endpoint*; otherwise x is called a *cutpoint*. A cutpoint of valence greater than two is called a *vertex*.

We have noted above that a singly degenerate Kleinian group with a locally connected limit set has a unique arc connecting any two given points. Thus the concepts defined above are applicable.

We first proceed to analyze the vertex set. Lemma 1 shows that each loxodromic fixed point is an endpoint. Parabolic fixed points are cutpoints of infinite valence, however there are two types. It is worthwhile examining them.

Suppose λ is the fixed point of some parabolic transformation γ in the singly degenerate group G. Let $f: \Delta \to \Omega(G)$ be the Riemann (conformal) map from Δ to Ω. Because f is a holomorphic equivalence, both domains are models for the holomorphic universal cover of $S = \Omega(G)/G$. $\eta = f^{-1} \circ \gamma \circ f$ is a möbius automorphism of Δ which covers the same homotopy class of closed curves as does γ. Since γ has infinite order, η must either be parabolic or loxodromic.[26] It is elementary to show that η is parabolic precisely when it covers a simple loop around a puncture. We then say that the isomorphism of G into $\mathcal{M}$öb preserves the type of η. It is possible that η be loxodromic—in which case we say that γ is an *accidental parabolic transformation* or an APT.

The two previous cases are distinguished by the homotopy class on S which γ covers. If γ covers a small loop C around a puncture, a component of the lift of C in Ω is a Jordan arc which may be completed to a Jordan curve by adjoining the fixed point of γ. It can be shown that C can be deformed so that the corresponding Jordan domain is a geometric disk in Ω. In the case of γ being an APT, there is no disk in Ω which is invariant under γ. Using a plane separation argument similar to that given in the previous Lemma, we can show that the arc $\overline{x\lambda}$ in $\Lambda(G)$, which connects an arbitrary point x to the fixed point λ of γ, must intersect its translates under powers of γ. However this intersection must consist solely of λ. The local structure of a locally connected limit set of a singly degenerate group near a parabolic fixed point is sketched in Figure 17.

Without proof, we note that any elliptic transformation γ has one fixed point λ in $\Lambda(G)$. If γ has order n, then λ has valence which is a multiple of n.

Selberg has shown that any finitely generated Kleinian group has a torsion-free subgroup of finite index. A finite index subgroup has the same limit set as the original group. We will therefore restrict our attention to torsion-free groups. The reintroduction of torsion then becomes a combinatorial problem.

Recall that $\Omega(G)$ is holomorphically equivalent to Δ via the map f. We

[26] It is common to include hyperbolic möbius transformations among the loxodromic ones. In this case, η must actually be hyperbolic.

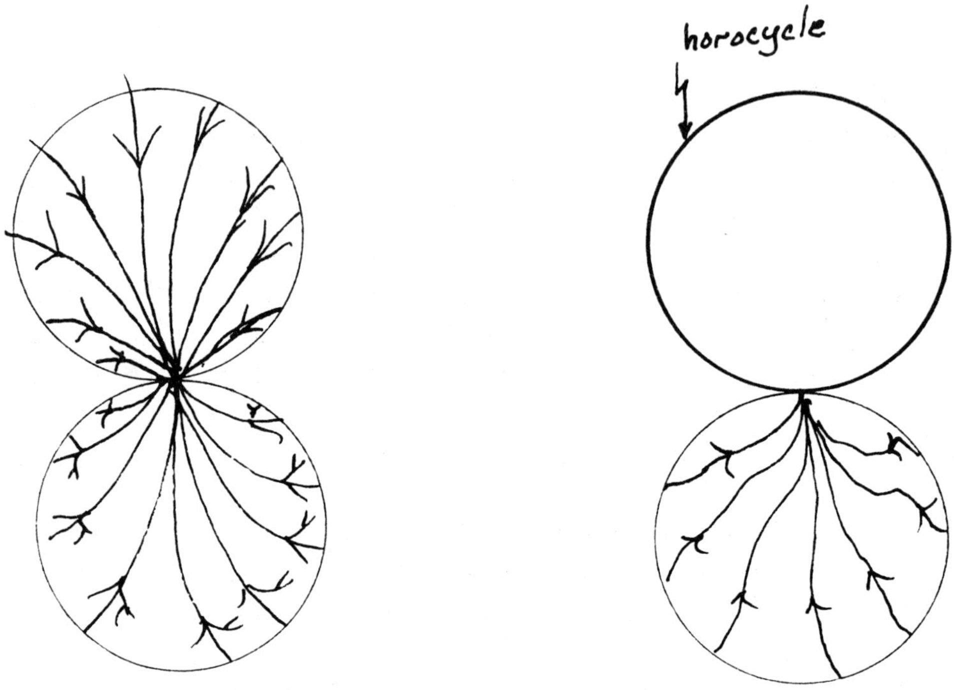

Figure 17: The local structure of the tree near the fixed point of an APT and a non-APT

can use f to push forward the hyperbolic geometry of the disk onto Ω. Since Λ is assumed locally connected, f extends to the closure of Δ and geodesics in Ω have endpoints on Λ.

Let S denote $\Omega(G)/G$. Suppose λ is a cutpoint of Λ of valence greater than two—we do not exclude the possibility that the valence is infinite. Further assume that λ is not a parabolic fixed point. On the circle there is a cyclic ordering to the preimages of λ under f. Assume first that there are finitely many preimages. Connect them cyclically by hyperbolic geodesics to form an asymptotic polygon P_1 in Δ. $P = f(P_1)$ is then an asymptotic polygon in the hyperbolic geometry of Ω. However it has the amazing property that λ is the one and only endpoint of each side of P. We claim that P is disjoint from all its translates.

Since λ is not a fixed point of any nontrivial transformation $\gamma \in G$, no side of P is mapped by γ to a side of P. Therefore, if a side σ of P has the property that $\gamma(\sigma)$ intersects the boundary of P at a side τ, the intersection must be transversal. We next show that $\gamma(\sigma)$ cannot intersect τ.

$\tau \cup \{\lambda\}$ (respectively, $\gamma(\sigma) \cup \gamma(\lambda)$) is a Jordan curve which is analytic except possibly at its unique limit point. The two curves do not intersect at their limit point and they have at least one transverse intersection point. Since the intersection number of any two curves on the sphere must be even,

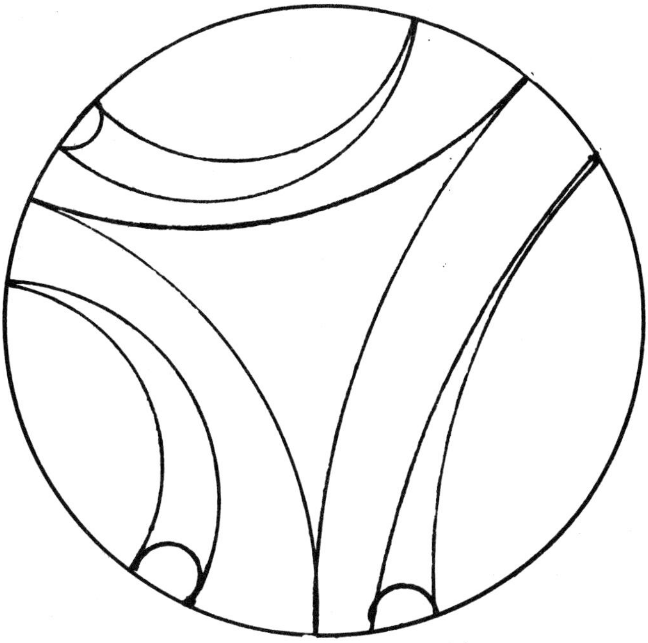

Figure 18: Asymptotic polygons with disjoint boundaries

there must be another intersection point and that point must lie in $\Omega(G)$. We have now reached the absurd conclusion that two hyperbolic geodesics in a model of the hyperbolic plane intersect in at least two points. We have proved that the boundary of P does not intersect any of its translates.

We have now shown that, when λ is not a fixed point, P is disjoint from its translates. It follows that P injects into S under the projection map from Ω. If P has k sides, its area is $(k-2)\pi$. But S has area $2\pi(2g+n)$ where g is the genus and n is the number of punctures on S. We therefore obtain a bound on the number of sides of P and a bound on the number of orbits under G of points, not parabolic fixed points, of valence greater than two. We can perform a similar construction at a point λ of valence two and we get a unique geodesic line instead of a polygon.

If λ is the fixed point of a parabolic transformation γ, then λ has infinite valence. Assume first that γ is not accidental. There is either one or two finite sided asymptotic polygons with the following property. Their translates under the powers of γ cover the interior of the infinite sided polygon obtained by cyclically connecting the preimages of λ in the boundary of the disk and pushing forward to Ω using f. On S, the corresponding puncture lies in the interior of a polygon P. If P has at least two sides, the two polygons are formed by connecting the puncture to two distinct vertices using geodesics. If P is a monogon, there is one polygon and it gets two extra sides from a geodesic arc connecting the single vertex to the puncture. Once we connect the puncture to one or more vertices of the polygon, the resulting lamination no longer has compact support.

If γ is an accidental parabolic transformation, then there is a geodesic line ℓ in $\Omega(G)$, with the property that λ is the unique endpoint of ℓ and ℓ covers a simple closed geodesic on S.

Notice that the family of sides of polygons and lines obtained from connecting point of valence two is a G-invariant lamination $\mathcal{L}$ of Ω. $\mathcal{L}$ therefore projects to a lamination of S. We have already noted that the leaves of a lamination of S have measure zero, hence so does its lift $\mathcal{L}$. Therefore the faces of $\mathcal{L}$ have full measure in $\Omega(G)$.

We have shown above that, under the locally connected limit set assumption, the limit set of a singly degenerate group may be viewed as the set obtained topologically by identifying to points, the finite sided closed faces and leaves of a lamination of the disk. This lamination must be invariant under a Fuchsian group. We also have to collapse the infinite sided faces left invariant by parabolic transformations. We will call those faces *parabolic*.

It is a consequence of a theorem of R. L. Moore (see Whyburn [Why42]) that the converse of this statement is also true in the topological category. Specifically, we take a G-invariant lamination of the unit disk Δ and identify to points its leaves, faces of finite area and parabolic faces. The resulting identifications of the sphere again form a sphere to which the group action may be pushed. The limit set is the continuous image of the unit circle hence is locally connected.

We have proved the following

Theorem 3 *Suppose G is a totally degenerate group. Then the limit set of G is locally connected if and only if there is a lamination $\mathcal{L}$ which collapses onto the limit set of G. If $\Lambda(G)$ is locally connected then*

1. *$\Lambda(G)\setminus\{endpoints\}$ is homeomorphic to the leaf space of the lamination $\mathcal{L}$ of Δ.*
2. *The only infinite valence vertices are at parabolic fixed points.*
3. *Loxodromic fixed points are endpoints.*
4. *Vertices of finite valence greater than two correspond to the faces of $\mathcal{L}$.*

The general setting for these locally connnected limit sets has been studied by Morgan and Shalen [MSb] among and others. Their technique for constructing these trees are quite different from the phenomenological analysis given above. A short description is given in the next subsection.

10.1 R-trees

Following Morgan and Shalen, we have the following

Definition 3 An **R**-tree is a metric space in which any two points are joined by a unique topological arc and any topological arc is isometric to an interval in **R**.

We have already shown the uniqueness of the arc connecting any two points in the limit set Λ of a singly degenerate group if Λ is locally connected. Further, Λ is the leaf space of a lamination $\mathcal{L}$ of the unit disk invariant under a Fuchsian group H. $\mathcal{L}$ projects to a lamination with compact support if we remove the finite number of leaves which terminate at punctures. Each such lamination admits a non-trivial transverse measure μ. μ may be integrated along any arc in Λ to give a distance function and an isometry with an interval. We therefore may restate part of the previous theorem as

Theorem 4 *Let G be a singly degenerate group then $\Lambda(G)$ is locally connected if and only if the set of non-endpoints (that is, interior points) of $\Lambda(G)$ is an* **R***-tree.*

The construction doesn't require any group invariance. We have really described a method for constructing measured laminations of simply connected domains D from **R**-trees embedded in the sphere as $\hat{\mathbf{C}} \setminus D$.

10.2 Some open questions

There are several important questions about degenerate groups with locally connected limit sets. We list some below. The first is the most important.

Question 1 *Does every singly degenerate group have a limit set which is locally connected?*

Question 2 *Can we explicitly construct a singly degenerate group G? Here explicitly means writing the generators of G as matrices with numeric entries.*

Question 3 *Suppose $\mathcal{L}$ is a lamination of the unit disk Δ which is invariant under a finitely generated Fuchsian group of the first kind and only has faces of finite area. Can the leaf space of $\mathcal{L}$ be realized as the limit set of a singly degenerate group?*

Question 4 *Suppose G_1 and G_2 are degenerate groups whose limit sets are projectively isometric* **R***-trees. Are the groups quasiconformally equivalent? Can the condition be weakened to the existence of a homeomorphism between the limit sets with the same conclusion valid?*

The last question is interesting because it starts to examine the notion of moduli for geometrically infinite groups. We know that the deformations of conformal structure of $\Omega(G)/G$ gives a class of deformations of G. Suppose we want to parametrize the degenerate groups H which have the same quotient surface $S = \Omega(G)/G$. Under the locally connected hypothesis, the limit sets are associated to measured laminations. The question asks the extent to which the laminations, measured or topological, determine the groups. In the next section we examine the extension of these ideas to more general Kleinian groups.

11 Formally Kleinian groups

This section contains an attempt to model the topology of all finitely generated Kleinian groups. We shall restrict our attention to torsion-free groups although the extension to groups with torsion is direct but linguistically clumsy. It generalizes the locally connnected hypothesis for the limit sets of singly degenerate groups. The generalization is the following:

Definition 4 Let H be a geometrically finite Kleinian group with no cusp annuli and let $\mathcal{L}$ be an invariant lamination of $\Omega(H)$. Assume X is the identification space formed by identifying the leaves of $\mathcal{L}$ to points and also identifying the closure of each finite area or parabolic face of $\mathcal{L}$ to a point. Denote by π the quotient map. The action of H pushes forward to a group G acting on X by the rule $\pi G = H\pi$. G is called a *formally Kleinian* *group* acting on X. The pair (X, G) is, by abuse of language, also called a formally Kleinian group and we say that it is *modelled* on the pair $(H, \mathcal{L})$ or its geometrically finite model is $(H, \mathcal{L})$.

Recall that Thurston has shown that the measured laminations form a boundary for the Teichmüller space of $\Omega(H)$. In the previous section, we showed that some, possibly all, singly degenerate groups lie in the boundary of the deformation space of a geometrically finite group—in fact, those groups were Fuchsian. There is a longstanding conjecture, first stated by Bers in a restricted case. The statement is

Conjecture 4 *Every torsion-free finitely generated Kleinian group (or, equivalently, its quotient manifold) is either geometrically finite or lies on the boundary of the deformation space of its non-singular geometrically finite model.*

It remains to define the non-singular geometrically finite model of a geometrically infinite group. Recall that every Kleinian manifold $\mathcal{M}$ with finitely generated fundamental group has a compact core $\mathcal{C}$. $\mathcal{C}$ is a compact submanifold whose inclusion is a homotopy equivalence. Without going into details, we think of $\mathcal{C}$ as the non-singular model for $\mathcal{M}$. If $\mathcal{C}$ admits a hyperbolic structure, then the cover group H is geometrically finite and all its parabolic transformations lie in rank two parabolic subgroups. It is a consequence of Thurston's uniformization theorem, the proof of which has started to appear, that $\mathcal{C}$ indeed admits a hyperbolic structure. McCullough, Miller and Swarup [DMS85] have shown that any two compact cores are homeomorphic. It then follows that a geometrically infinite group lies in the boundary of at most one non-singular model deformation space. To my knowledge, there is no geometrically finite group H which admits non-trivial deformations and for which the following is known to be true: Every geometrically infinite group whose non-singular model is H lies in the boundary of the deformation space of H.

The conjecture, however tenuous it is, motivates our definition of formally Kleinian groups for the following reasons. The process of moving to the boundary of a deformation space of a geometrically finite group implies that one or more of the border surfaces of the quotient manifold $\mathcal{M}$ move to the boundary of its Teichmüller space. That surface moves toward a lamination. We may think of that lamination as being collapsed—what else could it do?

While I have never seen an example of a singly degenerate Kleinian group with a locally connected limit set, our last theorem has the following easy consequence.

Theorem 5 *A singly degenerate Kleinian group has a locally connected limit set if and only if it is formally Kleinian.*

Cannon and Thurston [CT] have proved the following remarkable result. There are many doubly degenerate Kleinian groups which are known to be formally Kleinian. These are the fiber groups for hyperbolic 3-manifolds $\mathcal{M}$ which fiber over the circle. Equivalently, $\mathcal{M}$ is the mapping torus of an irreducible surface diffeomorphism f. These Kleinian groups are modelled on a Fuchsian group H and laminations of each of its components. The laminations are invariant under the diffeomorphism f (see Figure 19).

I should mention that regular b-groups are Formally Kleinian. They are geometrically finite but have cusp cylinders. The parabolic transformations covering the cylinders pull back to loxodromic elements in the model group. This is an alternative way of describing regular b-groups as having pinched loops—this was the original reason for studying them (see e.g. Abikoff [Abi75]). Here we are in the situation where the lamination $\mathcal{L}$ consists of simple closed curves on $\Omega(H)$ which determine non-trivial homotopy classes in $\pi_1\mathcal{M}(H)$. Formally Kleinian groups obtains by collapsing only simple closed curves on $\Omega(H)$ are called *rational.*

All the questions that we asked about singly degenerate groups may be rephrased in the context of formally Kleinian groups. In general we can expect, via R. L. Moore's theorem, that the underlying space X of a formally Kleinian group is a 2-sphere. We need only require that X be Hausdorff and that no point in X have a preimage which separates $\hat{\mathbf{C}}$.

Recently, several mathematicians have studied convergence groups. Informally, they are groups of homeomorphisms of a sphere which have many of the properties of Kleinian groups (see e.g. Gehring and Martin [GM]). It is not difficult to prove that a formally Kleinian group whose underlying space X is a 2-sphere is itself a convergence group. Martin and Skora [MSa] have shown that certain convergence groups are topologically conjugate to geometrically finite Kleinian groups.

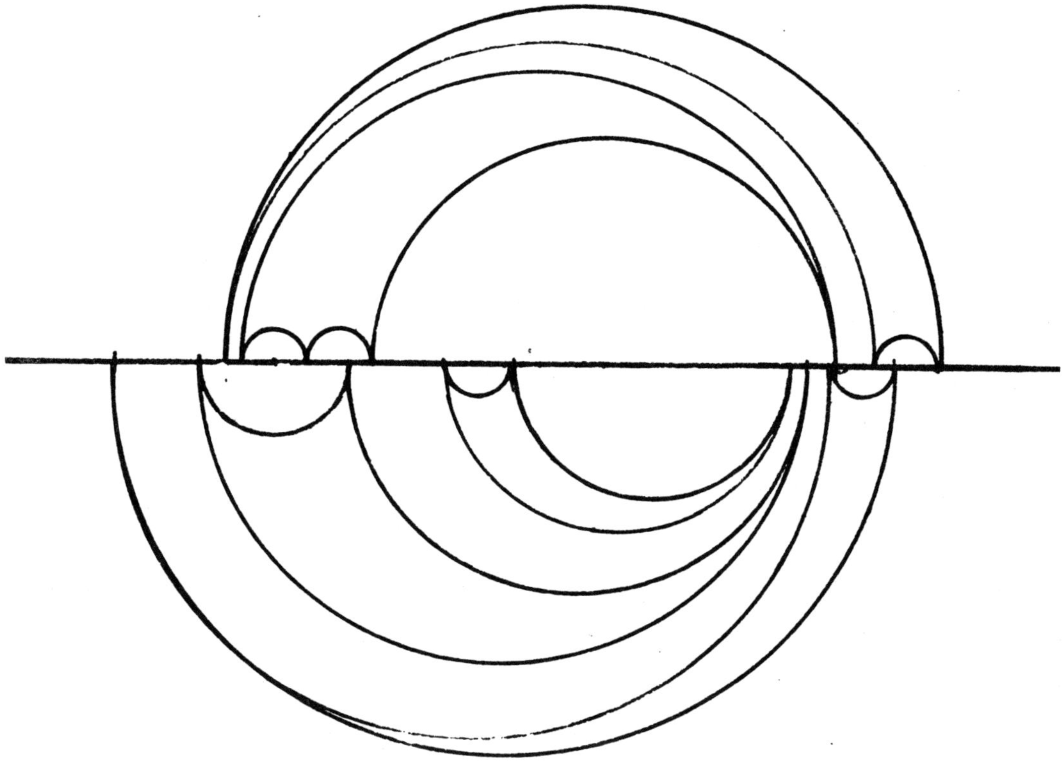

Figure 19: Laminations which collapse to form a doubly degenerate group

References

[Abi] W. Abikoff. To appear.

[Abi71] W. Abikoff. Some remarks on Kleinian groups. In *Advances in the Theorey of Riemann Surfaces*, pages 1–5, 1971.

[Abi75] W. Abikoff. On boundaries of teichmüller spaces and on Kleinian groups, iii. *Acta Math.*, 134:211–237, 1975.

[Abi76] W. Abikoff. Two theorems on totally degenerate kleinian groups. *Amer. J. of Math.*, 98:109–118, 1976.

[Abi80] W. Abikoff. *The real analytic theory of Teichmüller space.* Volume 820 of *Lecture Notes in Mathematics*, Springer-Verlag, 1980.

[Abi81] W. Abikoff. The uniformization theorem. *American Mathematical Monthly*, 88:574–592, 1981.

[Abi84] W. Abikoff. The bounded model for hyperbolic 3-space and a quaternionic uniformization theorem. *Math. Scand.*, 54:5–16, 1984.

[Abi86] W. Abikoff. The euler characteristic and inequalities for kleinian groups. *Proc. Amer. Math. Soc.*, 97:593–601, 1986.

[Ahl66a] L. Ahlfors. Fundamental polyhedrons and limit point sets of kleinian groups. *Proc. Nat. Acad. Sci. (USA)*, 55:251–254, 1966.

[Ahl66b] L. Ahlfors. *Lectures on quasiconformal mappings.* Van Nostrand, Princeton, 1966.

[Ahl66c] L. Ahlfors. Some remarks on kleinian groups. In *Proceedings of the Tulane Conference*, 1966.

[Bea83] A. Beardon. *Discrete Groups. Graduate Texts in Mathematics*, Springer-Verlag, 1983.

[Ber70] L. Bers. On boundaries of Teichmüller spaces and on Kleinian groups, I. *Ann. of Math.*, 91:570–600, 1970.

[Bon86] F. Bonahon. Bouts des variétés hyperboliques de dimension 3. *Ann. of Math.*, 124:71–158, 1986.

[Bro85] R. Brooks. Branched coverings of 3-manifolds which fiber over the circle. *Crelles J.*, 362:87–101, 1985.

[CT] J. Cannon and W. Thurston. To appear.

[DMS85] A. Miller D. McCullough and G. A. Swarup. Uniqueness of cores of non-compact 3-manifolds. *J. London Math. Soc.*, 32:548–556, 1985.

[FK97] R. Fricke and F. Klein. *Vorlesungen über die Theorie der automorphen Funktionen.* Volume I, B. G. Teubner Verlags, 1897.

[For51] L. R. Ford. *Autopmorphic Functions.* Chelsea Publishing Co., 1951.

[GM] F. Gehring and G. Martin. To appear.

[Gra84] J. J. Gray. Fuchs and the theory of differential equations. *Bulletin Amer. Math. Soc.*, 10:1–26, 1984.

[Gre66] L. Greenberg. Fundamental polyhedra for kleinian groups. *Ann. of Math.*, 84:433–441, 1966.

[Kle23] F. Klein. *Gesammelte mathematische Abhandlungen.* Volume II, Springer-Verlag, 1923.

[Kle63] F. Klein. *On Riemann's Theory of Algebraic Functions and their Integrals.* Dover Publications, 1963.

[Kra72] I. Kra. *Automorphic Forms and Kleinian Groups.* W. A. Benjamin, Inc., 1972.

[LV73] O. Lehto and K. Virtanen. *Quasiconformal mappings in the plane.* Springer-Verlag, 1973.

[Mar74a] A. Marden. Geometry of finitely generated kleinian groups. *Ann. of Math.*, 99:383–462, 1974.

[Mar74b] A. Marden. Schottky groups and circles. In *Contributions to Analysis*, pages 273–278, Academic Press, 1974.

[Mas71] B. Maskit. On poincaré's theorem for fundamental polygons. *Advances in Math.*, 7:219–230, 1971.

[Mas87] B. Maskit. *Kleinian Groups.* Springer-Verlag, 1987.

[MSa] G. Martin and R. Skora. To appear.

[MSb] J. Morgan and P. Shalen. To appear.

[Sco73] G. P. Scott. Compact submanifolds of 3-manifolds. *J. London Math. Soc.*, (2) 7:246–250, 1973.

[Sul81] D. Sullivan. On the ergodic theory at infinity of an arbitrary discrete group of hyperbolic motions. In *Ann. of Math. Studies, 97*, pages 465–496, 1981.

[Thu79] W. Thurston. *The geometry and topology of 3-manifolds.* Princeton University Math. Dept., 1979.

[Why42] G. Whyburn. *Analytic Topology.* Volume 28 of *Colloquium Publications*, Amer. Math. Soc., 1942.

Department of Mathematics
University of Connecticut
Storrs, CT 06268

Contemporary Mathematics
Volume **74**, 1988

THE REAL SPECTRUM COMPACTIFICATION OF TEICHMÜLLER SPACE

G. W. Brumfiel*

§1. Introduction. Let Σ denote a closed oriented surface of genus $g \geq 2$, and let T_g denote the Teichmüller space of marked hyperbolic structures on Σ. Points of T_g can be identified with equivalence classes of discrete, faithful representations $\rho : \pi_1(\Sigma) \to SL(2, \mathbf{R})$, where two such representations are equivalent if their induced $PSL(2, \mathbf{R}) = SL(2, \mathbf{R}) / \{\pm Id\}$ representations are conjugate.

It is known that T_g can be identified with a closed, semi-algebraic set in a real affine space, in fact, with a component of a real algebraic set [9], [15]. In modern real algebra, the notion of the real spectrum of a commutative ring, [1], leads to a natural compactification of any semi-algebraic set $W \subseteq \mathbf{R}^N$, that is, a compact Hausdorff space, which contains W as a dense subspace. If W is closed in $\mathbf{R}^N$ then W is open, as well as dense, in this compactification, which we will denote $\widetilde{W}$. In this paper, we study the compactification $\widetilde{T}_g$ of T_g.

Here are some properties of $\widetilde{T}_g$. First, there is a continuous surjection $\widetilde{T}_g \to \widehat{T}_g$, where $\widehat{T}_g$ is the Thurston compactification of T_g studied in [8]. This surjection is closely related to some work of J. Morgan and P. Shalen, [15]. The Morgan-Shalen paper, in fact, provided the motivation for all the work outlined in this paper, and will be discussed further below.

Secondly, the action of the outer automorphism group, $\mathrm{Out}(\pi_1(\Sigma))$, on T_g extends continuously to an action on $\widetilde{T}_g$, just as it does to an action on $\widehat{T}_g$, and the surjection $\widetilde{T}_g \to \widehat{T}_g$ is $\mathrm{Out}(\pi_1(\Sigma))$-equivariant. Moreover, each element of $\mathrm{Out}(\pi_1(\Sigma))$ has a fixed point in $\widetilde{T}_g$, so one can conclude the same about $\widehat{T}_g$, without first proving that the Thurston compactification of T_g is topologically a disk.

The continuous action of $\mathrm{Out}(\pi_1(\Sigma))$ on $\widetilde{T}_g$ is a consequence of the fact that the self-map of Teichmüller space T_g induced by an element of $\mathrm{Out}(\pi_1(\Sigma))$ is essentially a polynomial map, along with a routine functoriality property of the real spectrum of a ring. The fixed point property of $\widetilde{T}_g$ follows from a general fixed point theorem concerning self-maps of the real spectrum compactification of a semi-algebraic set, [2], [3], together with the fact that T_g is semi-algebraically a real affine space, $\mathbf{R}^{6g-6}$. Thus, certain desirable properties of $\widetilde{T}_g$

1980 **Mathematics Subject Classification** (1985 **Revision**), 32G15.

* Research supported by NSF grant DMS85–06816 at Stanford University.

hold for quite general reasons, and do not require any specific facts about representations or hyperbolic geometry to understand.

On the other hand, many interesting properties of $\widetilde{T}_g$ do require a closer look at the underlying representation theory. For example, it is possible to interpret points of the real spectrum boundary of Teichmüller space, that is, points of $\widetilde{B}_g = \widetilde{T}_g - T_g$, as equivalence classes of certain discrete, faithful representations $\rho : \pi_1(\Sigma) \to SL(2,F)$, where F is a real closed, non-Archimedean ordered field. The ordered field F can vary, which makes the precise definition of the appropriate equivalence relation a little complicated. Essentially, however, two such representations are equivalent if their induced $PSL(2,F)$ representations are conjugate. A precise statement is given in §6 below.

The nice fact to note here is that the compactification $\widetilde{T}_g$ is obtained from T_g by adding a boundary which consists of points very naturally related to the description of points of Teichmüller space itself. These points "at infinity" in $\widetilde{B}_g \subset \widetilde{T}_g$ correspond to certain representations $\rho : \pi_1(\Sigma) \to SL(2,F)$, where the trace in F of some matrix $\rho(\gamma) = \begin{pmatrix} a(\gamma) & b(\gamma) \\ c(\gamma) & d(\gamma) \end{pmatrix} \in SL(2,F)$, $\gamma \in \pi_1(\Sigma)$, is "infinitely large" relative to the real numbers $\mathbf{R} \subset F$. In the topology of $\widetilde{T}_g$, all such boundary points are limits of sequences of ordinary real representations $\rho_n : \pi_1(\Sigma) \to SL(2,\mathbf{R})$, where for some $\gamma \in \pi_1(\Sigma)$, $|\text{trace}(\rho_n(\gamma))| \to \infty$ as $n \to \infty$. (However, this convergence of sequences in T_g to points at infinity in the topology of $\widetilde{T}_g$ cannot be thought of as convergence of matrix entries in the fields $F \supset \mathbf{R}$.)

A representation $\rho : \pi_1(\Sigma) \to SL(2,F)$ can also be viewed as an action of $\pi_1(\Sigma)$ as a group of direct hyperbolic isometries of the hyperbolic plane HF^2 over the non-Archimedean ordered field F. The hyperbolic plane over F has as models both the upper half plane and the open unit ball in F^2, just as in classical hyperbolic geometry over $\mathbf{R}$. For a discrete, faithful representation, one can presumably make sense of something like a "fundamental domain" for the action of $\pi_1(\Sigma)$ on HF^2. This should be a finite-sided, geodesic polygon in HF^2 and should allow one to interpret boundary points of $\widetilde{T}_g$ as $(HF^2, PSL(2,F))$-geometric structures on a combinatorial surface Σ_F, built up using simplices over F, just as Σ is built up using $\mathbf{R}$-simplices. However, more work needs to be done to make this idea precise, and, hopefully, useful. For example, it definitely will not be the case that the $\pi_1(\Sigma)$-translates of such a "fundamental domain" will cover HF^2, and the orbit space $HF^2 / \pi_1(\Sigma)$ will not be as well-behaved as it is in the classical case over $\mathbf{R}$.

The relation between the real spectrum compactification, $\widetilde{T}_g$, and the Thurston compactification, $\widehat{T}_g$, of Teichmüller space arises from the following considerations. Let $S \subset \pi_1(\Sigma)$ be a subset which contains a representative of each conjugate class homotopic to a simple closed curve on Σ. Using lengths of simple closed geodesic curves on hyperbolic surfaces, Thurston, [8], studied an embedding of T_g in a variant of an infinite dimensional projective space over $\mathbf{R}$, specifically, an embedding of T_g in $\mathbf{P}^S = [0,\infty)^S - \vec{0} \big/ \sim$, where $\sim$ means all positive scaler multiples of non-zero S-tuples are identified. If $\gamma \in S$, then Thurston's geodesic curve length can be replaced by the closely related number $\log\,(|\text{trace}(\rho(\gamma))|)$, where $\rho: \pi_1(\Sigma) \to SL(2,\mathbf{R})$ represents a point of Teichmüller space T_g. That is, define $\theta: T_g \to \mathbf{P}^S$ by $\theta([\rho]) = (\log(|\text{trace}(\rho(\gamma))|))_{\gamma\in S} \in \mathbf{P}^S$. The Thurston compactification $\widehat{T}_g$ can be defined as the closure of $\theta(T_g)$ in $\mathbf{P}^S$.

Now, it turns out that for the non-Archimedean ordered fields F which arise in the description of the boundary points of the real spectrum compactification $\widetilde{T}_g$ of Teichmüller space, there is a very easily defined *real valued logarithm*, $\log: F^+ \to \mathbf{R}$, well-defined up to positive real scaler multiple. Moreover, if $\rho: \pi_1(\Sigma) \to SL(2,F)$ is a representation which defines a point of the boundary $\widetilde{B}_g = \widetilde{T}_g - T_g$ then the same formula, $\widetilde{\theta}([\rho]) = (\log(|\text{trace}(\rho(\gamma))|))_{\gamma\in S} \in \mathbf{P}^S$, extends the (perturbed) Thurston map $\theta: T_g \to \mathbf{P}^S$ above, *continuously* to $\widetilde{\theta}: \widetilde{T}_g \to \mathbf{P}^S$. It follows from compactness of $\widetilde{T}_g$ and density of T_g in $\widetilde{T}_g$ that $\widetilde{\theta}(\widetilde{T}_g) = \widehat{T}_g$.

In the paper of Morgan and Shalen, [15], which led to our study of $\widetilde{T}_g$, the Thurston boundary $\widehat{B}_g \subset \mathbf{P}^S$ of $\widehat{T}_g$ is shown to be a continuous image of a certain algebro-geometrically defined space of Krull valuations, defined on function fields associated to the real algebraic set X_g which contains Teichmüller space as a component. Our non-Archimedean fields F are exactly the real closures of function fields of subvarieties of X_g, on which an order is imposed. Moreover, the map $v: F^* \to \mathbf{R}$, $v(a) = -\log(|a|) \in \mathbf{R}$, where log is the non-Archimedean logarithm above, turns out to be a rank 1, Krull valuation, of essentially the same type as those which are constructed by Morgan and Shalen. It is certainly possible to translate many of our statements into the Morgan-Shalen context, and vice-versa. However, we will not do this, as it would add considerably to the length of the discussion. Despite the fact that our motivation was to understand the approach of Morgan and Shalen, our treatment can be followed without any knowledge of their work, or any expertise with valuations.

The surjection $\widetilde{\theta}: \widetilde{B}_g \to \widehat{B}_g$ defined above, between the two boundaries of Teichmüller space, is clearly a kind of forgetful map. Given an equivalence class of non-Archimedean representations $\rho: \pi_1(\Sigma) \to SL(2,F)$, the image $\widetilde{\theta}([\rho]) \in \widehat{B}_g \subset \mathbf{P}^S$ just keeps track of the projectivized S-tuple of real numbers $(\log(|\text{trace}(\rho(\gamma))|))_{\gamma\in S}$. Morgan and Shalen interpret these numbers as translation lengths assigned to elements of $\pi_1(\Sigma)$ acting as isometries of a certain metric $\mathbf{R}$-tree, [15]. In fact, these translation length numbers determine an action of $\pi_1(\Sigma)$ on a suitable $\mathbf{R}$-tree, up to conjugacy, so the Thurston boundary points can be interpreted as equivalence classes of certain actions of $\pi_1(\Sigma)$ on $\mathbf{R}$-trees. The Morgan-Shalen $\mathbf{R}$-trees arise from their valued function fields, by certain algebraic constructions.

In our approach, we obtain a more direct hyperbolic geometric intepretation of these same $\mathbf{R}$-trees. Specifically, we first define an F-valued, cross-ratio type "multiplicative distance" on our non-Archimedean hyperbolic planes, $D: HF^2 \times HF^2 \to F^+$. It is then easy to see that the formula $d(p,q) = \log(D(p,q)) \in \mathbf{R}$ defines a real valued *topological pseudo-metric* on HF^2. It turns out that the associated metric space quotient of HF^2 is a metric $\mathbf{R}$-tree, TF^2, [4]. (More precisely, TF^2 is a Λ-tree, where $\Lambda = \log(F^+) \subset \mathbf{R}$ is an additive subgroup.)

Thus, the familiar "log of the cross-ratio" distance formula of classical hyperbolic geometry has a very neat non-Archimedean analogue which yields **R**-trees. The map between boundaries of Teichmüller space, $\widetilde{\theta} : \widetilde{B}_g \to \widehat{B}_g$, can be interpreted as the forgetful map, which assigns to an equivalence class of actions of $\pi_1(\Sigma)$ as a group of hyperbolic isometries of a non-Archimedean hyperbolic plane HF^2, the induced action of $\pi_1(\Sigma)$ as a group of metric isometries of the quotient **R**-tree, TF^2.

It is plausible that our methods can be used to reprove certain results about the Thurston compactification of Teichmüller space. For example, if $B \subset \widetilde{B}_g$ is any dense subset, then the Thurston boundary $\widehat{B}_g \subset \mathbf{P}^S$ is obviously the closure of $\widetilde{\theta}(B) \subset \mathbf{P}^S$. A particularly easy subset of $\widetilde{B}_g$ to understand consists of the set of points represented by $\rho : \pi_1(\Sigma) \to SL(2, F)$, where F has transcendence degree one over **R**. Such points, defined by real curves, are easily seen to be dense in $\widetilde{B}_g$. The **R**-tree associated to the hyperbolic plane over an ordered function field of a real curve is an ordinary simplicial tree, that is, a **Z**-tree. This reproves one of the results about the Thurston compactification, also recovered by Morgan and Shalen by valuation theoretic methods, namely, that the *integral* points in $\widehat{B}_g \subset \mathbf{P}^S$ are dense. In the Thurston approach, these integral points correspond to a very simple kind of projective measured lamination on Σ, essentially just a finite system of curves.

The paper is organized as follows. In §2, we give the definition and basic properties of the real spectrum of a commutative ring. In §3, we specialize to affine coordinate rings of real algebraic sets, and define the relevant compactifications of semi-algebraic sets. Further properties of these compactifications, and a rather geometric interpretation of the added boundary points, are given in §4. These three sections contain few proofs, and thus do not really provide a course on the real spectrum. However, we do discuss more than the minimum necessary for the rest of the paper, because of our belief that the real spectrum looks like a promising tool for other problems in the general area of geometry of group representations.

In §5, we introduce non-Archimedean logarithms, and use them to map our real spectrum compactifications of semi-algebraic sets to various real projective spaces. Teichmüller space is interpreted as a semi-algebraic set in §6, so that all the previously developed theory of compactifications can be applied. We also give our interpretation of boundary points of Teichmüller space in §6. In §7, we compare the real spectrum compactification with the Thurston compactification, and discuss the action of the mapping class group. Finally, in §8, we outline how **R**-trees arise as quotients of non-Archimedean hyperbolic planes.

§2. Preliminaries on the real spectrum. Let A be a commutative ring with identity. In this section, we will summarize the definition and basic properties of a topological space, $\mathrm{Spec}_R(A)$, called the *real spectrum* of A. For details, we must refer to other sources, such as [1] and [12]. But we hope that our outline makes it clear how the real spectrum is a very natural tool in real algebraic geometry, quite analogous to the Zariski spectrum in algebraic geometry. The real spectrum has been introduced surprisingly recently, around 1978, by M.-F. Coste-Roy and M. Coste [5], [6], who were motivated by topos theoretic considerations. (A much earlier paper of Krivine, [11], contains a somewhat related construction.)

There are various descriptions of the points of $\text{Spec}_R(A)$. Perhaps the simplest is to define a point $\alpha \in \text{Spec}_R(A)$ to be a prime ideal $p(\alpha) \subset A$, together with a total order on the field of fractions, $K(\alpha)$, of $A/p(\alpha)$. (By this we mean, of course, an ordered field structure.) If $f \in A$, write $f(\alpha)$ for the image of f in $A/p(\alpha) \subseteq K(\alpha)$. Then exactly one of the statements $f(\alpha) > 0$, $f(\alpha) = 0$, $f(\alpha) < 0$, holds in the totally ordered field $K(\alpha)$. We define a topology on $\text{Spec}_R(A)$ by choosing as a subbasis for the open sets, all the sets $U(f) = \{\alpha \in \text{Spec}_R(A) \mid f(\alpha) > 0\}$, $f \in A$. This topology will be called the *real spectrum topology.*

Every ordered field K has a *real closure* F, which is unique up to unique order preserving isomorphism over K. Among the many characterizations of the real closure F of K is the following. F is an algebraic extension of K, with an order which extends the order on K. Moreover, every positive element of F is a square in F (so F has a unique ordered field structure), and every polynomial of odd degree over F has a root in F.

This leads to a second interpretation of a point $\alpha \in \text{Spec}_R(A)$ as a homomorphism $\alpha : A \to F(\alpha)$, where $F(\alpha)$ is a real closed field, algebraic over the subring $\alpha(A) \subseteq F(\alpha)$. Two such homomorphisms, $\alpha : A \to F$ and $\alpha' : A \to F'$, define the same point if there is an order preserving isomorphism $\varphi : F \xrightarrow{\sim} F'$ such that $\alpha' = \varphi \circ \alpha$. If φ exists, it is unique, since φ is determined by α' on $\alpha(A) \subseteq F$.

At this point, one can see clearly that if $X \subseteq \mathbf{R}^N$ is a real algebraic set, with affine coordinate ring $A(X) = \mathbf{R}[x_1 \ldots x_N]/I(X)$, then $X \subset \text{Spec}_R(A(X))$. Namely, a point $\vec{x} \in X \subseteq \mathbf{R}^N$ is the same as a homomorphism $A(X) \to \mathbf{R}$. Moreover, the real spectrum topology induces the usual real topology on X, since $X \cap U(f) = \{\vec{x} \in X \mid f(\vec{x}) > 0 \in \mathbf{R}\}$, for $f \in A(X)$.

The compactification of X which we are after will be the subspace of closed points, $\text{Spec}_R^m(A(X)) \subset \text{Spec}_R(A(X))$. In order to prove that for any ring A, $\text{Spec}_R(A)$ is quasi-compact and $\text{Spec}_R^m(A) \subset \text{Spec}_R(A)$ is compact and Hausdorff, it is very useful to have one additional interpretation of the points of $\text{Spec}_R(A)$. We will identify each point $\alpha \in \text{Spec}_R(A)$ with a subset of the ring A. This defines an embedding $\text{Spec}_R(A) \subset 2^A$, where 2^A is the power set of A. Then one checks that $\text{Spec}_R(A)$ is a *closed* subspace of 2^A in the product, or Tychonoff, topology on 2^A. The induced topology on $\text{Spec}_R(A)$, which is called the *constructible topology,* is thus compact and Hausdorff. It is an easy observation that the real spectrum topology on $\text{Spec}_R(A)$ defined above is weaker than the constructible topology, so $\text{Spec}_R(A)$ is quasi-compact in the real spectrum topology.

Before giving more details, we will point out that there are other examples of this same method for obtaining quasi-compact spaces in algebraic geometry. First, recall that the standard subbase for the open sets of the Tychonoff topology on 2^A is given by all the sets $W(f) = \{\alpha \subset A \mid f \in \alpha\}$ and $W'(f) = \{\beta \subset A \mid f \notin \beta\}$, $f \in A$. Two natural weaker topologies are thus defined by taking either of the families $\{W(f) \mid f \in A\}$ or $\{W'(f) \mid f \in A\}$ separately as a subbase.

The Zariski spectrum of a ring, $\mathrm{Spec}(A)$, whose points are the prime ideals $p \subset A$, can obviously be interpreted as a subset of 2^A. A subbase for the open sets in the Zariski topology on $\mathrm{Spec}(A)$ is given by the family of sets $D(f) = \{p \in \mathrm{Spec}(A) \mid f \notin p\}$, $f \in A$. These are obviously just the sets $W'(f) \cap \mathrm{Spec}(A)$.

As another example, if F is a field, the abstract Riemann surface, $S(F)$, of all valuation rings $\mathcal{O} \subset F$, can be interpreted as a subset of 2^F. The appropriate topology in this case, also called the Zariski topology, has as a subbase the family of sets $V(f) = \{\mathcal{O} \in S(F) \mid f \in \mathcal{O}\}$, $f \in F$. Clearly, these are just the sets $W(f) \cap S(F)$.

We will now define the embedding $\mathrm{Spec}_R(A) \subset 2^A$. To each point $\alpha : A \to F(\alpha)$ of $\mathrm{Spec}_R(A)$, we assign the subset of A consisting of all elements with non-negative image in $F(\alpha)$. To avoid introducing too much notation, we will take some liberties and also call this subset $\alpha \subset A$, that is, $\alpha = \{g \in A \mid g(\alpha) \geq 0 \in F(\alpha)\} \in 2^A$. Clearly, the subbasic open set $U(f) = \{\alpha \in \mathrm{Spec}_R(A) \mid f(\alpha) > 0\} \subset \mathrm{Spec}_R(A)$ is just $W'(-f) \cap \mathrm{Spec}_R(A)$, since $f(\alpha) > 0$ translates to $-f \notin \alpha \subset A$.

To see that $\mathrm{Spec}_R(A) \subset 2^A$ is closed in the Tychonoff topology, we need a list of properties of subsets $\alpha \subset A$ which characterize the points $\alpha \in \mathrm{Spec}_R(A)$.

Proposition 2.1. *A subset $\alpha \subset A$ belongs to $\mathrm{Spec}_R(A) \subset 2^A$ if and only if the following properties hold:*

(i) $x, y \in \alpha \Rightarrow x + y \in \alpha$ *and* $x \cdot y \in \alpha$

(ii) $-1 \notin \alpha$

(iii) $\alpha \cup -\alpha = A$

(iv) $\alpha \cap -\alpha \subset A$ *is a prime ideal.* ∎

One checks now that these properties are closed conditions on subsets $\alpha \subset A$ in the Tychonoff topology on 2^A. (Actually, it is easiest to check that the negation of each property is an open condition.) This same method can also be used to prove that the Zariski spectrum of a ring and the abstract Riemann surface of a field are compact in their constructible topology, which, as we have observed, is stronger than their respective Zariski topologies.

Next, we need to describe the closed points of $\mathrm{Spec}_R(A)$. We write $\alpha \to \beta$, and say β *specializes* α, if $\beta \in \overline{\{\alpha\}}$, that is, if β is in the closure of the point α in the real spectrum topology. It is a triviality to check that $\alpha \to \beta$ if and only if $\alpha \subseteq \beta$, when α and β are regarded as subsets of A, that is, if $g(\alpha) \geq 0$ implies $g(\beta) \geq 0$, for $g \in A$. The following proposition summarizes the basic facts about specializations and the subspace of closed points $\mathrm{Spec}_R^m(A) \subset \mathrm{Spec}_R(A)$.

Proposition 2.2. (a) *If $\alpha \to \beta$ and $\alpha \to \gamma$ in $Spec_R(A)$, then either $\beta \to \gamma$ or $\gamma \to \beta$. That is, the specializations of one point $\alpha \in Spec_R(A)$ form a totally ordered chain of specializations.*

(b) *For each $\alpha \in Spec_R(A)$, there exists a unique closed point $\beta \in Spec_R^m(A)$ with $\alpha \to \beta$. In particular, any open cover of $Spec_R^m(A)$ covers all of $Spec_R(A)$, hence $Spec_R^m(A)$ is quasi-compact.*

(c) *If $\alpha \not\to \alpha'$ and $\alpha' \not\to \alpha$ then there exists $f \in A$ with $f(\alpha) > 0$ and $f(\alpha') < 0$. In particular, $Spec_R^m(A)$ is Hausdorff.*

(d) *The map $r: Spec_R(A) \to Spec_R^m(A)$, defined by $r(\alpha) = \beta$, where β is the unique closed point with $\alpha \to \beta$, is a continuous retraction of $Spec_R(A)$ onto $Spec_R^m(A)$.*

(e) *A point $\alpha \in Spec_R(A)$ is closed if and only if, when α is regarded as a homomorphism $\alpha: A \to F(\alpha)$, then every element of the ordered field $F(\alpha)$ is bounded by an element of the subring $\alpha(A) \subseteq F(\alpha)$.* ∎

We will not give the proof of these important facts here, although they are not difficult. Details can be found in [1] and [12]. It is the remarkable property 2.2(a) which seems to be the main underlying reason for the "good" topological properties of $\mathrm{Spec}_R^m(A)$. Note that if $A = A(X)$, where $X \subseteq \mathbf{R}^N$ is a real algebraic set, then 2.2(e) implies that points of $X \subset \mathrm{Spec}_R(A)$ are closed points, since any homomorphism $A(X) = \mathbf{R}[x_1 \ldots x_N]/I(X) \to \mathbf{R}$ is clearly surjective.

The real spectrum is a contravariant functor from the category of commutative rings to the category of topological spaces. That is, if $\varphi: A \to B$ is a ring homomorphism and $\beta \subset B$ defines a point of $\mathrm{Spec}_R(B)$, then $\alpha = \varphi^{-1}(\beta) \subset A$ defines a point of $\mathrm{Spec}_R(A)$ and the induced map $\varphi^*: \mathrm{Spec}_R(B) \to \mathrm{Spec}_R(A)$ is easily seen to be continuous. In general, these maps φ^* do not preserve closed points. However, if $\varphi: A \xrightarrow{\sim} B$ is a ring isomorphism (e.g. an automorphism of one ring), then the induced map $\varphi^*: \mathrm{Spec}_R(B) \xrightarrow{\sim} \mathrm{Spec}_R(A)$ is a homeomorphism, hence preserves closed points. But, even more generally, the continuous retractions $r: \mathrm{Spec}_R(A) \to \mathrm{Spec}_R^m(A)$ of 2.2(d) can be composed with the maps $\varphi^*\big|_{\mathrm{Spec}_R^m(B)}$ induced by $\varphi: A \to B$, and, in this way, $\mathrm{Spec}_R^m(\cdot)$ is also a functor from rings to compact, Hausdorff spaces.

§3. **Compactifications of semi-algebraic sets.** If $X \subseteq \mathbf{R}^N$ is a real algebraic set and if $A(X)$ is the affine coordinate ring of X, then we have seen in the previous section that $X \subset \mathrm{Spec}_R^m(A(X)) \subset \mathrm{Spec}_R(A(X))$. Since $\mathrm{Spec}_R^m(A(X))$ is a compact, Hausdorff space, it follows that for any subspace $W \subseteq X$, the closure $\widetilde{W}$ of W in $\mathrm{Spec}_R^m(A(X))$ is a compact, Hausdorff space, which contains W as a dense subspace. In this section, we want to summarize some further properties of these compactifications $\widetilde{W}$, when W is a *closed, semi-algebraic subset* of X. (For more general semi-algebraic subsets $Y \subset X$ there is a "better" compactification $Y^m \subset \mathrm{Spec}_R(A(X))$, which we will also mention briefly below.)

We recall that a *semi-algebraic* subset of $\mathbf{R}^N$ is a subset which is a finite union of sets defined by finitely many polynomial inequalities $f_i(\vec{x}) > 0$ and equalities $g_j(\vec{x}) = 0$, $\vec{x} \in \mathbf{R}^N$, $f_i, g_j \in \mathbf{R}[x_1 \ldots x_N]$. Now, the key observation is that if $f, g \in A(X) = \mathbf{R}[x_1 \ldots x_N]/I(X)$, then an inequality $f > 0$ or an equality $g = 0$, which define subsets of X, also define subsets of $\text{Spec}_R(A(X))$, specifically $U(f) = \{\alpha \in \text{Spec}_R(A(X)) \mid f(\alpha) > 0\}$ and $Z(g) = \{\alpha \in \text{Spec}_R(A(X)) \mid g(\alpha) = 0\}$. Clearly, $X \cap U(f) = \{\vec{x} \in X \subset \mathbf{R}^N \mid f(\vec{x}) > 0\}$ and $X \cap Z(g) = \{\vec{x} \in X \subset \mathbf{R}^N \mid g(\vec{x}) = 0\}$.

If A is any ring, the subsets of $\text{Spec}_R(A)$ which are finite unions of finite intersections of the subsets $U(f)$ and $Z(g)$ of $\text{Spec}_R(A)$ are called ***constructible subsets*** of $\text{Spec}_R(A)$. It is easy to see from compactness of $\text{Spec}_R(A) \subset 2^A$ in the Tychonoff topology that the constructible sets are precisely the sets which are both closed and open (clopen) in the Tychonoff topology. In particular, all constructible sets are quasi-compact in the real spectrum topology on $\text{Spec}_R(A)$.

If $A = A(X)$ is the affine coordinate ring of a real algebraic set X then we have seen above that there is a natural map from the collection of constructible subsets of $\text{Spec}_R(A(X))$ *onto* the collection of semi-algebraic subsets of X. Namely, to $C \subseteq \text{Spec}_R(A(X))$, assign $X \cap C \subseteq X$. A theorem of Artin and Lang esentially asserts that this correspondence is *bijective*.

Theorem 3.1. (Artin-Lang) *Suppose* $\{f_i\}$ *and* $\{g_j\}$ *are finite collections of elements of* $A(X) = \mathbf{R}[x_1 \ldots x_N]/I(X)$. *If there exists some ordered field* F *and a homomorphism* $\alpha : A(X) \to F$ *such that* $f_i(\alpha) > 0$ *and* $g_j(\alpha) = 0$, *all* i, j, *then there exists a point* $\vec{x} \in X \subset \mathbf{R}^N$ *such that* $f_i(\vec{x}) > 0$ *and* $g_j(\vec{x}) = 0$, *all* i, j. ∎

To see that this theorem implies that the correspondence between constructible sets $C \subseteq \text{Spec}_R(A(X))$ and semi-algebraic sets $X \cap C \subseteq X$ is bijective, observe that this correspondence is obviously inclusion preserving, and preserves unions and intersections, so it suffices to prove that $C \neq \emptyset$ implies $X \cap C \neq \emptyset$. But this is *exactly* what the Artin-Lang Theorem says, since a point $\alpha \in \text{Spec}_R(A(X))$ is a homomorphism $\alpha : A(X) \to F(\alpha)$ and C is described by finite collections of inequalities $f_i > 0$ and equalities $g_j = 0$.

Another consequence of the Artin-Lang Theorem is the conclusion that $X \cap C$ is always *dense* in C, for any constructible set $C \subseteq \text{Spec}_R(A(X))$. Namely, a basic open neighborhood of $\alpha \in C$ is defined by a set $U\{f_i\} = \bigcap_{i=1}^{m} U(f_i)$, where $f_i(\alpha) > 0$. Then $C \cap U\{f_i\}$ is constructible, and the Artin-Lang Theorem implies that $X \cap C \cap U\{f_i\} \neq \emptyset$, since $\alpha \in C \cap U\{f_i\} \neq \emptyset$.

One proof of the Artin-Lang Theorem amounts to observing that it is a trivial corollary of the Tarski-Seidenberg Principle (elimination of quantifiers) for real closed fields, [7]. In his graduate algebra text, [13], Lang gives another proof by induction on the dimension of X, beginning with special arguments for algebraic curves.

Our goal for the remainder of this section is to establish the following description of the compactification $\widetilde{W}$ of a closed, semi-algebraic subset $W \subseteq X$, and its "boundary" $\widetilde{W} - W$.

Proposition 3.2. (a) *Let $W \subseteq X$ be a closed, semi-algebraic subset and let $C(W) \subseteq \mathrm{Spec}_R(A(X))$ be the associated constructible subset. Then the closure, $\widetilde{W}$, of W in $\mathrm{Spec}^m_R(A(X))$ is given by $\widetilde{W} = C(W) \cap \mathrm{Spec}^m_R(A(X))$.*

(b) *The points of the boundary $\widetilde{B}(W) = \widetilde{W} - W$ are exactly those points $\alpha : A(X) = \mathbf{R}[x_1 \dots x_N]\big/ I(X) \to F(\alpha)$ with $\alpha \in \widetilde{W}$, which satisfy $\Sigma x_i^2(\alpha) - r > 0 \in F(\alpha)$, for all $r \in \mathbf{R}$. (Equivalently, some $|x_i(\alpha)| \in F(\alpha)$ is infinitely large relative to $\mathbf{R} \subset F(\alpha)$.) In particular, W is open in $\widetilde{W}$.* ∎

The main result needed to establish 3.2(a) is a very useful refinement of the Artin-Lang Theorem, which is called the Finiteness Theorem. One form of the Finiteness Theorem is the assertion that the correspondence above between constructible subsets $C \subseteq \mathrm{Spec}_R(A(X))$ and semi-algebraic subsets $X \cap C \subseteq X$, also preserves open and closed sets. That is, a constructible set C is open in $\mathrm{Spec}_R(A(X))$ if and only if $X \cap C$ is open in X. Here is another formulation of the Finiteness Theorem, which explains the name.

Theorem 3.3. (Finiteness Theorem) *Suppose $V \subseteq X$ is a semi-algebraic subset which is open in X. Then V is a finite union of basic open sets, $V = \bigcup\limits_{j=1}^{n} X \cap U\{f_{ij}\}$, where $X \cap U\{f_{ij}\} = \{\vec{x} \in X \mid f_{ij}(\vec{x}) > 0,\ 1 \leq i \leq m_j\}$.* ∎

The relation between the two formulations is seen as follows. Let $C(V) \subseteq \mathrm{Spec}_R(A(X))$ be the constructible set corresponding to $V \subseteq X$. The Finiteness Theorem 3.3, along with the Artin-Lang Theorem 3.1, immediately implies that if $V \subseteq X$ is open then $C(V)$ is a finite union of basic open sets $U\{f_{ij}\} \subseteq \mathrm{Spec}_R(A(X))$, hence $C(V)$ is open. Conversely, if it is known that V open implies $C(V)$ is open, then quasi-compactness of $C(V)$ implies that $C(V)$ is a finite union of basic open sets $U\{f_{ij}\} \subseteq \mathrm{Spec}_R(A(X))$, and hence $V = X \cap C(V)$ is a finite union of basic open sets $X \cap U\{f_{ij}\}$.

The Finiteness Theorem is discussed in both [1] and [12]. There are several proofs, of one formulation or the other, which can be found in papers referred to in [1] and [12].

Proof of 3.2. (a) If $W \subseteq X$ is a closed, semi-algebraic subset, then, by the Finiteness Theorem, $C(W) \subseteq \mathrm{Spec}_R(A(X))$ is closed. Thus, $C(W) \cap \mathrm{Spec}^m_R(A(X))$ is closed in $\mathrm{Spec}^m_R(A(X))$. Since we already know that W is dense in $C(W)$, we conclude that $\widetilde{W} = C(W) \cap \mathrm{Spec}^m_R(A(X))$. ∎

Proof of 3.2. (b) Suppose $\alpha : A(X) = \mathbf{R}[x_1 \dots x_N]\big/ I(X) \to F(\alpha)$ represents a point $\alpha \in \widetilde{W}$, and suppose all $|x_i(\alpha)| \in F(\alpha)$, $1 \leq i \leq N$, are bounded by real numbers. Then every element of the subring $\alpha(A(X)) \subseteq F(\alpha)$ is bounded by a real number. But, since $\alpha \in \mathrm{Spec}^m_R(A(X))$ is a closed point, Proposition 2.2(e) then implies that every element in

$F(\alpha)$ is bounded by a real number, and $F(\alpha)$ would thus be an Archimedean ordered field containing $\mathbf{R}$. But this means $F(\alpha) = \mathbf{R}$, so $\alpha \in \widetilde{W}$ is actually a point of $X \cap C(W) = W$. ∎

Remark 3.4. Let us summarize our principle conclusions. If $X \subseteq \mathbf{R}^N$ is a real algebraic set and if $W \subseteq X$ is a closed semi-algebraic subset, then the Finiteness Theorem (applied to the open set $X - W$) implies that W is a finite union of basic closed sets, $W = \bigcup_{j=1}^{n} X \cap W\{g_{ij}\}$, where $X \cap W\{g_{ij}\} = \{\vec{x} \in X \mid g_{ij}(\vec{x}) \geq 0, \quad 1 \leq i \leq m_j\}$. The compactification $\widetilde{W}$ is the union $\bigcup_{j=1}^{n} \widetilde{W}_j$, where $W_j = X \cap W\{g_{ij}\}$. The points of $\widetilde{W}_j \subset \mathrm{Spec}_R(A(X))$ are exactly the homomorphisms $\alpha :\ A(X) = \mathbf{R}[x_1 \dots x_N] \big/ I(X) \to F(\alpha)$ with the following properties: (i) $g_{ij}(\alpha) \geq 0, \quad 1 \leq i \leq m_j$, and (ii) either $F(\alpha) = \mathbf{R}$ or for some x_k, $1 \leq k \leq N$, $|x_k(\alpha)|$ is infinitely large relative to $\mathbf{R}$ and every element of $F(\alpha)$ is bounded by some power of $|x_k(\alpha)|$. The points of the boundary $\widetilde{B}(W) = \widetilde{W} - W$ are those points $\alpha \in \widetilde{W}$ such that some $|x_k(\alpha)|$ is infinitely large relative to $\mathbf{R}$, that is, such that $F(\alpha)$ is non-Archimedean. The space W is open and dense in the compact, Hausdorff space $\widetilde{W}$.

Remark 3.5. Suppose $Y \subset X$ is a semi-algebraic subset which is not closed. Then $\widetilde{Y}$, the closure of Y in $\widetilde{X} = \mathrm{Spec}_R^m(A(X))$, is certainly a compact, Hausdorff space, containing Y as a dense subset. But there is a better compactification, Y^m, of Y. As above, let $C(Y) \subset \mathrm{Spec}_R(A(X))$ denote the constructible subset corresponding to $Y \subset X$. Define $Y^m \subset C(Y)$ to be the set of points of $C(Y)$ which are *closed in* $C(Y)$. Such points need not be closed in $\mathrm{Spec}_R(A(X))$. However, the arguments behind the proof of Proposition 2.2, which shows that $\mathrm{Spec}_R^m(A(X))$ is a compact, Hausdorff space, can be extended to prove that $Y^m \subset \mathrm{Spec}_R(A(X))$ is a compact, Hausdorff space for any semi-algebraic $Y \subset X$. If $W \subseteq X$ is closed, then $W^m = \widetilde{W}$, since $C(W) \subseteq \mathrm{Spec}_R(A(X))$ is closed. We will see in the next section that for all semi-algebraic sets Y, the compact, Hausdorff space Y^m depends only on the semi-algebraic homeomorphism type of Y, not on the particular embeddings $Y \subset X \subset \mathbf{R}^N$. ∎

Remark 3.6. If $W \subseteq \mathbf{R}^N$ is a closed, semi-algebraic set then the algebraic topology of the compactification $\widetilde{W}$ is essentially the same as the algebraic topology of W. For example, the connected components of W and $\widetilde{W}$ coincide, [1]. More generally, such results as the Artin-Lang Theorem 3.1, the Finiteness Theorem 3.3, and the existence of triangulations of finite collections of semi-algebraic sets, can be used to prove that the Čech homology of $\widetilde{W}$, coincides with the homology of W. Also, the Čech homology of $\widetilde{B}(W) = \widetilde{W} - W$ coincides with the homology of $W \cap (\mathbf{R}^N - B_r) = W - W \cap B_r$ for all sufficiently large r, where $B_r = \{\vec{x} \in \mathbf{R}^N \mid \|\ \vec{x}\ \| < r\}$ is the open ball of radius r at the origin in $\mathbf{R}^N$. Similar statements hold for the K-theory of $\widetilde{W}$ and $\widetilde{B}(W)$. The point is that finite, constructible, open covers of $\widetilde{W}$, which are cofinal among all open covers, are the same thing as finite, semi-algebraic, open covers of W. ∎

§4. Ultrafilters of semi-algebraic sets. In this section, we give another interpretation of points of $\mathrm{Spec}_R(A(X))$, where $X \subseteq \mathbf{R}^N$ is a real algebraic set. This interpretation is more geometric than the descriptions of points of $\mathrm{Spec}_R(A)$ given for general rings A in §2, hence provides added insight into the real spectrum compactifications of semi-algebraic sets. For example, we use the new interpretation to show that the assignment of the constructible sets $C(Y) \subseteq \mathrm{Spec}_R(\mathbf{R}[x_1 \dots x_N])$ to affine semi-algebraic sets $Y \subseteq \mathbf{R}^N$ is actually a *functor* from the category of affine semi-algebraic sets and continuous semi-algebraic maps to the category of quasi-compact spaces. In particular, the space $C(Y)$, and the compact, Hausdorff space of closed points $Y^m \subset C(Y)$, mentioned in Remark 3.5, depend only on the semi-algebraic homeomorphism type of Y, and not on the particular affine embedding $Y \subseteq \mathbf{R}^N$. We also describe a useful dense set of points in the boundary $\widetilde{B}(W) = \widetilde{W} - W$ when W is a closed semi-algebraic set, in terms of branches of real algebraic curves at infinity in W.

If $\mathcal{C}$ is a family of subsets of some fixed set S, then a *filter* in $\mathcal{C}$ is a non-empty subfamily $\mathcal{F} \subset \mathcal{C}$ such that (i) $\emptyset \notin \mathcal{F}$, (ii) $Y_1, Y_2 \in \mathcal{F}$ implies $Y_1 \cap Y_2 \in \mathcal{F}$, (in particular, $Y_1 \cap Y_2 \neq \emptyset$), and (iii) $Y_1 \in \mathcal{F}$, $Y_1 \subseteq Y_2$ implies $Y_2 \in \mathcal{F}$. An *ultrafilter* in $\mathcal{C}$ is a filter which is not properly contained in any other filter, that is, a maximal filter. Zorn's Lemma implies that every filter is contained in an ultrafilter. If the complement $Y' = S - Y$ of each set $Y \in \mathcal{C}$ also belongs to $\mathcal{C}$, then a filter $\mathcal{F}$ is an ultrafilter if and only if for each $Y \in \mathcal{C}$, either $Y \in \mathcal{F}$ or $Y' \in \mathcal{F}$. Also in this case, if $y \in S$ is a single element then the family of subsets $\mathcal{F}_y = \{Y \in \mathcal{C} \mid y \in Y\}$ is clearly an ultrafilter in $\mathcal{C}$. The ultrafilters $\mathcal{F}_y$ are called ***principal ultrafilters***.

If A is a ring, let $\mathcal{C}$ denote the family of all constructible subsets of $\mathrm{Spec}_R(A)$. A routine consequence of compactness of $\mathrm{Spec}_R(A)$ in the constructible topology is that all ultrafilters in $\mathcal{C}$ are principal ultrafilters. In fact, the correspondence which assigns to a point $\alpha \in \mathrm{Spec}_R(A)$, the principal ultrafilter $\mathcal{F}_\alpha = \{C \in \mathcal{C} \mid \alpha \in C\}$ is bijective.

Now, suppose that $A = A(X)$ is the affine coordinate ring of a real algebraic set X. The Artin-Lang Theorem 3.1 identifies the Boolean algebra of constructible subsets of $\mathrm{Spec}_R(A(X))$ with the Boolean algebra of semi-algebraic subsets of X. Thus, from the above paragraph, we conclude that *points* of $\mathrm{Spec}_R(A(X))$ correspond bijectively with *ultrafilters* of semi-algebraic subsets of X.

The inverse of the above correspondence can be specifically defined as follows. Regard elements $f \in A(X)$ as real valued functions $f : X \to \mathbf{R}$, and regard points of $\mathrm{Spec}_R(A(X))$ as subsets $\alpha \subset A(X)$, as in Proposition 2.1.

Proposition 4.1. *Suppose $\mathcal{F}$ is an ultrafilter of semi-algebraic subsets of X. Then $\alpha = \alpha(\mathcal{F}) = \{f \in A(X) \mid \text{for some } Y \in \mathcal{F},\ f|_Y \geq 0\} \subset A(X)$ is a point of $\mathrm{Spec}_R(A(X))$. Moreover, this correspondence from ultrafilters of semi-algebraic subsets of X to points of $\mathrm{Spec}_R(A(X))$ is bijective.* ∎

It is relatively easy to directly check the conditions 2.1(i)–(iv) for the subset $\alpha = \alpha(\mathcal{F}) \subset A(X)$ defined in Proposition 4.1. In particular, $p(\alpha) = \alpha \cap -\alpha \subset A(X)$ is a prime ideal which corresponds to an irreducible subvariety $Z \subset X$, which is called the *support* of α and

$\mathcal{F}$. Geometrically, one should think of the ultrafilter $\mathcal{F}$ associated to α as a kind of localized germ of Zariski dense, semi-algebraic subsets of Z. A function $f(\alpha) \in A(X)/p(\alpha) = A(Z)$ is non-negative in the order on $A(Z)$ if $f \in A(X)$ is non-negative on this germ, that is, on some sufficiently small subset of Z representing the germ.

We will give now a rather specific example of this ultrafilter construction, which, in fact, describes the simplest possible points in the boundaries $\widetilde{B}(W) = \widetilde{W} - W$ of our compactifications of closed semi-algebraic subsets $W \subseteq X$. Suppose $Z \subset X$ is a real algebraic curve, and let $Z_0 \subset Z \cap W$ denote a "branch of Z at infinity in W". This means that Z_0 is a subset of $Z \cap W$ such that for some coordinate half-axis $L_+ \subset \mathbf{R}^N$ and for all sufficiently large $r \in \mathbf{R}$, orthogonal projection of $\mathbf{R}^N$ onto L carries $Z_0 \cap (\mathbf{R}^N - B_r)$ bijectively onto a ray in L_+, where $B_r = \{\vec{x} \in \mathbf{R}^N \mid \|\vec{x}\| < r\}$.

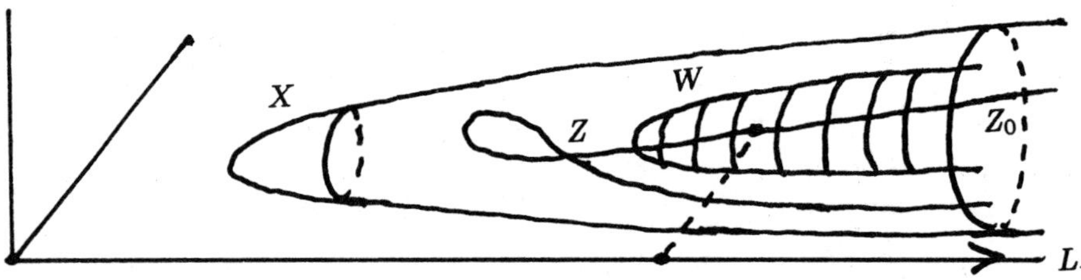

Then the sets $Z_0 \cap (\mathbf{R}^N - B_r)$, as $r \to \infty$, determine an ultrafilter of semi-algebraic subsets of W, with support Z, and a point $\alpha_0 \in C(W) - W$. For $f \in A(X)$, we have $f(\alpha_0) \geq 0$ if $f\big|_{Z_0 \cap (\mathbf{R}^N - B_r)} \geq 0$ for sufficiently large r.

Proposition 4.2. *The points $\alpha_0 \in C(W) - W$ represented by branches Z_0 of real algebraic curves at infinity in W are closed points of $C(W)$, hence belong to $\widetilde{B}(W) = \widetilde{W} - W$. The set of such points is dense in $\widetilde{B}(W)$.*

Proof. Say $x_1 \in A(X) = R[x_1 \ldots x_N]/I(X)$ is a coordinate such that $x_1 \geq 0$, $x_i = 0$, $2 \leq i \leq N$, corresponds to a half-axis $L_+ \subset \mathbf{R}^N$ which "parametrizes Z_0 at infinity". Then the field of fractions $K(\alpha_0)$ of $A(X)/p(\alpha_0) = A(Z)$ is an algebraic extension of $\mathbf{R}(x_1)$. The induced order on $\mathbf{R}(x_1)$ is described as follows. First, $x_1 > 0$ and x_1 is infinitely large relative to $\mathbf{R} \subset \mathbf{R}(x_1)$, because a function $x_1 - s$, $s \in \mathbf{R}$, is non-negative on $Z_0 \cap (\mathbf{R}^N - B_r)$ for sufficiently large r. But now, there is only one order on $\mathbf{R}(x_1)$ such that x_1 is positive and infinitely large relative to $\mathbf{R}$. Namely, in any ordered field, $f/g > 0$ if and only if $fg > 0$, so it suffices to consider polynomials $f(x_1)$. Write $f(x_1) = a_0 x_1^n + a_1 x_1^{n-1} + \ldots + a_n = x_1^n(a_0 + a_1/x_1 + \ldots + a_n/x_1^n)$, with $a_i \in \mathbf{R}$, $a_0 \neq 0$. Since the higher terms a_j/x_1^j are infinitesimally small relative to $\mathbf{R}$, we must have sign $(f(x_1)) = \text{sign}(a_0)$. It follows easily that in this order on $\mathbf{R}(x_1)$ every element is bounded by some power of x_1, hence every element of the algebraic extension $K(\alpha_0)$ of $\mathbf{R}(x_1)$, and of its real closure $F(\alpha_0)$, is also bounded by some power of x_1. Thus, from Proposition 2.2(e), we see that $\alpha_0 \in \text{Spec}_R(A(X))$ is a closed point.

Next, we must show that every point $\alpha \in \widetilde{B}(W) = \widetilde{W} - W$ can be approximated by such points α_0. Suppose $f_i(\alpha) > 0$, $1 \leq i \leq m$, so that $U\{f_i\} \cap \widetilde{W}$ defines a basic open neighborhood of α in $\widetilde{W}$. It follows from the Artin-Lang Theorem 3.1 that $U\{f_i\} \cap W$ is an ***unbounded*** semi-algebraic set in $\mathbf{R}^N$. Namely, by Proposition 3.2(b), $\alpha \in \widetilde{W} - W$ satisfies $\Sigma x_i^2(\alpha) - r > 0$, for all $r \in \mathbf{R}$. Thus, $\alpha \in C(U\{f_i\} \cap W \cap (\mathbf{R}^N - B_r)) \neq \emptyset$, so $U\{f_i\} \cap W \cap (\mathbf{R}^N - B_r) \neq \emptyset$, for all real r.

We now apply the Curve Selection Lemma of real algebraic geometry, [14], to deduce the existence of a branch Z_0 of a curve at infinity in $U\{f_i\} \cap W$. Then Z_0 represents a point α_0 in the neighborhood $U\{f_i\} \cap \widetilde{W}$ of α, which proves the set of all such points α_0 is dense in $\widetilde{B}(W)$. More specifically, the map $I: \mathbf{R}^N - (\vec{0}) \to \mathbf{R}^N - (\vec{0})$ defined by $I(\vec{x}) = \vec{x} \big/ \| \vec{x} \|^2$ sends the unbounded semi-algebraic set $U\{f_i\} \cap W$ to a semi-algebraic set with $\vec{0}$ in its closure. The Curve Selection Lemma gives a branch of a curve $Y_0 \subset I(U\{f_i\} \cap W)$ with $\vec{0} \in \overline{Y_0}$. Then take $Z_0 = I^{-1}(Y_0)$. ∎

We will include here one further application of the ultrafilter interpretation of points of $\mathrm{Spec}_R(A(X))$ given in Proposition 4.1. If $Y \subset X$ is a fixed semi-algebraic subset then the constructible set $C(Y) \subset \mathrm{Spec}_R(A(X))$ can be identified under the correspondence of 4.1 with the set of ultrafilters of semi-algebraic subsets of X with $Y \in \mathcal{F}$, or equivalently, with the set of all ultrafilters of semi-algebraic subsets of Y. Note this description of points of $C(Y)$ is independent of X.

It is a consequence of the Tarski-Seidenberg Principle that images of semi-algebraic sets under semi-algebraic functions are still semi-algebraic sets. A semi-algebraic function $\varphi : Y \to Y'$ between two affine semi-algebraic sets is a function whose graph in $Y \times Y'$ is a semi-algebraic set. If $\mathcal{F}$ is an ultrafilter of semi-algebraic subsets of Y then $\widetilde{\varphi}(\mathcal{F}) = \{V' \subset Y' \mid V'$ semi-algebraic and $\varphi(V) \subset V'$ for some $V \in \mathcal{F}\}$ is easily seen to be an ultrafilter of semi-algebraic subsets of Y'. Thus, $\varphi : Y \to Y'$ induces a map $\widetilde{\varphi} : C(Y) \to C(Y')$ which depends functorially on φ. This map $\widetilde{\varphi}$ is always continuous in the *constructible* topologies on $C(Y)$ and $C(Y')$, regardless of whether $\varphi : Y \to Y'$ is continuous. Moreover, a careful look at the real spectrum topologies, combined with the Finiteness Theorem 3.3 leads to the following.

Proposition 4.3. If $\varphi : Y \to Y'$ is a continuous semi-algebraic mapping then $\widetilde{\varphi} : C(Y) \to C(Y')$ is continuous in the real spectrum topologies. In particular, if $\varphi : Y \xrightarrow{\sim} Y'$ is a semi-algebraic homeomorphism, then $\widetilde{\varphi} : C(Y) \xrightarrow{\sim} C(Y')$ is a homeomorphism. ∎

§5. Non-Archimedean logarithms. Let F be an ordered field. We call a positive element $b \in F$ a ***big element*** if for all $a \in F$ there exists an integer $m \geq 1$ such that $a < b^m$. Given a big element $b \in F$, we define a real valued logarithm $\log_b : F^+ \to \mathbf{R}$ as follows. (There is only one possible definition.) If $a > 0$ in F then $\log_b(a) \in \mathbf{R}$ is the Dedekind cut of $\mathbf{Q}$ defined by

(5.1) $$\frac{m'}{n} \leq \log_b(a) \leq \frac{m}{n} \quad \text{if } b^{m'} \leq a^n \leq b^m,\ n, m, m' \in \mathbf{Z},\ n > 0$$

We will summarize in the next proposition the basic properties of this logarithm, all of which are quite easy to prove from (5.1).

Proposition 5.2. (a) $\log_b(b^m) = m$ *for all* $m \in \mathbf{Z}$.

(b) $\log_b(aa') = \log_b(a) + \log_b(a')$, *for all* $a, a' \in F^+$

(c) *If* $0 < a < a'$ *then* $\log_b(a) \leq \log_b(a')$.

(d) *If* b *and* b' *are two big elements and* $a \in F^+$ *then* $\log_{b'}(a) = \log_{b'}(b)\log_b(a)$ *and* $\log_{b'}(b) > 0$.

(e) *If* F *is a non-Archimedean ordered field then* $\log_b(r) = 0$ *for all positive rational numbers* $r \in \mathbf{Q}^+ \subset F^+$.

(f) *If* F *is a non-Archimedean ordered field then* $\log_b(|a+a'|) \leq \text{Max}(\log_b(|a|), \log_b(|a'|))$, *for all* $a, a' \in F$ *(Define* $\log_b(0) = -\infty$, *so that this makes sense in all cases.)*

(g) *If* F *is a non-Archimedean ordered field then* $v_b : F^* \to \mathbf{R}$, *defined by* $v_b(a) = -\log_b(|a|)$, *is a non-trivial, rank 1, Krull valuation of* F. ∎

The question arises, which ordered fields contain big elements. A simple induction shows that any ordered field F of finite transcendence degree over a subfield K which contains a big element, also contains a big element. The starting point is the observation that there is only one extension of the order on K to an order on the transcendental extension $K(x)$, such that x is positive and infinitely large relative to K. Moreover, in this order, $x \in K(x)$ is a big element. (The proof of this for $\mathbf{R}(x)$, given in the course of the proof of Proposition 4.2, works for any K.) Now, if K contains a big element and if F is Archimedean over K, then a big element in K is also big in F. If F is not Archimedean over K, there exists an element $x \in F$ which is positive and infinitely large relative to K. Necessarily, x is transcendental over K, and F has lower transcendence degree over $K(x)$ than over K.

Suppose $X \subseteq \mathbf{R}^N$ is a real algebraic set with affine coordinate ring $A(X)$. Suppose $W \subseteq X$ is a closed semi-algebraic subset and suppose $S \subset A(X)$ is a family of elements with the following properties: (i) S contains generators of the ring $A(X)$ over $\mathbf{R}$, (ii) for all $\vec{x} \in W$ and all $f \in S$, $|f(\vec{x})| \geq 1$, (iii) for all $\vec{x} \in W$, $|f(\vec{x})| > 1$ for some $f \in S$. Then there is a map $\theta : W \to \mathbf{P}^S = [0,\infty)^S - (\vec{0})/\sim$, defined by $\theta(\vec{x}) = (\log(|f(\vec{x})|))_{f \in S}$. Our main result in this section is that θ extends continuously to the real spectrum compactification $\widetilde{W} \subset C(W) \subseteq \text{Spec}_R(A(X))$, by exactly the same formula.

Proposition 5.3. *The formula* $\tilde{\theta}(\alpha) = (\log(|f(\alpha)|))_{f \in S}$, $\alpha \in \widetilde{W}$, *defines a continuous map* $\tilde{\theta} : \widetilde{W} \to \mathbf{P}^S$.

Proof. Recall that $f(\alpha)$ means the image of $f \in A(X)$ under $\alpha : A(X) \to F(\alpha)$. If $|f(\vec{x})| \geq 1$ for all $\vec{x} \in W$ then $|f(\alpha)| \geq 1$ for all $\alpha \in \widetilde{W}$, so $\log(|f(\alpha)|) \geq 0$. Next, the choice of big base element for the logarithm at any point is unimportant, since, by Proposition 5.2 (d), change of base element just multiplies all logarithms by a positive real scaler. To complete the proof that $\widetilde{\theta}(\alpha)$ is defined for all $\alpha \in \widetilde{W} - W$, it is necessary to check that for some $f \in S$, $\log(|f(\alpha)|) \neq 0$. In the non-Archimedean case, this is not the same as $|f(\alpha)| > 1$, but, instead, amounts to showing that some $|f(\alpha)| \in F(\alpha)$ is a big element. This follows from the fact that S contains a set of generators for $A(X)$ over $\mathbf{R}$ and the fact that $\alpha \in \widetilde{W} - W$ is a closed point at infinity in $\mathrm{Spec}_R(A(X))$. Namely, if $A(X) = \mathbf{R}[x_1 \ldots x_N]/I(X)$ and $\alpha : A(X) \to F(\alpha)$ is a closed point of $\mathrm{Spec}_R(A(X))$ at infinity, then Proposition 2.2(e) and 3.2(b) imply that some $|x_k(\alpha)| \in F(\alpha)$ is a big element. (See also the summary of properties of points $\alpha \in \widetilde{B}(W) = \widetilde{W} - W$ given in Remark 3.4.)

Now we will prove that $\widetilde{\theta} : \widetilde{W} \to \mathbf{P}^S$ is continuous. Let $\{x_i\} \subset S$ be a finite set of generators for $A(X)$ over $\mathbf{R}$ and set $r = 2 + \Sigma x_i^2 \in A(X)$. The argument above showing that $(\log(|f(\alpha)|))_{f \in S} \not\equiv \vec{0}$ also shows that $\log(r(\alpha)) > 0$ for all $\alpha \in \widetilde{W}$. We can then define a lifting of $\widetilde{\theta} : \widetilde{W} \to \mathbf{P}^S$ to $\widetilde{\Theta} : \widetilde{W} \to [0, \infty)^S$, by $\widetilde{\Theta}(\alpha) = (\log(|f(\alpha)|) \big/ \log(r(\alpha)))_{f \in S}$. Note that the quotient of two logarithms is independent of the base, so $\widetilde{\Theta}$ is well-defined.

It suffices to show that $\widetilde{\Theta}$ is continuous. Let $f \in S$ be a "coordinate" and consider $\widetilde{\Theta}_f : \widetilde{W} \to \mathbf{R}$ where $\widetilde{\Theta}_f(\alpha) = \log(|f(\alpha)|) \big/ \log(r(\alpha))$. Then, for $s \geq 0 \in \mathbf{R}$, one can show from (5.1) that

$$\widetilde{\Theta}_f^{-1}([0, s]) = \bigcap_{\frac{m}{n} > s} \left\{ \alpha \in \widetilde{W} \;\middle|\; |f^n(\alpha)| \leq r^m(\alpha) \right\}$$

$$\widetilde{\Theta}_f^{-1}(([s, \infty)) = \bigcap_{\frac{m}{n} < s} \left\{ \alpha \in \widetilde{W} \;\middle|\; r^m(\alpha) \leq |f^n(\alpha)| \right\} .$$

Since the sets on the right are closed in $\widetilde{W}$, it follows that $\widetilde{\Theta}_f^{-1}((s, \infty))$ and $\widetilde{\Theta}_f^{-1}([0, s))$ are open in $\widetilde{W}$, and hence $\widetilde{\Theta}_f$ is continuous. ∎

§6. **Teichmüller space as a semi-algebraic set.** In this section, we review briefly how the Teichmüller space T_g associated to a closed surface Σ of genus $g \geq 2$ is a component of a real algebraic set, [9]. Our outline is taken directly from [15].

Let V_g denote the real algebraic set $\mathrm{Hom}(\pi_1(\Sigma), SL(2, \mathbf{R}))$. If $\pi_1(\Sigma) = \langle \gamma_i \mid 1 \leq i \leq n \rangle$ and we write $\rho(\gamma_i) = \begin{pmatrix} a_i(\rho), & b_i(\rho) \\ c_i(\rho) & d_i(\rho) \end{pmatrix}$ for $\rho \in V_g$, then $A(V_g) = \mathbf{R}[a_i, b_i, c_i, d_i] \big/ I(V_g)$. The group $SL(2, \mathbf{R})$ acts on V_g by conjugation, inducing an action of $SL(2, \mathbf{R})$ as ring automorphisms of $A(V_g)$. The subring of invariant elements in $A(V_g)$ is intimately related to the conjugate classes of $SL(2, \mathbf{R})$ representations of $\pi_1(\Sigma)$.

It is well-known that this subring $\mathbf{R}[\mathrm{trace}(\rho(\gamma)) \mid \gamma \in \pi_1(\Sigma)] = A(V_g)^{SL(2,\mathbf{R})}$ is finitely generated. Specifically, a set of generators is given by the elements $\left\{ \mathrm{trace}(\rho(\gamma_{i_1} \cdots \gamma_{i_r})) \;\middle|\; i_1 < \ldots < i_r \right\}$, [16]. Let X_g denote the real algebraic set with $A(X_g) = A(V_g)^{SL(2,\mathbf{R})}$. The inclusion $A(X_g) \to A(V_g)$ induces a map $\pi : V_g \to X_g$, and conjugate representations of $\pi_1(\Sigma)$ in $SL(2, \mathbf{R})$ have the same image in X_g. Points of X_g are called *real characters* of representations and X_g is called the *character variety*, or, more precisely, X_g is the set of real points of the character variety.

The projection $\pi : V_g \to X_g$ is not surjective. This is because $SU(2,\mathbf{C})$ representations of $\pi_1(\Sigma)$ have real characters, but are not generally conjugate to $SL(2,\mathbf{R})$ representations inside $SL(2,\mathbf{C})$. However, let $DV_g \subset V_g$ denote the subspace of discrete, faithful representations and let $DX_g = \pi(DV_g) \subset X_g$. Then $SL(2,\mathbf{R})$ acts freely on $DV_g \subset V_g$, the projection $\pi \big|_{DV_g} : DV_g \to DX_g$ is a principal $SL(2,\mathbf{R})$ fibre bundle, $DV_g \subset V_g$ is both closed and open, and $DX_g \subset X_g$ is both closed and open. Since any real algebraic set has only finitely many components, each of which is a semi-algebraic subset, one concludes that $DX_g \subset X_g$ is a union of components of X_g, and is thus a closed semi-algebraic subset. All these results are discussed in [15].

It is possible to describe $DX_g \subset X_g$ and $DV_g \subset V_g$ in terms of (infinitely many) algebraic inequalities. A theorem of Jørgensen, [10], implies that a representation $\rho : \pi_1(\Sigma) \to SL(2,\mathbf{R})$ is discrete and faithful if and only if for each pair $\gamma, \gamma' \in \pi_1(\Sigma)$ which do not commute, the following inequality holds.

$$\left|\text{trace}([\rho(\gamma),\rho(\gamma')]) - 2\right| + \left|\text{trace}^2(\rho(\gamma)) - 4\right| \geq 1\,. \tag{6.1}$$

The points of Teichmüller space T_g are actually conjugate classes of discrete, faithful $PSL(2,\mathbf{R})$ representations of $\pi_1(\Sigma)$, rather than $SL(2,\mathbf{R})$ representations. Since $PSL(2,\mathbf{R}) = SL(2,\mathbf{R})\big/\{\pm\text{Id}\}$ and since $\pi_1(\Sigma)$ has a presentation with only one relation, which is a product of commutators, it is easy to see that all $PSL(2,\mathbf{R})$ representations of $\pi_1(\Sigma)$ lift to $SL(2,\mathbf{R})$ representations and that $\text{Hom}(\pi_1(\Sigma), \{\pm Id\}) = H^1(\Sigma, \mathbf{Z}/2)$ acts freely on V_g, by $\sigma \cdot \rho(\gamma) = \sigma(\gamma)\rho(\gamma)$, where $\sigma : \pi_1(\Sigma) \to \{\pm Id\}$ and $\rho : \pi_1(\Sigma) \to SL(2,\mathbf{R})$. Clearly the discrete, faithful representations $DV_g \subset V_g$ are preserved by this action. Moreover, the action of $\text{Hom}(\pi_1(\Sigma), \{\pm Id\})$ on $DV_g \subset V_g$ descends to an action on $DX_g \subset X_g$. From all this, and the fact that Teichmüller space T_g is topologically $\mathbf{R}^{6g-6}$, which is contractible, it follows that $\text{Hom}(\pi_1(\Sigma), \{\pm Id\})$ acts freely as a transitive permutation group on the components of DX_g and that any one component of DX_g projects bijectively to Teichmüller space T_g. In other words, the projection $DX_g \to T_g$ is a topologically trivial principal $(\mathbf{Z}/2)^{2g}$-covering, where $(\mathbf{Z}/2)^{2g} = H^1(\Sigma, \mathbf{Z}/2) = \text{Hom}(\pi_1(\Sigma), \{\pm Id\})$.

The above discussion, along with the general theory of the compactifications $\widetilde{W} \subset \text{Spec}_R(A(X))$ of closed semi-algebraic subsets of real varieties $W \subseteq X$, now characterizes the real spectrum compactification $\widetilde{T}_g$ of T_g as one of the components of the compactification $\widetilde{DX}_g \subset \text{Spec}_R(A(X_g))$ of $DX_g \subset X_g$. Somewhat more invariantly, we can also describe $\widetilde{T}_g$ as the orbit space of a free action of $\text{Hom}(\pi_1(\Sigma), \{\pm Id\})$ on $\widetilde{DX}_g$. Namely, $\text{Hom}(\pi_1(\Sigma), \{\pm\, Id\})$ acts algebraically on X_g, by sign changes of the coordinates $\text{trace}(\rho(\gamma)) \in A(X_g)$, $\gamma \in \pi_1(\Sigma)$. Thus, $\text{Hom}(\pi_1(\Sigma), \{\pm\, Id\})$ acts on $\text{Spec}_R(A(X_g))$, and preserves $\widetilde{DX}_g$. The components of $\widetilde{DX}_g$ are thus permuted by this action in exactly the same way as the components of DX_g, since, from Remark 3.6, components of DX_g and $\widetilde{DX}_g$ correspond bijectively.

We now wish to give a representation theoretic characterization of boundary points of $\widetilde{T}_g$. If F is a real closed field containing $\mathbf{R}$, denote by $V_g(F)$ and $X_g(F)$ the F-valued points of the affine schemes associated to the $\mathbf{R}$-algebras $A(V_g)$ and $A(X_g) = A(V_g)^{SL(2,\mathbf{R})}$, respectively.

Thus, $V_g(F) = \mathrm{Hom}_{\mathbf{R}}(A(V_g), F)$ and $X_g(F) = \mathrm{Hom}_{\mathbf{R}}(A(X_g), F)$, and the inclusion $A(X_g) \subset A(V_g)$ induces a map of F-affine algebraic sets $\pi_F : V_g(F) \to X_g(F)$. Obviously, $V_g(F) = \mathrm{Hom}_{\mathbf{R}}(A(V_g), F) = \mathrm{Hom}(\pi_1(\Sigma), SL(2, F))$ and conjugate $SL(2, F)$ representations of $\pi_1(\Sigma)$ have the same image in $X_g(F)$.

We have seen that $\widetilde{T}_g$ is a quotient of $\widetilde{DX}_g \subset \mathrm{Spec}_R(A(X_g))$, where $DX_g \subset X_g = X_g(\mathbf{R})$ is a certain closed, semi-algebraic set. Let $DX_g(F) \subset X_g(F)$ denote the F-affine semi-algebraic set defined by the same finite Boolean combination of inequalities which defines $DX_g \subset X_g$ in real affine space. This makes sense because $\mathbf{R} \subset F$. Of course, there is not a unique such system of inequalities, but the Tarski Principle (or the Artin-Lang Theorem 3.1) implies that any two systems which define the same real affine semi-algebraic set also define the same F-affine semi-algebraic set, for any real closed field F containing $\mathbf{R}$. Let $DV_g(F) = \pi_F^{-1}(DX_g(F)) \subset V_g(F)$. Note that we are *not* defining $DV_g(F) \subset V_g(F)$ as the set of all discrete, faithful representations $\pi_1(\Sigma) \to SL(2, F)$. The relation between these two sets is unclear.

Now, let $[\alpha] \in \widetilde{B}_g = \widetilde{T}_g - T_g$ be a boundary point of the real spectrum compactification of T_g, where $\alpha \in \widetilde{B}(DX_g) = \widetilde{DX}_g - DX_g$ is represented by $\alpha : A(X_g) \to F$, with F a non-Archimedean, real closed field. From Proposition 3.2 and Remark 3.4, we have $\alpha \in DX_g(F)$. Also, F is algebraic and Archimedean over the subring $\alpha(A(X_g)) \subset F$. The following result shows that α actually is the character of a representation $\pi_1(\Sigma) \to SL(2, F)$.

Proposition 6.2. *For all real closed fields F containing $\mathbf{R}$, the map $\pi_F : DV_g(F) \to DX_g(F)$ is surjective, and is a semi-algebraic, principal, $SL(2, F)$ fibre bundle. In particular, all points of $DX_g(F)$ are characters of $SL(2, F)$-representations of $\pi_1(\Sigma)$, and two representations in $DV_g(F) \subset V_g(F)$ have the same image in $DX_g(F)$ if and only if they are conjugate as $SL(2, F)$-representations.*

Proof. First, we clarify that by a *semi-algebraic fibre bundle*, we mean a bundle such that the base is covered by finitely many open, semi-algebraic sets, over each of which the bundle is trivial, and such that the change of coordinate functions for this cover are semi-algebraic. We know the proposition is true when $F = \mathbf{R}$. Then the Tarski Principle implies the result for any F, since the algebraic morphism $\pi_F : V_g(F) \to X_g(F)$ is just obtained from $\pi : V_g(\mathbf{R}) \to X_g(\mathbf{R})$ by the base extension $\mathbf{R} \subset F$. ∎

Our next goal is to understand the representations $\rho : \pi_1(\Sigma) \to SL(2, F)$ in $DV_g(F)$ which lift boundary points of $\widetilde{T}_g$. The first comment is that we do not want or need *all* points "at infinity" in $DV_g(F)$, as F varies over non-Archimedean ordered fields containing $\mathbf{R}$. We only want points which lie over points "at infinity" in $DX_g(F)$. This means some $|\mathrm{trace}(\rho(\gamma))| \in F$, $\gamma \in \pi_1(\Sigma)$, must be infinitely large relative to $\mathbf{R} \subset F$, not just some matrix entry of $\rho(\gamma) \in SL(2, F)$. But also, since $\dim(V_g) = \dim(X_g) + 3$, if $\alpha : A(V_g) \to F$ is the ring homomorphism associated to some $\rho : \pi_1(\Sigma) \to SL(2, F)$, then, in general, F will have *positive* transcendence degree over the subring $\alpha(A(X_g)) \subset F$. Proposition 6.2 allows us to *start* with $\alpha : A(X_g) \to F$, so that F is algebraic and Archimedean over $\alpha(A(X_g))$, and lift α to $\rho : \pi_1(\Sigma) \to SL(2,F)$, or, equivalently, to $\alpha : A(V_g) \to F$, with the same F. Summarizing, all boundary points of $\widetilde{T}_g$ lift to representations $\rho : \pi_1(\Sigma) \to SL(2,F)$ where F is a non-Archimedean, real closed field containing $\mathbf{R}$, such that some element $|\text{trace}(\rho(\gamma))| \in F$ is a big element of F, $\gamma \in \pi_1(\Sigma)$, and such that F is algebraic over the real function field generated by all elements trace$(\rho(\gamma))$, $\gamma \in \pi_1(\Sigma)$.

In the geometric language of the real spectrum and the map $i^* : \text{Spec}_R(A(V_g)) \to \text{Spec}_R(A(X_g))$ induced by the inclusion $i : A(X_g) \to A(V_g)$, the meaning of the above paragraph is that all points of $\widetilde{B}(DX_g) = \widetilde{DX}_g - DX_g \subset \text{Spec}_R(A(X_g))$ are of the form $i^*(\alpha)$, where $\alpha \in \text{Spec}_R(A(V_g))$ satisfies dim(Support(α)) = dim(Support$(i^*(\alpha))$). Here Support$(\alpha) \subset V_g$ and Support$(i^*(\alpha)) \subset X_g$ denote the irreducible subvarieties of V_g and X_g corresponding to α and $i^*(\alpha)$, respectively, as discussed in §4. It follows easily that any such α belongs to $\widetilde{B}(DV_g) = \widetilde{DV}_g - DV_g \subset \text{Spec}_R(A(V_g))$, hence α is in the closure of $DV_g = DV_g(\mathbf{R})$, the discrete, faithful $SL(2,\mathbf{R})$ representations of $\pi_1(\Sigma)$.

There are several characterizations of $DV_g(\mathbf{R})$. For example, in addition to being the set of all discrete, faithful, $SL(2,\mathbf{R})$ representations of $\pi_1(\Sigma)$, $DV_g(\mathbf{R})$ can be characterized as the set of all representations $\rho : \pi_1(\Sigma) \to SL(2,\mathbf{R})$ which satisfy the Jørgensen inequalities (6.1), or as the set of all *totally hyperbolic* representations, that is, all $\rho : \pi_1(\Sigma) \to SL(2,\mathbf{R})$ such that $|\text{trace}(\rho(\gamma))| > 2$ for all $\gamma \neq Id \in \pi_1(\Sigma)$. Also, $DV_g(\mathbf{R})$ consists of all representations $\rho : \pi_1(\Sigma) \to SL(2,\mathbf{R})$ so that the Euler class $\chi(\rho)$ of the associated 2-plane bundle over Σ satisfies $|\chi(\rho)| = 2g-2$. It is unclear whether any of these four characterizations also describe $DV_g(F)$ for non-Archimedean F. Another problem is to produce a concrete semi-algebraic description of $DV_g(\mathbf{R})$, and hence also $DV_g(F)$. It seems highly likely that the Euler class condition has an analogue for $SL(2,F)$ representations, and it also seems likely that the Euler class condition can be translated to an explicit semi-algebraic formula for DV_g.

In a positive direction, we point out that all representations $\rho : \pi_1(\Sigma) \to SL(2,F)$ which correspond to points of $\widetilde{DV}_g \subset \text{Spec}_R(A(V_g))$ are totally hyperbolic (hence faithful) and satisfy the Jørgensen inequalities (6.1). This is simply because any algebraic inequality which holds on $DV_g = DV_g(\mathbf{R})$, also holds on $\widetilde{DV}_g = C(DV_g) \cap \text{Spec}_R^m(A(V_g))$, by the Artin-Lang Theorem 3.1. A less trivial fact is that such representations are also discrete. This is due to Charles Frohman.

Proposition 6.3. (Frohman) *Any representation $\rho : \pi_1(\Sigma) \to SL(2,F)$ which corresponds to a point of $\widetilde{DV}_g \subset \text{Spec}_R(A(V_g))$ is discrete.*

Proof. We use the fact that $\rho \in \widetilde{DV}_g$ is in the closure of the discrete, faithful $SL(2,\mathbf{R})$ representations of $\pi_1(\Sigma)$, along with a result of Margulis. The Margulis result implies that there is a real constant $\epsilon > 0$, such that if $\Gamma \subset SL(2,\mathbf{R})$ is a discrete subgroup and if $\gamma, \gamma' \in \Gamma$ are within ϵ of the identity, (in the standard Euclidean metric on $SL(2,\mathbf{R}) \subset \mathbf{R}^4$), then γ and γ' commute, [18, Chapter 5]. For this same constant $\epsilon \in \mathbf{R} \subset F$, we claim that if $\gamma, \gamma' \in \pi_1(\Sigma)$ and if $\rho(\gamma)$, $\rho(\gamma') \in SL(2,F)$ are within ϵ of the identity, then γ and γ' commute. Assuming this claim, it follows that the subgroup of $\pi_1(\Sigma)$ generated by all such elements γ is abelian, hence cyclic. Choose a generator γ. If $\gamma \neq Id$, then $|\text{trace}(\rho(\gamma))| > 2 \in F$, and, for $n \in \mathbf{Z}$, $|\text{trace}(\rho(\gamma^n))| \geq |\text{trace}(\rho(\gamma))|$. (To see this, write $|\text{trace}(\rho(\gamma))| = \lambda + \lambda^{-1}$ with $\lambda > 1 \in F$. Then $|\text{trace}(\rho(\gamma^n))| = \lambda^n + \lambda^{-n}$.) Thus, $\rho(\pi_1(\Sigma)) \subset SL(2, F)$ is a discrete subgroup, since, if $\delta = |\text{trace}(\rho(\gamma))| - 2 \in F$, then for all $\gamma' \neq Id \in \pi_1(\Sigma)$ either $\rho(\gamma') \in SL(2, F)$ is at least distance ϵ from the identity or $|\text{trace}(\rho(\gamma'))| - 2 \geq \delta$, which also bounds $\rho(\gamma')$ away from the identity.

To complete the proof, we must show that if $\rho(\gamma)$ and $\rho(\gamma') \in SL(2, F)$ are within ϵ of the identity then γ and γ' commute. If $[\gamma, \gamma'] \neq Id \in \pi_1(\Sigma)$ then $|\text{trace}(\rho([\gamma, \gamma']))| > 2$ since ρ is totally hyperbolic. Since $DV_g(\mathbf{R})$ is dense in $\widetilde{DV}_g$, there would exist a discrete, faithful real representation $\rho_0 : \pi_1(\Sigma) \to SL(2, \mathbf{R})$ such that $|\text{trace}(\rho_0([\gamma, \gamma']))| > 2$, and such that $\|\rho_0(\gamma) - Id\| < \epsilon$ and $\|\rho_0(\gamma') - Id\| < \epsilon$. But this contradicts the Margulis result. ∎

To complete our representation theoretic description of points of $\widetilde{B}_g = \widetilde{T}_g - T_g$, we will clarify when two suitable representations $\rho : \pi_1(\Sigma) \to SL(2, F)$ and $\rho' : \pi_1(\Sigma) \to SL(2, F')$, with F and F' non-Archimedean, real closed fields, determine the same point of $\widetilde{B}_g$. Essentially, there are three things to take into account. First, conjugations will not change the images in $DX_g(F)$ and $DX_g(F')$. Secondly, a possible ordered field isomorphism (even if $F = F'$) is built into the definition of when two homomorphisms $A(X_g) \to F$ and $A(X_g) \to F'$ represent the same point of $\widetilde{DX}_g \subset \text{Spec}_R(A(X_g))$. Finally, we must take into account the action of $\text{Hom}(\pi_1(\Sigma), \{\pm Id\})$, since $\widetilde{T}_g = \widetilde{DX}_g / \text{Hom}(\pi_1(\Sigma), \{\pm Id\})$. This last can be formulated by simply passing to the induced $PSL(2)$ representations. Summarizing, we have the following description of the real spectrum boundary of Teichmüller space.

Proposition 6.4. (a) *All points of $\widetilde{B}_g = \widetilde{T}_g - T_g$ lift to representations $\rho : \pi_1(\Sigma) \to SL(2, F)$, where F is a non-Archimedean, real closed field containing $\mathbf{R}$, such that $\rho \in DV_g(F) \subset V_g(F)$, and such that F is algebraic and Archimedean over the (finitely generated) subring $\mathbf{R}[\text{trace}(\rho(\gamma)) \mid \gamma \in \pi_1(\Sigma)] \subset F$. All such representations ρ are discrete, faithful, and totally hyperbolic.*

(b) *Two representations $\rho : \pi_1(\Sigma) \to SL(2, F)$ and $\rho' : \pi_1(\Sigma) \to SL(2, F')$ with the properties in (a) determine the same point of $\widetilde{B}_g$ if and only if there is an ordered field isomorphism $\varphi : F \xrightarrow{\sim} F'$ such that the diagram below commutes up to conjugacy in $PSL(2, F')$.*

$$\begin{array}{ccc} & & PSL(2,F) \\ & \overset{\rho}{\nearrow} & \\ \pi_1(\Sigma) & & \downarrow\wr\ \varphi \\ & \underset{\rho'}{\searrow} & \\ & & PSL(2,F') \end{array}$$

∎

§7. Comparison with the Thurston compactification. Suppose $\gamma \in \pi_1(\Sigma)$ is represented by a simple closed curve on the surface Σ. Then given a hyperbolic structure on Σ, there exists a curve of shortest length homotopic to γ, and this curve is a simple, closed geodesic on Σ. The length of this geodesic is computed as follows. Suppose the hyperbolic structure on Σ is determined by the discrete, faithful representation $\rho : \pi_1(\Sigma) \to SL(2,\mathbf{R})$. Then $\rho(\gamma) \in SL(2,\mathbf{R})$ is a hyperbolic motion of $H\mathbf{R}^2$, hence is conjugate to $\begin{pmatrix} \lambda & 0 \\ 0 & \lambda^{-1} \end{pmatrix} \in SL(2,\mathbf{R})$, for some $|\lambda| > 1$. The curve length of the geodesic on Σ homotopic to γ is $l_\gamma(\rho) = \log(\lambda^2)$, which is the same as the translation distance of the hyperbolic motion $\rho(\gamma)$ along its axis in $H\mathbf{R}^2$.

Set $t = \lambda + \lambda^{-1} = \text{trace}(\rho(\gamma))$. Then $\lambda^2 - \lambda t + 1 = 0$, so $\lambda, \lambda^{-1} = \frac{t \pm \sqrt{t^2-4}}{2}$. Assuming λ and t are positive, we get $\lambda > 1$, $t > 2$, and λ and t are related by the formulas

$$\lambda = t\left(\frac{1+\sqrt{1-4/t^2}}{2}\right), \quad t = \lambda(1+1/\lambda^2)\,. \tag{7.1}$$

Let $S \subset \pi_1(\Sigma)$ be a subset containing an element of each conjugate class represented by a simple closed curve on Σ. (More precisely, S should be interpreted as a set of conjugate classes of elements of $\pi_1(\Sigma)$.) The curve lengths $l_\gamma(\rho)$, $\gamma \in S$, define an embeding $\theta' : T_g \to \mathbf{P}^S$, namely, $\theta'([\rho]) = (l_\gamma(\rho))_{\gamma \in S}$, [8]. We will work instead with the map $\theta : T_g \to \mathbf{P}^S$ defined by $\theta([\rho]) = (\log(|\text{trace}(\rho(\gamma))|))_{\gamma \in S}$. The formulas (7.1) above relating λ and t, along with the discussion in §5, show that both θ and θ' extend to continuous maps $\widetilde{T}_g \to \mathbf{P}^S$. In fact, the extensions $\widetilde{\theta}$ and $\widetilde{\theta}'$ of θ and θ' actually *coincide* on the boundary $\widetilde{B}_g = \widetilde{T}_g - T_g$, since $1 < \lambda < t < 2\lambda \in F$ implies $\log(\lambda) = \log(t)$ for one of our non-Archimedean logarithms $\log : F^+ \to \mathbf{R}$.

The Thurston compactification $\widehat{T}_g$ of T_g can be defined as the closure of $\theta(T_g)$ in $\mathbf{P}^S$, [8]. Since $\widetilde{T}_g$ is compact, $T_g \subset \widetilde{T}_g$ is dense, and $\widetilde{\theta} : \widetilde{T}_g \to \mathbf{P}^S$ is continuous, we have proved

Proposition 7.2. *The map $\widetilde{\theta} : \widetilde{T}_g \to \mathbf{P}^S$ defines a continuous surjection $\widetilde{\theta} : \widetilde{T}_g \to \widehat{T}_g$, extending the embedding $\theta : T_g \to \mathbf{P}^S$.* ∎

The automorphism group $\text{Aut}(\pi_1(\Sigma))$ acts on $V_g = \text{Hom}(\pi_1(\Sigma), SL(2,\mathbf{R}))$ by precomposition, that is $\mu\rho = \rho \circ \mu^{-1}$, $\mu \in \text{Aut}(\pi_1(\Sigma))$. This action clearly preserves the discrete, faithful representations $DV_g \subset V_g$. Moreover, the action can also be described as an action of $\text{Aut}(\pi_1(\Sigma))$ as ring automorphisms of $A(V_g)$.

Since inner automorphisms are conjugations, there is an induced action of the outer automorphism group, $\text{Out}(\pi_1(\Sigma))$, on $A(X_g) = A(V_g)^{SL(2,\mathbf{R})}$, hence an action of $\text{Out}(\pi_1(\Sigma))$ on $\text{Spec}_R(A(X_g))$. This action preserves $DX_g \subset X_g$, hence also $\widetilde{DX}_g$. Moreover, it can be seen easily that the action of $\text{Out}(\pi_1(\Sigma))$ on DX_g descends to an action on $T_g = DX_g \big/ \text{Hom}(\pi_1(\Sigma), \{\pm Id\})$. Thus, $\text{Out}(\pi_1(\Sigma))$ acts continuously on the real spectrum compactification of Teichmüller space, $\widetilde{T}_g = \widetilde{DX}_g \big/ \text{Hom}(\pi_1(\Sigma), \{\pm Id\})$.

One of the features of the Thurston compactification, $\widehat{T}_g$, is also that $\text{Out}(\pi_1(\Sigma))$ acts continuously on $\widehat{T}_g$, [8]. This action is described as follows. Since $\text{Out}(\pi_1(\Sigma))$ coincides with the mapping class group of the surface Σ, the set of homotopy classes of simple closed curves is preserved by automorphisms of $\pi_1(\Sigma)$. Thus, $\mathrm{Out}(\pi_1(\Sigma))$ acts on $\mathcal{S}$, and hence on $\mathbf{P}^{\mathcal{S}}$. It is clear that $\theta(T_g) \subset \mathbf{P}^{\mathcal{S}}$ is $\mathrm{Out}(\pi_1(\Sigma))$-invariant. In fact, the restricted action on $\theta(T_g)$ coincides with the action on T_g defined directly above, since the functions $\{|\mathrm{trace}(\rho(\gamma))|,\ \gamma \in \mathcal{S}\}$ on T_g are just being permuted by $\mathrm{Out}(\pi_1(\Sigma))$. Moreover, since $A(X_g)$ is generated by the functions $\{\mathrm{trace}(\rho(\gamma)) \mid \gamma \in \mathcal{S}\}$, the action of $\mathrm{Out}(\pi_1(\Sigma))$ on $\mathrm{Spec}_R(A(X_g))$ also is determined by the induced permutations of $\mathcal{S}$. We thus conclude

Proposition 7.3. *With respect to the actions of* $\mathrm{Out}(\pi_1(\Sigma))$ *on* $\widetilde{T}_g$ *and* $\widehat{T}_g$ *described above, the map* $\widetilde{\theta} : \widetilde{T}_g \to \widehat{T}_g$ *is* $\mathrm{Out}(\pi_1(\Sigma))$*-equivariant.* ∎

We mentioned previously the well-known fact that Teichmüller space T_g is topologically $\mathbf{R}^{6g-6}$. Now, actually, this identification can be chosen to be a *semi-algebraic* equivalence. Since each element of $\mathrm{Out}(\pi_1(\Sigma))$ acts as a semi-algebraic homeomorphism of T_g, we can apply the semi-algebraic Brouwer fixed point theorem for the real spectrum compactification $\widetilde{\mathbf{R}}^{6g-6}$ of $\mathbf{R}^{6g-6}$, which was proved in [2] and [3]. The conclusion is

Proposition 7.4. *Each element of the mapping class group* $\mathrm{Out}(\pi_1(\Sigma))$ *fixes at least one point of* $\widetilde{T}_g \simeq \widetilde{\mathbf{R}}^{6g-6}$. *Since* $\widetilde{\theta} : \widetilde{T}_g \to \widehat{T}_g$ *is* $\mathrm{Out}(\pi_1(\Sigma))$*-equivariant, the same conclusion holds for* $\widehat{T}_g$. ∎

Of course, the Thurston compactification $\widehat{T}_g$ is topologically a closed disk D^{6g-6}. However, our proof of the fixed point property for $\widehat{T}_g$ is independent of this rather difficult fact.

It should be useful to have a good understanding of what it means for a boundary point in $\widetilde{B}_g = \widetilde{T}_g - T_g$ to be fixed by an outer automorphism of $\pi_1(\Sigma)$. From Proposition 6.4(b) we deduce

Proposition 7.5. *Suppose* $\rho : \pi_1(\Sigma) \to PSL(2,F)$ *represents a point* $[\rho] \in \widetilde{B}_g = \widetilde{T}_g - T_g$, *where* F *is a non-Archimedean, real closed field, and suppose* $\mu \in Out(\pi_1(\Sigma))$. *Then* $\mu[\rho] = [\rho]$ *if and only if there exists an ordered field automorphism* $\varphi : F \xrightarrow{\sim} F$ *such that the diagram below commutes up to conjugacy in* $PSL(2,F)$.

$$\begin{array}{ccc} \pi_1(\Sigma) & \xrightarrow{\rho} & PSL(2,F) \\ \mu \downarrow \wr & & \downarrow \wr \ \varphi \\ \pi_1(\Sigma) & \xrightarrow{\rho} & PSL(2,F) \end{array}$$ ∎

Remark 7.6. Associated to a fixed point of $\mu \in \mathrm{Out}(\pi_1(\Sigma))$ in the Thurston boundary $\widehat{B}_g = \widehat{T}_g - T_g$, there is a certain constant, called the *multiplier*. The multiplier can be interpreted either in the Thurston context of measured laminations on Σ, or in the Morgan-Shalen context of actions of $\pi_1(\Sigma)$ on metric $\mathbf{R}$-trees. In the Thurston context, two measured laminations represent the same boundary point if the underlying laminations are isotopic and the measure of one is a scalar multiple of the measure of the other. In the Morgan-Shalen context, boundary points are suitable conjugate classes of actions of $\pi_1(\Sigma)$ as a group of isometries of metric $\mathbf{R}$-trees, modulo a scaling factor on the tree metric. Thus, in both contexts, to a boundary fixed point of $\mu \in \mathrm{Out}(\pi_1(\Sigma))$, there is associated a real scaling factor. In our context, say in the situation of Proposition 7.5, the multiplier associated to the fixed point $[\rho]$ of μ is $\log_b(\varphi(b)) > 0 \in \mathbf{R}$, where $b \in F$ is a big element and $\varphi : F \xrightarrow{\sim} F$ is the order preserving automorphism which occurs in 7.5. (It is easy to see that $\log_b(\varphi(b))$ is independent of b, say from Proposition 5.2(d).) ∎

§8. The connection with R-trees. In §6, we saw that a boundary point of $\widetilde{T}_g$ could be interpreted as an equivalence class of certain representations $\rho : \pi_1(\Sigma) \to SL(2,F)$, where F is a non-Archimedean, real closed field. For some $\gamma \in \pi_1(\Sigma)$, $|\mathrm{trace}(\rho(\gamma))| \in F$ is a big element. Let $\log : F^+ \to \mathbf{R}$ be the corresponding non-Archimedean logarithm studied in §5. From Proposition 5.2(g), the map $v : F^* \to \mathbf{R}$, $v(a) = -\log(|a|)$, is a Krull valuation. Let $\Lambda = \log(F^+) \subset \mathbf{R}$ denote the value group.

In another paper [4], we studied some elementary hyperbolic geometry over F. The main result was that the hyperbolic plane HF^2 admits a natural topological pseudo-metric, $d(p,q) = \log(D(p,q))$, such that the associated metric space quotient TF^2 is a Λ-tree. (We will define $D(p,q)$ below.) The group $SL(2,F)$ acts as hyperbolic isometries of the hyperbolic plane HF^2 and also as metric isometries of the underlying tree TF^2.

In the paper of Morgan and Shalen, [15], it is shown how any valuation $v : F^* \to \Lambda$ leads to a Λ-tree on which $SL(2,F)$ acts. The points of this tree are homothety classes of $\mathcal{O}$-lattices in F^2, where $\mathcal{O} \subset F$ is the valuation ring of v. This construction leads to ordinary simplicial trees when $\Lambda = \mathbf{Z}$, and this case is thoroughly studied in Serre's book, [17].

Since a real closed field, F, contains all positive integer roots of its positive elements, the value group $\Lambda = \log(F^+) \subset \mathbf{R}$ is always a *divisible* group. However, the field of definition of $\rho : \pi_1(\Sigma) \to SL(2,F)$ is really some finitely generated function field $K \subset F$, and F is the real closure of K. The Λ-tree TF^2 contains a Λ_K-subtree, TK^2, where $\Lambda_K = \log(K^+) \subset \Lambda$. The tree TK^2 is a quotient of the hyperbolic plane HK^2 over K, and $\pi_1(\Sigma)$ acts on TK^2, via $\rho : \pi_1(\Sigma) \to SL(2,K) \subset SL(2,F)$. If, for example, F has transcendence degree one over $\mathbf{R}$, then K is the function field of a real algebraic curve, and necessarily $\Lambda_K = \mathbf{Z}$ and $\Lambda = \mathbf{Q}$. In Proposition 4.2, we showed such points are dense in the boundary $\widetilde{B}_g = \widetilde{T}_g - T_g$.

We also showed in [4] that the quotient tree TF^2 of the hyperbolic plane HF^2 coincides with the Morgan-Shalen tree associated to the valuation $v : F^* \to \Lambda$, where $v(a) = -\log(|a|)$, and the analogous result for the Λ_K-tree quotient of HK^2, $K \subset F$. However, all the desirable properties of these trees and the actions of surface groups $\pi_1(\Sigma)$ induced by discrete, faithful representations $\pi_1(\Sigma) \to SL(2,K) \subset SL(2,F)$, can be directly established in the context of non-Archimedean hyperbolic geometry. We will very briefly sketch here the main step in the construction of the Λ-tree quotient of the hyperbolic plane HF^2, in order to give the flavor of our approach.

Interpret the hyperbolic plane HF^2 as the upper half plane of the field $C = F[i]$, that is, $HF^2 = \{x + iy \mid x, y \in F,\ y > 0\}$. Define a cross-ratio type, F-valued, "multiplicative distance" on HF^2 by the formula

$$D(p,q) = \frac{1+t}{1-t} \geq 1 \in F,\ t = \frac{\|p-q\|}{\|\bar{p}-q\|},\ p,q \in HF^2\ . \tag{8.1}$$

Here, if $z = x + iy \in C = F[i]$ then $\bar{z} = x - iy \in C$ and $\|z\| = \sqrt{x^2+y^2} \in F$, as usual. We call D a "multiplicative distance" because of the triangle inequality $D(p,r) \leq D(p,q)D(q,r)$, which holds for all $p,q,r \in HF^2$, with equality if and only if p,q,r are hyperbolically collinear. The group $SL(2,F)$ acts on $HF^2 \subset C$ by the usual formula $\begin{pmatrix} a & b \\ c & d \end{pmatrix} z = \frac{az+b}{cz+d}$, and the cross-ratio D is $SL(2,F)$-invariant.

Of course, the same cross-ratio formula occurs in classical hyperbolic geometry over $\mathbf{R}$. In fact, the hyperbolic distance on $H\mathbf{R}^2$ is given by $d(p,q) = \log(D(p,q))$. Now, this same formula defines a function $d : HF^2 \times HF^2 \to \Lambda \subset \mathbf{R}$, for any F. Moreover, the multiplicative triangle inequality for D implies an additive triangle inequality for d. Thus, $d(p,q) = \log(D(p,q))$ is a topological pseudo-metric on HF^2. Set $TF^2 = HF^2 \big/ \sim$, where $p \sim q$ means $d(p,q) = 0$. Then TF^2 is a Λ-metric space. Since D is $SL(2,F)$-invariant, we also see that $SL(2,F)$ acts as metric isometries of the space TF^2.

The key idea in proving that the metric space TF^2 is actually a Λ-tree, is to show that if $p,q,r \in HF^2$ are three hyperbolically non-collinear points with distinct images in TF^2, then the image in TF^2 of the two-dimensional hyperbolic triangle $\Delta pqr \subset HF^2$ is one of the following one-dimensional pictures.

(8.2)

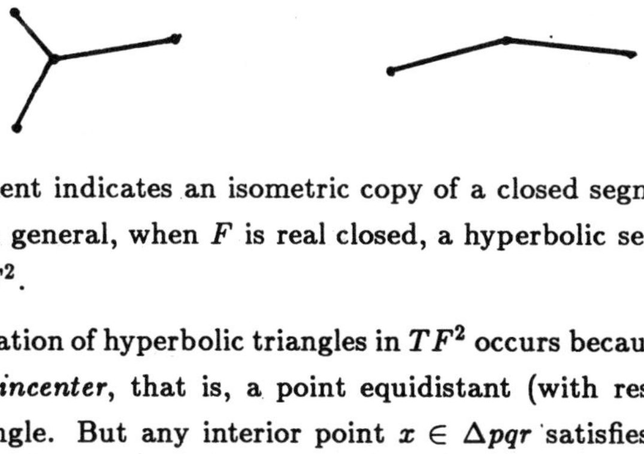

Here, a segment indicates an isometric copy of a closed segment in the ordered abelian group $\Lambda \subset \mathbf{R}$. In general, when F is real closed, a hyperbolic segment in HF^2 has image a Λ-segment in TF^2.

This degeneration of hyperbolic triangles in TF^2 occurs because every hyperbolic triangle in HF^2 has an *incenter*, that is, a point equidistant (with respect to D) from all three sides of the triangle. But any interior point $x \in \Delta pqr$ satisfies $D(x, \partial\Delta) < \sqrt{3}$, and our non-Archimedean logarithm satisfies $\log(\sqrt{3}) = 0$. Thus, if $o \in \Delta pqr$ is the incenter of a hyperbolic triangle then there are points p', q', r' on sides qr, pr, and pq, respectively, with $d(o,p') = d(o,q') = d(o,r') = 0$.

(8.3)

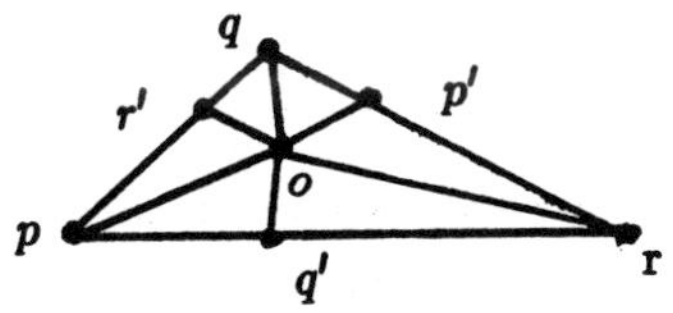

It is then relatively easy to prove that each of the small triangles in $\Delta pqr \subset HF^2$ in the Figure (8.3) has image a single segment in TF^2. If the image $[o] \in TF^2$ is distinct from

$[p], [q], [r] \in TF^2$, then $\Delta pqr \subset HF^2$ degenerates to three Λ-segments in TF^2, with a common endpoint $[o]$, as pictured on the left in Figure (8.2). If $[o]$ coincides with the image of one of the vertices, then the image of Δpqr in TF^2 consists of just two Λ-segments, as pictured on the right in Figure (8.2).

References

1. E. Becker, "On the real spectrum of a ring and its application to semi-algebraic geometry", *Bull. Amer. Math. Soc.* **15** (1986), 19–60.
2. G. W. Brumfiel, "A semi-algebraic Brouwer fixed point theorem for real affine space", preprint, Stanford University, 1987.
3. ———, "A Hopf fixed point theorem for semi-algebraic maps", preprint, Stanford University, 1987.
4. ———, "The tree of a non-Archimedean hyperbolic plane", preprint, Stanford University, 1987.
5. M. Coste and M.-F. Coste-Roy, "Topologies for real algebraic geometry", in *Topos Theoretic Methods in Geometry* (A. Kock, ed.), Various Publ. Series 30, Aarhus Universitet, 1979.
6. M.-F. Coste-Roy, "Spectre réel d'un anneau et topos étale réel", Thèse, Univ. Paris-Nord, 1980.
7. D. Dubois and G. Efroymson, "Algebraic theory of real varieties", in *Studies and Essays Presented to Yu-Why Chen on his 60th Birthday*, Taipei 1970, 107–135.
8. A. Fathi, F. Laudenbach, and V. Poenaru, "Travaux de Thurston sur les surfaces", *Astérisque* 66–67, (1979).
9. H. Helling, "Diskrete Untergruppen von $SL_2(\mathbf{R})$", *Inventiones Math.* **17** (1972), 217–229.
10. T. Jørgensen, "On discrete groups of Möbius transformations", *Amer. J. Math.* **98** (1976), 739–749.
11. J. L. Krivine, "Anneaux Préordonnés", *Journal D' Analyse Mathematique* **12** (1964), 307–326.
12. T. Y. Lam, "An introduction to real algebra", *Rocky Mountain J. Math.* **14** (1984), 767–814.
13. S. Lang, *Algebra*, 2nd ed., Addison-Wesley, Reading, Mass, 1967.
14. J. Milnor, *Singular points of complex hypersurfaces, Annals of Mathematics Studies 61,* Princeton University Press, Princeton, N. J., 1968.
15. J. Morgan and P. Shalen, "Valuations, trees, and degenerations of hyperbolic structures, I", *Ann. of Math.* **120** (1984), 407–476.

16. C. Procesi, "The invariant theory of $n \times n$ matrices", *Adv. in Math.* **19** (1976), 306–381.

17. J. P. Serre, *Trees*, Springer Verlag, Berlin-Heidelberg- New York, 1980.

18. W. Thurston, "The geometry and topology of 3-manifolds" Notes, Princeton University.

MATHEMATICS DEPARTMENT
STANFORD UNIVERSITY
STANFORD, CA 94305-2125

Contemporary Mathematics
Volume **74**, 1988

A SEMI-ALGEBRAIC BROUWER FIXED POINT THEOREM FOR REAL AFFINE SPACE

G. W. Brumfiel*

Introduction. It would seem that there is no such theorem. For example, the map $\varphi(x) = x + 1$ of the real line to itself has no fixed points. However, if the real line is compactified by adding two points, at $+\infty$ and $-\infty$, then the map φ extends continuously and fixes the boundary points.

In semi-algebraic geometry, real affine n-space means the real spectrum of a polynomial ring, $\mathbf{A}^n = \operatorname{Spec}_R(\mathbf{R}[x_1 \cdots x_n])$. If A is a commutative ring, a point α of the real spectrum $\operatorname{Spec}_R(A)$ is given by a prime ideal $p(\alpha) \subset A$ together with a total order on the integral domain $A/p(\alpha)$. Equivalently, the field of fractions of $A/p(\alpha)$ is given the structure of an ordered field. If $f \in A$ then $f(\alpha) > 0$ makes sense, namely, the image of f in $A/p(\alpha)$ is either positive, negative, or zero. A topology on $\operatorname{Spec}_R(A)$ is defined by taking as a subbasis for the open sets, all sets $U(f) = \{\alpha \in \operatorname{Spec}_R(A) \mid f(\alpha) > 0\}, \quad f \in A$.

In general, $\operatorname{Spec}_R(A)$ is a quasi-compact space, but points need not be closed. We will summarize further properties of the real spectrum in the next section, referring to the articles [1], [2], [6] for details. Note that if $A = \mathbf{R}[x_1 \cdots x_n]$, then $\mathbf{R}^n \subset \mathbf{A}^n = \operatorname{Spec}_R(\mathbf{R}[x_1 \cdots x_n])$, because a point of $\mathbf{R}^n$ is a homomorphism in $\mathbf{R}[x_1 \cdots x_n] \to \mathbf{R}$.

The subspace of closed points of $\mathbf{A}^1 = \operatorname{Spec}_R(\mathbf{R}[x])$ is exactly the two point compactification $\mathbf{R} \cup \pm\infty$. In general, the subspace $\mathbf{A}^n_c \subset \mathbf{A}^n$ of closed points of $\mathbf{A}^n$ is a compact, Hausdorff space, which contains $\mathbf{R}^n$ as a dense open set, (a compactification of $\mathbf{R}^n$). For all n, the boundary $\partial\mathbf{A}^n_c = \mathbf{A}^n_c - \mathbf{R}^n$ is "at infinity". That is, if $\alpha \in \partial\mathbf{A}^n_c$ then the image of some x_i is infinitely large relative to $\mathbf{R}$ in the ordered domain $\mathbf{R}[x_1 \cdots x_n]/p(\alpha)$. This real spectrum boundary of $\mathbf{R}^n$ is a Čech homology $(n-1)$-sphere, but setwise it is very complicated if $n > 1$.

A semi-algebraic map is a continuous function whose graph is a semi-algebraic set. Any semi-algebraic map $\varphi : \mathbf{R}^n \to \mathbf{R}^n$ has a unique continuous extension $\widetilde{\varphi} : \mathbf{A}^n \to \mathbf{A}^n$. The map $\widetilde{\varphi}$ does not necessarily take closed points to closed points if φ is not a homeomorphism, nor does $\widetilde{\varphi}$ necessarily take points at infinity to points at infinity if φ is not proper. The purpose of this note is to prove the following fixed point theorem.

1980 **Mathematics Subject Classification** (1985 **Revision**), 55M20.

* Research supported by NSF grant DMS85-06816 at Stanford University.

Theorem. If $\varphi : \mathbf{R}^n \to \mathbf{R}^n$ is any semi-algebraic map, then either φ has a fixed point or $\widetilde{\varphi} : \mathbf{A}^n \to \mathbf{A}^n$ has a closed fixed point at infinity.

Before turning to the proof, we make a few remarks. Our proof of the theorem is based on the property of $\mathbf{A}^n$ that finite open covers of $\mathbf{A}^n$ correspond to nice open covers of $\mathbf{R}^n$, in fact, they correspond essentially to finite, open, semi-algebraic covers of $\mathbf{R}^n$. Use of such covers allows one to compute the Čech homology of various spaces associated to the real spectrum $\mathbf{A}^n$. For example, the pairs $(\mathbf{A}^n, \mathbf{A}^n_\infty)$ and $(\mathbf{A}^n_c, \partial\mathbf{A}^n_c)$ have the same Čech homology sequence as (D^n, S^{n-1}). Here, $\mathbf{A}^n_\infty \subset \mathbf{A}^n$ is the closed subspace of all points at infinity, $\mathbf{A}^n_c \subset \mathbf{A}^n$ is the subspace of closed points, $\partial\mathbf{A}^n_c = \mathbf{A}^n_c \cap \mathbf{A}^n_\infty$ as defined earlier, D^n is the closed n-disc in $\mathbf{R}^n$, and S^{n-1} is its sphere boundary. These homology computations by themselves do not seem quite enough to deduce the fixed point theorem above. However, a Čech homology proof can be given, analogous to a seldom seen proof of the classical Brouwer fixed point theorem, exploiting the diagonal copy of $\mathbf{R}^n$ and the graph of $\varphi : \mathbf{R}^n \to \mathbf{R}^n$ which both lie in $\mathbf{R}^n \times \mathbf{R}^n$. In this paper we give a more elementary proof, by reducing the theorem to the classical Brouwer fixed point theorem for the disc. Of course, something like homology is usually used to prove that.

The next remark is that the theorem extends to any real closed ground field k. That is, a semi-algebraic map $\varphi : k^n \to k^n$ extends to $\widetilde{\varphi} : \mathbf{A}^n_k \to \mathbf{A}^n_k$, where $\mathbf{A}^n_k = \mathrm{Spec}_R(k[x_1 \cdots x_n])$, and $\widetilde{\varphi}$ will always fix either a point of $k^n \subset \mathbf{A}^n_k$ or a closed point at infinity relative to k. If $k \neq \mathbf{R}$, there are closed points in $\mathbf{A}^n_k$ which are not in k^n and yet which are also not at infinity relative to k. This is because k admits non-trivial Dedekind cuts. Thus, the theorem is superficially a little stronger than just asserting that $\widetilde{\varphi} : \mathbf{A}^n_k \to \mathbf{A}^n_k$ has a closed fixed point. If $k = \mathbf{R}$, all closed points of $\mathbf{A}^n$ are automatically either in $\mathbf{R}^n$ or at infinity, as previously mentioned.

The extension of the theorem to arbitrary real closed fields will be treated in a later paper. This is not trivial, and depends first on the existence of basic semi-algebraic topology, including things like triangulations and homology over arbitrary ground fields, as developed in [3], [4]. We will not say anything more about ground fields $k \neq \mathbf{R}$ in this paper.

A key tool in the proof of the fixed point theorem we give here is the existence of triangulations of finite collections of semi-algebraic sets, [5]. This means that given a semi-algebraic set V and a finite collection of semi-algebraic subsets $V_j \subset V$, there exists a finite simplicial complex K and a semi-algebraic homeomorphism from V to some union of (open) simplices of K, carrying each V_j also to a union of simplices. Since V need not be compact and $V_j \subset V$ need not be closed, V and the V_j are definitely not closed subcomplexes of K in general, just unions of simplices.

In the next section, we summarize the facts about real spectra which we need. Following that, we give the proof of the fixed point theorem. In the final section of the paper, we briefly outline an application to compactifications of Teichmüller spaces.

Preliminaries on the real spectrum. If A is a commutative ring, it follows immediately from the definition of the topology on $\mathrm{Spec}_R(A)$ given in the introduction that if

$\alpha, \beta \in \mathrm{Spec}_R(A)$ then β is in the closure of α, (written $\alpha \to \beta$ and pronounced "β specializes α"), if and only if for all $f \in A$, $f(\beta) > 0$ implies $f(\alpha) > 0$, that is, $\beta \in U(f)$ implies $\alpha \in U(f)$. As a consequence, suppose $\beta \not\to \gamma$ and $\gamma \not\to \beta$. Then there exists $f \in A$ with $f(\beta) > 0$ and $f(\gamma) < 0$. Namely, $\beta \not\to \gamma$ implies there is $g \in A$ with $g(\gamma) > 0$ and $g(\beta) \leq 0$ and, similarly, $\gamma \not\to \beta$ implies there is $h \in A$ with $h(\beta) > 0$ and $h(\gamma) \leq 0$. Now take $f = h - g$.

A somewhat less elementary result is that given $\alpha \in \mathrm{Spec}_R(A)$, there exists a unique β such that β is closed and $\alpha \to \beta$. It follows that any collection of open subsets of the quasi-compact space $\mathrm{Spec}_R(A)$ which covers the closed points, actually covers all of $\mathrm{Spec}_R(A)$ and hence has a finite subcover. The reader not familiar with these basic properties of the real spectrum, for example, quasi-compactness, can find them developed in [1], [2], [6].

We now turn to more specific properties of $\mathbf{A}^n = \mathrm{Spec}_R(\mathbf{R}[x_1 \cdots x_n])$. If $f \in \mathbf{R}[x_1 \cdots x_n]$ then $U(f) \cap \mathbf{R}^n = \{\bar{x} \in \mathbf{R}^n \mid f(\bar{x}) > 0\}$. In general, intersection with $\mathbf{R}^n \subset \mathbf{A}^n$ defines a correspondence from the Boolean algebra of subsets of $\mathbf{A}^n$ generated by the $U(f)$, (the constructible sets), onto the Boolean algebra of semi-algebraic subsets of $\mathbf{R}^n$. The Artin-Lang theorem asserts exactly that this correspondence is a bijection, [1]. Moreover, another theorem, known as the finiteness theorem, asserts that this bijection of Boolean algebras also corresponds open constructible sets in $\mathbf{A}^n$ bijectively with open semi-algebraic sets in $\mathbf{R}^n$, [2]. This is important in studying the extension $\widetilde{\varphi} : \mathbf{A}^n \to \mathbf{A}^n$ of $\varphi : \mathbf{R}^n \to \mathbf{R}^n$, which we now discuss.

Each point $\alpha \in \mathbf{A}^n$ determines an ultrafilter of constructible subsets of $\mathbf{A}^n$, namely, the principal ultrafilter of all constructible sets which contain α. The ultrafilter theorem asserts that this correspondence is, in fact, a bijection from $\mathbf{A}^n$ to the set of all ultrafilters of constructible sets, [6]. Therefore, because of the Artin-Lang correspondence, points of $\mathbf{A}^n$ also correspond to ultrafilters of semi-algebraic sets in $\mathbf{R}^n$.

Given a semi-algebraic map $\varphi : \mathbf{R}^n \to \mathbf{R}^n$ and an ultrafilter $\mathbf{S} = \{S_i\}$ of semi-algebraic sets in $\mathbf{R}^n$, we claim that the filter $\widetilde{\varphi}(\mathbf{S})$ generated by the images $\{\varphi(S_i) \mid S_i \in \mathbf{S}\}$, is, in fact, an ultrafilter. (Images and inverse images of semi-algebraic sets under semi-algebraic maps are semi-algebraic, by the Tarski theorem.) Namely, if $Y \subset \mathbf{R}^n$ is any semi-algebraic set, then either $\varphi^{-1}(Y)$ or $\varphi^{-1}(\mathbf{R}^n - Y)$ belongs to $\mathbf{S}$. Therefore, since $Y \supset \varphi\varphi^{-1}(Y)$ and $\mathbf{R}^n - Y \supset \varphi\varphi^{-1}(\mathbf{R}^n - Y)$, we conclude that either Y or $\mathbf{R}^n - Y$ belongs to $\widetilde{\varphi}(\mathbf{S})$.

The map $\widetilde{\varphi} : \mathbf{A}^n \to \mathbf{A}^n$ is then just the map which assigns the ultrafilter $\widetilde{\varphi}(\mathbf{S})$ defined above to an ultrafilter $\mathbf{S}$. The continuity of $\widetilde{\varphi}$ is a consequence of the fact that $\widetilde{\varphi}^{-1}(U(f))$ identifies (via the various correspondences above) with the constructible subset of $\mathbf{A}^n$ corresponding to the open semi-algebraic set $\varphi^{-1}(U(f) \cap \mathbf{R}^n) \subset \mathbf{R}^n$. Thus, by the finiteness theorem, $\widetilde{\varphi}^{-1}(U(f))$ is open.

We now have two descriptions of points $\alpha \in \mathbf{A}^n$, first as prime ideals $p(\alpha) \subset \mathbf{R}[x_1 \cdots x_n]$ together with an order on $\mathbf{R}[x_1 \cdots x_n]/p(\alpha)$, and, secondly, as ultrafilters $\mathbf{S}_\alpha$ of semi-algebraic sets in $\mathbf{R}^n$. The relation between the two descriptions is as follows. Let $V_\alpha \subset \mathbf{R}^n$ denote the subvariety corresponding to the prime ideal $p(\alpha) \subset \mathbf{R}[x_1 \cdots x_n]$. Then $V_\alpha \in \mathbf{S}_\alpha$, and for all $S \in \mathbf{S}_\alpha$, $S \cap V_\alpha$ is Zariski-dense in V_α, that is, the only polynomials which vanish on $S \cap V_\alpha$

are the polynomials in the prime ideal $p(\alpha)$. Moreover, for any $f \in \mathbf{R}[x_1 \cdots x_n]$, $f(\alpha) > 0$ if and only if $f \mid S > 0$ for some $S \in \mathbf{S}_\alpha$.

If $\alpha \in \mathbf{A}^n$, let $\dim(\alpha) = \dim(V_\alpha) = \dim_{\mathbf{R}}(\mathbf{R}[x_1 \cdots x_n]/p(\alpha))$. If $\alpha \to \beta$ then $p(\alpha) \subseteq p(\beta)$, since $f(\alpha) = 0$ implies $f(\beta) = 0$, and therefore $\dim(\alpha) \geq \dim(\beta)$, with equality only if $\alpha = \beta$. A semi-algebraic map $\varphi : \mathbf{R}^n \to \mathbf{R}^n$ cannot increase the dimension of any subset, that is, for any semi-algebraic set $S \subset \mathbf{R}^n$, $\dim(\varphi(S)) \leq \dim(S)$. We then see from the ultrafilter description of the map $\widetilde{\varphi} : \mathbf{A}^n \to \mathbf{A}^n$ that it is impossible that $\widetilde{\varphi}(\beta) \to \beta$ unless $\widetilde{\varphi}(\beta) = \beta$.

If $\beta \in \mathbf{A}^n$ is closed then it is also impossible that $\beta \to \widetilde{\varphi}(\beta)$ unless $\widetilde{\varphi}(\beta) = \beta$. We conclude that if $\beta \in \mathbf{A}^n$ is a closed point which is not fixed by $\widetilde{\varphi}$, then $\beta \not\to \widetilde{\varphi}(\beta)$ and $\widetilde{\varphi}(\beta) \not\to \beta$, hence there exists $f_\beta \in \mathbf{R}[x_1 \cdots x_n]$ with $f_\beta(\beta) > 0$ and $f_\beta(\widetilde{\varphi}(\beta)) < 0$. Consider the open set $U_\beta = U(f_\beta) \cap \widetilde{\varphi}^{-1}(U(-f_\beta))$. Then $\beta \in U_\beta$ and for all $\alpha \in U_\beta$ we have $\widetilde{\varphi}(\alpha) \notin U_\beta$. We will now prove the fixed point theorem stated in the introduction, by exploiting these neighborhoods U_β of closed points β which are not fixed by $\widetilde{\varphi}$.

Proof of the theorem. The points at infinity, $\mathbf{A}^n_\infty \subset \mathbf{A}^n$, form a closed, hence quasi-compact, subspace of $\mathbf{A}^n$. Namely, $\mathbf{A}^n_\infty$ is the intersection of the complements of the open "balls" $U(r^2 - \sum x_i^2) \subset \mathbf{A}^n$, $r \in \mathbf{R}$. Suppose $\widetilde{\varphi} : \mathbf{A}^n \to \mathbf{A}^n$ has no closed fixed point in $\mathbf{A}^n_\infty$. Then for each closed $\beta \in \mathbf{A}^n_\infty$, find a neighborhood U_β of β such that for all $\alpha \in U_\beta$, $\widetilde{\varphi}(\alpha) \notin U_\beta$, as constructed in the preceeding section. Since $\mathbf{A}^n_\infty$ is quasi-compact, we may choose a finite subcover, say $\mathbf{A}^n_\infty \subset U = \cup U_i$.

Let $V_i = U_i \cap \mathbf{R}^n$, an open semi-algebraic set in $\mathbf{R}^n$. Then $V = \cup V_i$ is a neighborhood of infinity in $\mathbf{R}^n$, that is, V contains the complement of some ball. This is because the closed set $\mathbf{A}^n - U \subset \mathbf{A}^n$ is quasi-compact and is covered by the "balls" $U(r^2 - \sum x_i^2) \subset \mathbf{A}^n$, hence in fact, $\mathbf{A}^n - U \subset U(r^2 - \sum x_i^2)$ for some $r \in \mathbf{R}$.

Since $V_i \subset U_i$, we know that for all $\vec{x} \in V_i$, $\varphi(\vec{x}) \notin V_i$. We will prove that $\varphi : \mathbf{R}^n \to \mathbf{R}^n$ has a fixed point. In fact, we will find a large closed ball $B^n \subset \mathbf{R}^n$, whose interior $\mathring{B}^n$ contains the complement of V, and a retraction $r : \mathbf{R}^n \to B^n$, which maps $\mathbf{R}^n - \mathring{B}^n$ onto ∂B^n and which preserves the decomposition $\mathbf{R}^n - \mathring{B}^n = \cup(V_i \cap (\mathbf{R}^n - \mathring{B}^n))$. That is, $\vec{x} \in V_i \cap (\mathbf{R}^n - \mathring{B}^n)$ if and only if $r(\vec{x}) \in V_i \cap \partial B^n$.

The point is, $r\varphi : \mathbf{R}^n \to \mathbf{R}^n \to B^n$ must have a fixed point in B^n by the classical Brouwer fixed point theorem. If $\vec{x} \in \partial B^n$ and $\varphi(\vec{x}) \notin \mathring{B}^n$, suppose $\vec{x} \in V_i$ and $\varphi(\vec{x}) \in V_j - V_i$. Then also $r\varphi(\vec{x}) \in V_j - V_i$, so $\vec{x} \neq r\varphi(\vec{x})$. The conclusion is, any fixed point of $r\varphi$ must be a point $\vec{x} \in \mathring{B}^n$. But $r\varphi(\vec{x}) = \vec{x}$ then implies $\varphi(\vec{x}) \in \mathring{B}^n$ also, so $r\varphi(\vec{x}) = \varphi(\vec{x})$ and hence $\varphi(\vec{x}) = \vec{x}$.

It remains to construct the retraction $r : \mathbf{R}^n \to B^n$. Choose a triangulation of $\mathbf{R}^n$ by smoothly embedded simplices so that all the open sets V_i are subcomplexes, that is, unions of simplices. Choose a big ball B^n so that all bounded simplices of the triangulation are contained in $\mathring{B}^n$. Also choose B^n large enough so that for each simplex σ which meets $\mathbf{R}^n - \mathring{B}^n$, the map $\sum x_i^2 : \sigma \cap (\mathbf{R}^n - \mathring{B}^n) \to \mathbf{R}$ has no critical points. This can always be done because the set of radii for which σ is somewhere tangent to the sphere of that radius is a 0-dimensional semi-algebraic set, hence finite.

For each simplex σ which meets $\mathbf{R}^n - \overset{\circ}{B^n}$, we now want to parametrize $\sigma \cap (\mathbf{R}^n - \overset{\circ}{B^n})$ as $(\sigma \cap \partial B^n) \times [b^2, \infty)$, so that projection onto $[b^2, \infty)$ is the map $\sum x_i^2$. (Here, b is the radius of B^n.) We also want to do this continuously at faces of simplices, that is, we want $\bar{\sigma} \cap (\mathbf{R}^n - \overset{\circ}{B^n})$ parametrized as $(\bar{\sigma} \cap \partial B^n) \times [b^2, \infty)$, so that two such parametrizations for $\bar{\sigma}$ and $\bar{\tau}$ coincide on $\bar{\sigma} \cap \bar{\tau} \cap (\mathbf{R}^n - \overset{\circ}{B^n})$. This is easy in differential topology, where one can integrate a suitable vector field, but it probably can also be done semi-algebraically. The retraction $r : \mathbf{R}^n \to B^n$ is then defined as projection $(\sigma \cap \partial B^n) \times [b^2, \infty) \to \sigma \cap \partial B^n$ on each simplex $\sigma \cap (\mathbf{R}^n - \overset{\circ}{B^n})$. Since the open sets V_i covering infinity are unions of simplices of the triangulation, it is clear that the retraction r preserves the decomposition $\mathbf{R}^n - \overset{\circ}{B^n} = \cup V_i \cap (\mathbf{R}^n - \overset{\circ}{B^n})$ as desired.

An application to Teichmüller theory. The various Teichmüller spaces of marked Riemann surfaces, possibly with finitely many punctures and boundary curves, typically have semi-algebraic structures homeomorphic to real affine spaces [7]. Moreover, the modular group acts as semi-algebraic homeomorphisms of Teichmüller space. Therefore, the modular group acts continuously on the real spectrum compactification of Teichmüller space, and, by the theorem proved in this paper, each element of the modular group must have a fixed point in this compactification.

In another paper, we will show how the Thurston compactification of Teichmüller space is a continuous image of the real spectrum compactification, equivariant with respect to the action of the modular group. (This construction was motivated by, and is similar to, the methods developed in [7].) Therefore, one can conclude that each element in the modular group has a fixed point in the Thurston compactification of Teichmüller space, without first proving that the Thurston compactification is, in fact, a closed disc.

Although it is true that, by itself, the existence of fixed points is not sufficient to draw strong conclusions about elements of the modular group, the author hopes that further study of the dynamics of self-homeomorphisms of real spectra will prove to be very interesting and useful in the theory of group representations.

References

1. E. Becker, *On the real spectrum of a ring and its application to semialgebraic geometry,* Bull. A. M. S. **15** (1986), 19-60.

2. M. Coste and M. F. Coste-Ray, *La topologie du spectre réel,* Contemporary math **8** (1982), 27-60.

3. H. Delfs, *Kohomologie affiner semialgebraischen Räume,* Diss. Univ. Regensburg, 1980.

4. H. Delfs and M. Knebusch, *Semialgebraic topology over real closed fields II; Basic theory of semialgebraic spaces,* Math. Z. **178** (1981), 175-213.

5. H. Hironaka, *Triangulations of algebraic sets,* Algebraic Geometry, Arcata 1974, A. M. S. Proc. Symp. Pure Math. **29** (1975), 165-184.

6. T. Y. Lam, *An introduction to real algebra,* Rocky Mountain J. Math. **14** (1984), 767-814.

7. J. Morgan and P. Shalen, *Valuations, trees, and degenerations of hyperbolic structures, I*, Ann. of Math. **120** (1984), 407-476.

MATHEMATICS DEPARTMENT
STANFORD UNIVERSITY
STANFORD, CA 94305–2125, USA

Contemporary Mathematics
Volume **74**, 1988

THE TREE OF A NON-ARCHIMEDEAN HYPERBOLIC PLANE

G. W. Brumfiel*

Introduction.

In this paper, we study the hyperbolic plane, HF^2, over a non-Archimedean ordered field F. Actually, for the main results, we work only with microbial ordered fields F (all definitions are given in the text below), but the class of microbial ordered fields is large enough to include all ordered fields of finite transcendence degree over a subfield of the real numbers, $\mathbf{R}$. We define a real valued topological pseudo-metric on HF^2, $d(p,q) = \log D(p,q) \in \mathbf{R}$, where p, q are points of HF^2, $D(p,q) \geq 1$ is an algebraically defined cross-ratio, taking values in an algebraic extension field $\overline{F}$ of F, and $\log(D)$ is a real valued logarithm, defined on the positive elements of $\overline{F}$ by comparing powers of positive elements with powers of a sufficiently large fixed element of F.

The main theorem is that the metric space quotient TF^2, of HF^2, determined by the pseudometric d, is a (generalized) tree. The group $PGL_+(2,F)$ of direct hyperbolic isometries of HF^2 acts as metric isometries of the tree TF^2. Thus all subgroups of $PGL_+(2,F)$ act on HF^2 and TF^2.

In another paper, [1], we will interpret boundary points of the real spectrum compactification of the Teichmüller space of a surface Σ, as certain (equivalence clases of) representations $\pi_1(\Sigma) \to SL(2,F)$, where F is a non-Archimedean ordered field. Thus, a boundary point of the real spectrum compactification of the Teichmüller space of Σ can be interpreted as a certain type of representation of $\pi_1(\Sigma)$ as hyperbolic isometries of a non-Archimedean hyperbolic plane. (This fits very nicely with the interpretation of ordinary points of Teichmüller space as actions of $\pi_1(\Sigma)$ on the classical real hyperbolic plane.) On the other hand, the paper of John Morgan and Peter Shalen, [2], interprets a boundary point of the Thurston compactification of Teichmüller space as a certain type of representation of $\pi_1(\Sigma)$ as metric isometries of a (generalized) tree. The real spectrum compactification maps naturally onto the Thurston compactification, [1]. On the boundary this map turns out to be exactly the passage from the action of $\pi_1(\Sigma)$ as hyperbolic isometries of planes HF^2 to the induced action of $\pi_1(\Sigma)$ as metric isometries of the underlying trees TF^2.

The Morgan-Shalen work, which inspired our approach, exploits the trees associated to valued fields, as studied, say, in Serre's book, [3]. Our non-Archimedean ordered fields admit

1980 **Mathematics Subject Classification** (1985 **Revision**), 51M10.
* Research supported by NSF grant DMS85–06816 at Stanford University

natural, order compatible, rank 1 valuations, (essentially the real valued logarithm mentioned above), and this provides the very close connection between the two approaches. In this sense, our approach is simply a reformulation of the work of Morgan and Shalen. Nonetheless, it was perhaps a bit of a mystery why valuations of fields and groups of metric isometries of trees should be so closely related to the Thurston compactification of Teichmüller space. We believe our approach clarifies these relationships.

If every positive element of F is a square in F, (that is, if F is a Euclidean field), then the study of hyperbolic geometry over F and the study of the tree TF^2 is particularly easy. Probably, this case is adequate for all the applications to Teichmüller theory, since any ordered field admits canonical Euclidean extension fields. The study of the tree TF^2 in the general case is quite a bit more delicate. It seems to require use of the Cauchy completion of the ordered field F to establish the desired properties of TF^2. Much of the length of this paper is due to our decision to develop the theory for arbitrary (microbial) ordered fields F. Essentially, the added difficulty is caused by the fact that the group $PGL_+(2, F)$ does not act transitively on hyperbolic lines in HF^2, if F is not Euclidean.

Although the real valued logarithm, $\log : F^+ \to \mathbf{R}$, is defined without reference to any hyperbolic geometry, there is an interesting historical connection. In axiomatic, coordinate-free synthetic geometry, the method of Eudoxus for "comparing the lengths" of two segments applies to non-Euclidean and non-Archimedean geometries, as well as to classical Euclidean geometry. That is, one measures the ratio of two segments by comparing iterated translations of one along a line with iterated translations of the other. If seven translations of one segment lies between three and four translations of the second, then the ratio of lengths is between $3/7$ and $4/7$. Translations along a line are defined synthetically, without reference to coordinates. If an ordered field, F, of coordinates is introduced in a hyperbolic geometry, then the group of translations along a line is essentially the *multiplicative group*, F^+. The Eudoxus ratio of two segment lengths, which is a Dedekind cut of $\mathbf{Q}$, is exactly the ratio of two real logarithms of elements of F^+.

Another bit of history related to the ideas of this paper is the fact that non-Archimedean ordered fields were considered long ago by Hilbert and others in the context of the foundations of geometry. That is, in thinking about the axioms of synthetic geometry and the introduction of a field of coordinates in both Euclidean and hyperbolic geometries, Hilbert and others thought about which theorems could be proved without Archimedes' axiom, and studied non-Archimedean models of axiom systems for geometry.

The metric tree quotient of a non-Archimedean hyperbolic plane studied in this paper can be generalized to hyperbolic n-space, although we do not consider higher dimensions here. Another interesting direction of generalization would be a study of Tits buildings associated to Lie groups of rank greater than one over a non-Archimedean ordered field.

Hyperbolic planes over ordered fields. Let F be an arbitrary ordered field and let $GL_+(2,F) \subset GL(2,F)$ denote the subgroup of invertible 2×2 matrices over F with positive determinant. We set $C = F[i]$, $i^2 = -1$, and consider the upper half plane over F,

$$HF^2 = \left\{ z \in C \;\middle|\; \mathrm{Im}(z) > 0 \right\} \cong \left\{ (x,y) \in F^2 \;\middle|\; y > 0 \right\}$$

The group $GL_+(2,F)$ acts on HF^2 by

$$\begin{pmatrix} a & b \\ c & d \end{pmatrix} z = \frac{az+b}{cz+d}, \quad z \in HF^2 \subset C. \tag{1}$$

The center subgroup F^* of diagonal matrices in $GL_+(2,F)$ acts trivially on HF^2, hence $PGL_+(2,F) = GL_+(2,F)\big/F^*$ acts on HF^2. This action is clearly transitive since $\begin{pmatrix} a & b \\ 0 & 1 \end{pmatrix} i = ai + b$.

Let $B = \left\{ z \in C \;\middle|\; \|z\| < 1 \right\} \cong \left\{ (x,y) \in F^2 \;\middle|\; x^2 + y^2 < 1 \right\}$ denote the unit disk over F. We can identify B and HF^2 via the Möbius transformations

$$\begin{aligned} t &: HF^2 \xrightarrow{\sim} B & t(z) &= \frac{z-i}{z+i} \\ t^{-1} &: B \xrightarrow{\sim} HF^2 & t^{-1}(w) &= i\left(\frac{1+w}{1-w}\right). \end{aligned} \tag{2}$$

Let $HS^2 = \left\{ (x,y,z) \in F^3 \;\middle|\; x^2 + y^2 + z^2 = 1, \quad z > 0 \right\}$ denote the upper hemisphere of $S^2 \subset F^3$. We can identify HS^2 and HF^2 via the "rational parametrization of the sphere" given by

$$\begin{aligned} u &: HF^2 \xrightarrow{\sim} HS^2 & u(x,y) &= \left(\frac{1-(x^2+y^2)}{1+x^2+y^2}, \frac{-2x}{1+x^2+y^2}, \frac{2y}{1+x^2+y^2} \right) \\ u^{-1} &: HS^2 \xrightarrow{\sim} HF^2 & u^{-1}(x,y,z) &= \left(\frac{-y}{1-x}, \frac{z}{1-x} \right). \end{aligned} \tag{3}$$

Let $B_0 = \left\{ (x,y) \in F^2 \;\middle|\; x^2 + y^2 + z^2 = 1, \text{ some } z > 0 \in F \right\}$. Clearly B_0 and HS^2 are identified via projection

$$\begin{aligned} v &: HS^2 \xrightarrow{\sim} B_0 & v(x,y,z) &= (x,y) \\ v^{-1} &: B_0 \xrightarrow{\sim} HS^2 & v^{-1}(x,y) &= \left(x, y, \sqrt{1-x^2-y^2} \right). \end{aligned} \tag{4}$$

We thus have four models for the points of the hyperbolic plane over F, namely HF^2, B, HS^2, and B_0. It is to be emphasized that in general $B_0 \subsetneq B$ since $(x,y) \in B_0$ if and only if $1 - (x^2 + y^2)$ is a positive *square* in F, while $(x,y) \in B$ whenever $1 - (x^2 + y^2)$ is positive. Nonetheless, the transformations t, u, v above identify all four models with one another. The figure below, which portrays some stereographic projections in F^3, interprets the transformations t, u, and v geometrically.

(5)

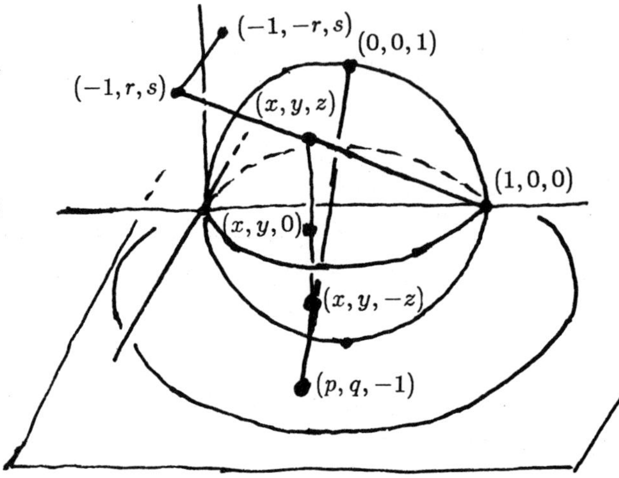

$$(x,y,z) \in HS^2$$
$$v(x,y,z) = (x,y) \in B_0$$
$$u^{-1}(x,y,z) = 1/2(-r,s) \in HF^2$$
$$tu^{-1}(x,y,z) = 1/2(p,q) \in B$$

The hyperbolic plane over F has a boundary circle in each of the four models. In the models B, HS^2, and B_0, the boundary circle is just the standard unit circle $S^1 = \left\{(x,y) \in F^2 \,\middle|\, x^2+y^2=1\right\}$. In the model HF^2, the boundary "circle" is $F \bigcup \infty$, where $F = \left\{z \in C \,\middle|\, \mathrm{Im}(z) = 0\right\}$ is the "real axis" in C. The transformations t, u, v above which identify the four models extend naturally to the boundary circles. This can be seen either in the formulas (2), (3), (4), or in the figure (5). The action of $PGL_+(2,F)$ on HF^2 extends to the boundary circle by formula (1). We let $PGL_+(2,F)$ act on the other three models B, HS^2, and B_0 via the identifications with HF^2. We will refer to HF^2 and B as the *Poincarè models* and to HS^2 and B_0 as the *Klein models* of the hyperbolic plane over F.

Hyperbolic planes over Euclidean fields. An ordered field F is said to be a *Euclidean field* if every positive element of F is a square in F. Given any ordered field F there exist algebraic extensions $F \subset \overline{F}$ such that $\overline{F}$ is a Euclidean field whose (unique) order extends the given order on F. For example, the real closure of F is such a Euclidean field. Inside the real closure of F there is a unique minimal Euclidean extension of F, called the *Euclidean closure* of F. In what follows, $\overline{F}$ will denote an arbitrary Euclidean algebraic extension of F, and $H\overline{F}^2$, $\overline{B}$, $H\overline{S}^2$, and $\overline{B}_0$ will denote the corresponding four models of the hyperbolic plane over $\overline{F}$. Of course, now $\overline{B} = \overline{B}_0$ as point-set, but the identification $v\,u\,t : \overline{B} \xrightarrow{\sim} H\overline{F}^2 \xrightarrow{\sim} H\overline{S}^2 \xrightarrow{\sim} \overline{B}_0$ is certainly not the identity.

The advantage of Euclidean fields is that *hyperbolic lines* are well-behaved. Specifically, in the Poincaré model $\overline{B}$, the hyperbolic lines are either diameters or arcs of circles in $\overline{F}^2$ which intersect the boundary circle of $\overline{B}$ orthogonally. In the upper half plane model $H\overline{F}^2$, the hyperbolic lines are either vertical lines or half-circles with diameter on the real axis boundary $\overline{F}$ of $H\overline{F}^2$. In the Klein model $\overline{B}_0$, the hyperbolic lines are the chords of the unit disc, which correspond under the transformation v of (4) to arcs of circles on $H\overline{S}^2$ obtained by intersecting $H\overline{S}^2$ with vertical planes in $\overline{F}^3$.

The transformations t and u of (2) and (3) which identify $H\overline{F}^2$, $\overline{B}$, and $H\overline{S}^2$ preserve lines. Also the group $PGL_+(2,\overline{F})$ acting on the hyperbolic plane preserves lines. Every line is determined by its two ideal endpoints on the boundary circle $\overline{S}^1$ of $\overline{B}$, and the group $PGL_+(2,\overline{F})$ acts transitively on lines.

The main point here is that in the *Euclidean* plane $\overline{F}^2$, the elementary continuity principle holds: a Euclidean line and a circle, or a pair of circles, intersect whenever they ought too. This elementary continuity principle does not hold in a general ordered field plane F^2 since one can't always solve the necessary quadratic equations rationally. For example, a hyperbolic line through two rational points in B_0 or B need not intersect the unit circle $S^1 \subset F^2$. (Example: the diameter through $(0,0)$ and $(1/3, 2/3)$ in $\mathbf{Q}^2$.) In general, $PGL_+(2,F)$ does not act transitively on rational hyperbolic lines. Despite this bad hyperbolic geometry in HF^2, we will have good reason to keep track of the rational points $HF^2 \subset H\overline{F}^2$ in what follows.

Let $SL(2,F) \subset GL_+(2,F)$ denote the subgroup of matrices with determinant 1. In the Euclidean field case, $SL(2,\overline{F}) \to PGL_+(2,\overline{F})$ is clearly surjective, with Kernel$= \{\pm Id\}$. In general, $SL(2,F) \to PGL_+(2,F)$ is not surjective. For example, the determinant (more precisely, the discriminant) in $F^+\big/(F^+)^2$ is an obstruction, where $(F^+)^2 \subset F^+ \subset F^*$ are the multiplicative subgroups of squares and positive elements respectively, in the non-zero elements $F^* \subset F$. In fact, $SL(2,F)$ does not even act transitively on HF^2 in general. One can check that if $z, z' \in HF^2 \subset C = F[i]$, then $\begin{pmatrix} a & b \\ c & d \end{pmatrix} z = z'$ for some $\begin{pmatrix} a & b \\ c & d \end{pmatrix} \in SL(2,F)$ if and only if $Im(z')\big/Im(z)$ is a sum of two squares in F. Thus the orbits of the action of $SL(2,F)$ on HF^2 correspond to elements of the quotient group $F^+\big/(\Sigma^2 F)^+$, where $(\Sigma^2 F)^+$ denotes the multiplicative subgroup of F^+ consisting of elements which are sums of two squares.

Let $RO(2,F) = \left\{ \begin{pmatrix} a & -b \\ b & a \end{pmatrix} \,\middle|\, a^2 + b^2 \neq 0 \right\} \subset PGL_+(2,F)$. Clearly, from (1), $RO(2,F)$ is the subgroup of $PGL_+(2,F)$ which fixes the point $i \in HF^2$. If $p = a + bi \in C$ and $R = \begin{pmatrix} a & -b \\ b & a \end{pmatrix} \in RO(2,F)$ is conjugated to a transformation of the unit disk B via (2), then R becomes the rotation multiplication by $w = \bar{p}^2/p\bar{p}$, where $\bar{p} = a - bi$. That is, R becomes the linear rotation $\begin{pmatrix} \frac{a^2-b^2}{a^2+b^2}, & \frac{2ab}{a^2+b^2} \\ \frac{-2ab}{a^2+b^2}, & \frac{a^2-b^2}{a^2+b^2} \end{pmatrix}$ of B. Conversely, the linear rotation $\begin{pmatrix} c & -d \\ d & c \end{pmatrix}$ of B, that is, multiplication by $w = c + di$, where $c^2 + d^2 = 1$, conjugates to the transformation $S = \begin{pmatrix} -d & c-1 \\ 1-c & -d \end{pmatrix} \in RO(2, F)$ on HF^2, at least if $w \neq 1$. If the equation $w = c + di = (a - bi)^2 = \bar{p}^2$ can be solved for $p = a + bi \in C$, which is always the case for a Euclidean field $\overline{F}$, then $S = \begin{pmatrix} -d & c-1 \\ 1-c & -d \end{pmatrix} \equiv \begin{pmatrix} a & -b \\ b & a \end{pmatrix} \in RO(2, F)$. But, in any case, $S = \begin{pmatrix} -d & c-1 \\ 1-c & -d \end{pmatrix} \in RO(2, F)$ is a rational Möbius transformation for any rational rotation $\begin{pmatrix} c & -d \\ d & c \end{pmatrix} \neq \begin{pmatrix} 1 & 0 \\ 0 & 1 \end{pmatrix} \in SO(2, F)$ of B.

Since $PGL_+(2, F)$ acts transitively on points of HF^2, the isotropy group of any point $z \in HF^2$ is a conjugate of the isotropy group $RO(2, F)$ of $i \in HF^2$, hence is isomorphic to the unit circle (of rotations) $SO(2, F)$. By way of contrast, the subgroup of $PGL_+(2, F)$ which leaves invariant the imaginary axis (that is, which fixes 0 and ∞) is the subgroup $TR(2, F) = \left\{ \begin{pmatrix} a & 0 \\ 0 & 1 \end{pmatrix} \middle| \, a > 0 \right\}$ of homotheties, $\begin{pmatrix} a & 0 \\ 0 & 1 \end{pmatrix} z = az, \quad z \in HF^2 \subset C$. Clearly $TR(2, F) \simeq F^+$. The invariant group of any line with two *rational* endpoints on the boundary $F \cup \infty$ of HF^2 will be a conjugate of $TR(2, F)$ in $PGL_+(2, F)$, but it is unclear what the invariant group of other lines in HF^2 looks like. Of course, for a Euclidean field $\overline{F}$, the invariant group of any line in $H\overline{F}^2$ is conjugate to $TR(2, \overline{F}) \subset PGL_+(2, \overline{F})$, and is isomorphic to the subgroup $\left\{ \begin{pmatrix} \alpha & 0 \\ 0 & \alpha^{-1} \end{pmatrix} \middle| \, \alpha > 0 \right\}$ of $SL(2, \overline{F})$.

We next consider a multiplicative hyperbolic distance in the Euclidean field case. In the Klein model $\overline{B}_0$ and the Poincaré model $\overline{B}$, we have cross-ratio formulas which are purely algebraic.

$\overline{B}_0 =$ 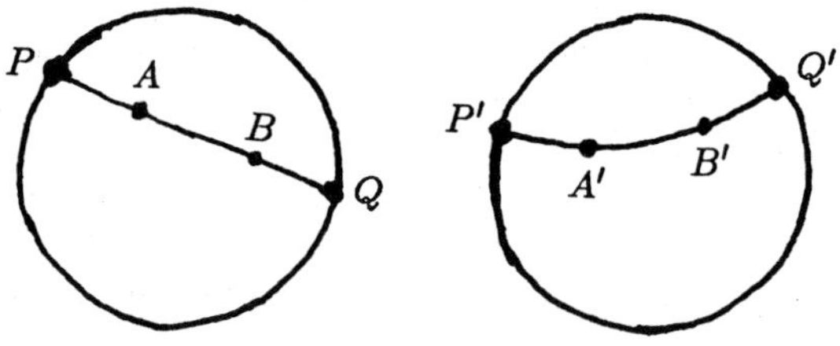 $= \overline{B}$

(6)
$$D(A, B) = \sqrt{\frac{|PB|\,|QA|}{|PA|\,|QB|}} \in \overline{F}$$
$$D'(A', B') = \frac{|P'B'|\,|Q'A'|}{|P'A'|\,|Q'B'|} \in \overline{F}$$

Here, if $U, V \in \overline{F}^2$, say $U = (x_1, y_1)$, $V = (x_2, y_2)$, then by $|UV|$ we mean the Euclidean distance $\sqrt{(x_1 - x_2)^2 + (y_1 - y_2)^2} \in \overline{F}$. If (P, A, B, Q) corresponds to (P', A', B', Q') under our identification of models $\overline{B}_0 \simeq \overline{B}$, then $D(A, B) = D'(A', B')$. This is a well-known computation in hyperbolic geometry which works for any Euclidean ground field $\overline{F}$.

We also have the following formula for D' in terms of complex coordinates $z_1' = x_1' + iy_1'$, $z_2' = x_2' + iy_2' \in \overline{B} \subset \overline{C} = \overline{F}[i]$.

(7) $$D'(z_1', z_2') = \frac{1+t'}{1-t'} \text{ where } t' = \|z_1' - z_2'\| \Big/ \|1 - \bar{z}_1' z_2'\|$$

If we translate D' over to the upper half plane model $H\overline{F}^2$ we obtain the following formula for two points $z_1'' = x_1'' + iy_1''$, $z_2'' = x_2'' + iy_2''$, $y_1'', y_2'' > 0$.

(8) $$D''(z_1'', z_2'') = \frac{1+t''}{1-t''} \text{ where } t'' = \|z_1 - z_2\| \Big/ \|\bar{z}_1 - z_2\|$$

We will simplify notation and just write D for all these formulas for D, D', and D''. It is then necessary to keep track of which model is being used in any specific discussion.

Note that even if we begin with a pair of rational points R, S in $B_0 \subset \overline{B}_0$, $B \subset \overline{B}$, or $HF^2 \subset H\overline{F}^2$, the number $D(R,S)$ belongs to $\overline{F}$, and is not generally in F. Here are the important properties of D.

(9) $$D(A,B) \geq 1, \text{ with equality if and only if } A = B.$$

(10) $$D(A,B)\, D(B,C) \leq D(A,C),$$ with equality if and only if A, B, C are (hyperbolically) collinear.

(11) $$D \text{ is } PGL_+(2,\overline{F})\text{-invariant.}$$

Microbial fields and real valued logarithms. An ordered field F is said to be *microbial* if there exists a positive element $s \in F$ such that $\lim_{n\to\infty} s^n = 0$. Such an element s is called a *microbe*. By a *big element* $b \in F$, we mean the reciprocal of a microbe. Thus $b \in F$ is big if b is positive and if for all $a \in F$ there exists an integer $m \geq 1$ such that $a < b^m$.

Given a big element $b \in F$, we define a *real valued logarithm* $\log_b : F^+ \to \mathbf{R}$ as follows. Given $a > 0$ in F, $\log_b(a) \in \mathbf{R}$ is the Dedekind cut of $\mathbf{Q}$ defined by

(12) $$\frac{m'}{n} \leq \log_b(a) \leq \frac{m}{n} \quad \text{if } b^{m'} \leq a^n \leq b^m,\ m', m, n \in \mathbf{Z},\ n > 0.$$

Here are the basic properties of this logarithm.

(13) $$\text{If } a, a' > 0 \text{ then } \log_b(aa') = \log_b(a) + \log_b(a').$$

(14) $$\text{If } 0 < a \leq a' \text{ then } \log_b(a) \leq \log_b(a').$$

(15) $$\text{If } b \text{ and } b' \text{ are two big elements, then } \log_{b'}(a) = \log_{b'}(b) \cdot \log_b(a).$$

If F is an Archimedean ordered field, that is, if $F \subset \mathbf{R}$, then $\log_b$ is the usual logarithm. If F is non-Archimedean and $b \in F$ is big then $1/b < r^n < b$ for positive rational numbers

$r \in \mathbf{Q} \subset F$ and integers n. It follows that $\log_b(r) = 0$. Since for all $a, a' \in F$ we have $|a + a'| \leq 2\ \mathrm{Max}(|a|, |a'|)$ where $|\ |$ means absolute value, we conclude from (13) and (14) and the fact that $\log_b(2) = 0$ the following.

(16) $$\log_b(|a + a'|) \leq \mathrm{Max}(\log_b(|a|), \log_b(|a'|)).$$

We set $\log_b(0) = -\infty$, so that (16) holds for all $a, a' \in F$.

Combining (13) and (16), we see that $v_b(a) = -\log_b(|a|)$ is a Krull valuation $v_b : F^* \to \mathbf{R}$. The valuation ring is $\mathcal{O} = \left\{a \in F \,\middle|\, \log_b(|a|) \leq 0\right\}$ with maximal ideal $\mathcal{M} = \left\{a \in F \,\middle|\, \log_b(|a|) < 0\right\}$. Formula (15) shows that replacing one big element, b, by another, b', just multiplies v_b by a positive real scalar and does not change $\mathcal{O}$ and $\mathcal{M}$. In other words, v_b and $v_{b'}$ are equivalent valuations. Formula (14) shows that the valuation v_b is compatible with the order on F. That is, $0 < a \leq a'$ implies $v_b(a) \geq v_b(a')$.

In general, a valuation ring in an ordered field, $\mathcal{O} \subset F$, is said to be *convex* or *order-compatible* if $0 < a \leq a'$ and $a' \in \mathcal{O}$ implies $a \in \mathcal{O}$. This is equivalent to convexity of the maximal ideal $\mathcal{M} \subset \mathcal{O}$, that is, $0 < a \leq a'$ and $a' \in \mathcal{M}$ implies $a \in \mathcal{M}$. Another formulation is that the residue field $k = \mathcal{O}/\mathcal{M}$ inherits an order so that projection $\mathcal{O} \to k$ is order preserving. If $\mathcal{O}$ and $\mathcal{O}'$ are two convex valuation rings in F, then either $\mathcal{O} \subset \mathcal{O}'$ or $\mathcal{O}' \subset \mathcal{O}$. In other words, the order compatible valuation rings in F form a chain under inclusion. It follows that there is at most one order compatible, rank 1 valuation ring $\mathcal{O} \in F$. (A valuation ring is of *rank 1* if the value group $\Lambda = F^* \big/ \mathcal{O}^*$, where $\mathcal{O}^* \subset \mathcal{O}$ denotes the units, is an Archimedean ordered group, that is, order isomorphic to an additive subgroup of $\mathbf{R}$. Recall that if $\lambda, \lambda' \in \Lambda$ are images of $a, a' \in F^*$, then $\lambda \leq \lambda'$ means $a'/a \in \mathcal{O}$.) Thus, if F is a microbial ordered field, the valuation $v_b : F^* \to \mathbf{R}$ constructed above, $v_b(a) = -\log_b(|a|)$, represents the unique rank 1, order compatible Krull valuation on F.

We now point out that the Cauchy completion of F with respect to the order topology and the valuation completion of F with respect to the valuation v_b are one and the same. Namely, since F is microbial, the order Cauchy completion is defined in terms of the sequence of neighborhoods of $0 \in F$ given by the intervals $\{(-s^n, s^n),\ n \geq 1\}$, where $s = 1/b$ is a microbe. The valuation completion is defined in terms of the neighborhoods of $0 \in F$ given by the additive subgroups $\mathcal{M}_n = \left\{a \in F \,\middle|\, v_b(a) \geq n\right\}$. But $-s^n < a < s^n$ implies $v_b(a) = -\log_b(|a|) \geq n$, which in turn implies $-s^{n-1} < a < s^{n-1}$, so the two neighborhood systems yield the same completion. We will denote this completion of F by $\widehat{F}$.

Let $\Lambda = \log_b(F^+) \subset \mathbf{R}$, the value group of v_b. Then Λ is a rank 1 ordered group, but not in general discrete, that is, Λ is not necessarily isomorphic to $\mathbf{Z}$. If $\Lambda \cong \mathbf{Z}$, then $\mathcal{O} \subset F$ is a discrete valuation ring and $\mathcal{M}_n = \mathcal{M}^n \subset \mathcal{O}$. In general, if $\lambda \in \Lambda$, $\lambda > 0$, set $\mathcal{M}_\lambda = \left\{a \in F \,\middle|\, v_b(a) \geq \lambda\right\}$. Then, since Λ is of rank 1,

$$\hat{\mathcal{O}} = \varprojlim_{\lambda \in \Lambda^+} \mathcal{O}/\mathcal{M}_\lambda = \varprojlim_{n \in \mathbf{Z}^+} \mathcal{O}/\mathcal{M}_n$$

and $\hat{\mathcal{O}}$ is a complete valuation ring, with field of fractions $\hat{F}$. The valuation $v_b : F^* \to \Lambda$ extends to $\hat{v}_b : \hat{F} \to \Lambda$, with valuation ring $\hat{\mathcal{O}} \subset \hat{F}$. The order on F also extends to $\hat{F}$, $b \in F \subset \hat{F}$ is still a big element, and $\hat{v}_b(\hat{a}) = -\log_b(|\hat{a}|)$. In particular, $\hat{v}_b$ is order compatible on $\hat{F}$.

We will describe an example. Let $F = \mathbf{Q}(X)$, the field of rational functions in one indeterminate over $\mathbf{Q}$. We can interpret a rational function $r(X) = f(X)/g(X)$ as a germ of a function on intervals $(0, \epsilon) \subset \mathbf{Q}$, $\epsilon > 0$, since for sufficiently small ϵ, $g(X)$ will have no roots in $(0, \epsilon)$. Order $\mathbf{Q}(X)$ by defining $r(X) < s(X)$ if $r(X) < s(X)$ as functions on some $(0, \epsilon)$. Then $0 < X$ and X is a microbe, infinitesimal relative to $\mathbf{Q}$. The associated valuation $v : \mathbf{Q}(X) \to \mathbf{Z}$, $v(r(X)) = -\log_{X^{-1}}(|r(X)|)$, is clearly the X-adic valuation. It is not difficult to show that this order on $\mathbf{Q}(X)$ is the unique order such that X is a positive infinitesimal.

The construction above works for $K(X)$ where K is any ordered field. That is, there is a unique order on $K(X)$ which extends the order on K in which X is positive and infinitesimal relative to K. Moreover, in this order on $K(X)$, X is a microbe.

Some elements of $F = \mathbf{Q}(X)$ pictured as graphs of germs of functions on $(0, \epsilon)$.

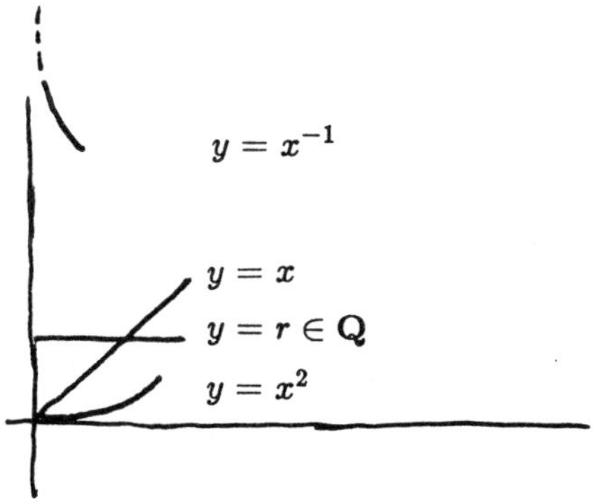

We conclude this section by pointing out that the class of microbial ordered fields is closed under ordered field extensions of finite transcendence degree. Namely, suppose $K \subset F$ is such an extension with K microbial. If F is Archimedean over K, then any microbe in K is a microbe in F. If F is not Archimedean over K, choose $X \in F$ positive and infinitesimal relative to K. Necessarily X is transcendental over K. Then, by the comments above, X is a microbe in $K(X)$ and the transcendence degree of $K(X) \subset F$ is less than that of $K \subset F$. By induction, F is microbial.

As a special case of the above discussion, we see that any ordered function field F over a subfield $K \subset \mathbf{R}$ is microbial, as is any Euclidean algebraic extension $\overline{F}$ of F. Note that if $\overline{F}$ is Euclidean [resp. real closed] and $\overline{v} : \overline{F}^* \to \mathbf{R}$ is a valuation, then $\overline{\Lambda} = \overline{v}(\overline{F}^*)$ is a 2-divisible [resp. divisible] subgroup of $\mathbf{R}$. This is because $\overline{F}$ contains roots of its positive elements. In fact, if $\overline{F}$ is the Euclidean closure of F then $\overline{\Lambda}$ is the 2-divisible hull of Λ in $\mathbf{R}$, and if $\overline{F}$ is the real closure of F then $\overline{\Lambda}$ is the divisible hull of Λ in $\mathbf{R}$, where $\Lambda = \overline{v}(F^*) \subset \mathbf{R}$.

The hyperbolic pseudo-metric on HF^2. We fix a microbial non-Archimedean ordered field F and a big element in F, with associated logarithm $\log : F^+ \to \Lambda \subset \mathbf{R}$. A different choice of big element just changes log by a positive real scaler multiple, which is unimportant in what follows.

If $P, Q \in HF^2$, we have the cross-ratio $D(P,Q) \geq 1 \in \overline{F}$. In general, $D(P,Q) \notin F$. However, $\log D(P,Q) \in \Lambda = \log(F^+)$. This can be seen from the formulas (7) and (8). For example, in the formulas (7) and (8), use the fact that if $0 < t < 1$ then $\log(1+t) = 0$, so

$$\log\left(\frac{1+t}{1-t}\right) = \log\left(\frac{(1+t)^2}{1-t^2}\right) = \log\left(\frac{1}{1-t^2}\right),$$

hence if $t^2 \in F$ then $\log\left(\frac{1+t}{1-t}\right) \in \Lambda$.

Given $P, Q \in HF^2$, we define

$$d(P,Q) = \log D(P,Q) \in \Lambda \subset \mathbf{R}. \tag{17}$$

Combining properties (9), (10), (11) of D with properties (13), (14) of log, we see

$$d(P,Q) \geq 0 \text{ and } d(P,Q) + d(Q,R) \leq d(P,R)\ . \tag{18}$$

Thus, $d(P,Q)$ is a pseudo-metric on HF^2. Also, $d(P,Q)$ is $PGL_+(2,F)$ invariant.

If $P, Q \in HF^2$, define $P \sim Q$ if $d(P,Q) = 0$. Then the triangle inequality in (18) implies that $\sim$ is an equivalence relation. We define

$$TF^2 = HF^2 / \sim,$$

the set of equivalence classes. The triangle inequality also implies that $d : TF^2 \times TF^2 \to \Lambda \subset \mathbf{R}$ is well-defined and gives, in fact, a topological metric on TF^2. The group $PGL_+(2,F)$ acts on TF^2 as a group of isometries.

We will give some explicit formulas for $d(P,Q)$ in the various models HF^2, B, and HS^2 of the hyperbolic plane over F. First, if $P = (x,y)$, $Q = (x',y') \in HF^2$, write $w = x+iy$, $w' = x'+iy' \in C$. Then

$$\begin{aligned} d(P,Q) &= \log\left(\frac{(x-x')^2 + y^2 + (y')^2}{yy'}\right) \\ &= \operatorname{Max}\left(\log\left(\frac{(x-x')^2}{yy'}\right), \log\left(\frac{y}{y'}\right), \log\left(\frac{y'}{y}\right)\right) \\ &= \log\left(\|\bar{w} - w'\|^2 / yy'\right) \\ &= \operatorname{Max}\left(0, \log\left(\|w - w'\|^2 / yy'\right)\right). \end{aligned} \tag{19}$$

Next, suppose $w = x+iy$, $w' = x'+iy' \in B$, which is the unit disk in C. Then

$$d(w,w') = \log\left(\frac{\|1 - \bar{w}w'\|^2}{(1-\|w\|^2)(1-\|w'\|^2)}\right). \tag{20}$$

Finally, suppose $R = (x, y, z)$, $S = (x', y', z') \in HS^2$, that is, $x^2 + y^2 + z^2 = (x')^2 + (y')^2 + (z')^2 = 1$, $z, z' > 0$. Then

$$d(R, S) = \log\left(\frac{1 - (xx' + yy')}{zz'}\right). \tag{21}$$

We will leave the verification of the formulas (20) and (21), as well as the equalities in (19) as exercises, since we do not explicitly need them below. The nice point is that the formulas for $d = \log D$ are rather simpler than the formulas for D itself.

We will make use of a variant of (19), more precisely, a special case. Let $i = (0, 1) \in HF^2$ and let $M = \begin{pmatrix} a & b \\ c & d \end{pmatrix} \in GL_+(2, F)$. Then

$$d(i, Mi) = \text{Max}\left(\log(a^2), \log(b^2), \log(c^2), \log(d^2)\right) - \log\left(\det(M)\right). \tag{22}$$

Here is a proof of (22), which can be easily extended to prove (19). Let $D(i, Mi) = \alpha \geq 1 \in \overline{F}$, and let

$$z = Mi = x + iy = \frac{ai + b}{ci + d} = \frac{(ac + bd) + (ad - bc)i}{c^2 + d^2}.$$

Then, from hyperbolic geometry, we know that z lies on the Euclidean circle in $H\overline{F}^2$ with diameter the segment $[\alpha^{-1}i, \alpha i]$. Thus,

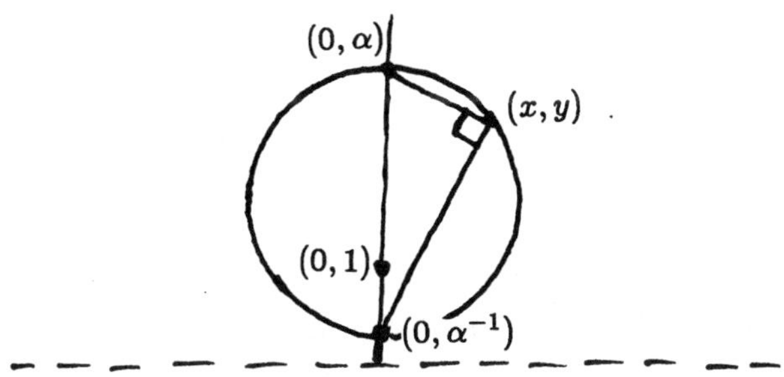

$$\begin{aligned} 0 = (x, y - \alpha) \cdot (x, y - \alpha^{-1}) &= x^2 + y^2 + 1 - (\alpha + \alpha^{-1})y \\ &= \frac{a^2 + b^2 + c^2 + d^2}{c^2 + d^2} - (\alpha + \alpha^{-1})\frac{\det(M)}{c^2 + d^2}. \end{aligned}$$

It follows that $(\alpha + \alpha^{-1})\det(M) = a^2 + b^2 + c^2 + d^2$, hence

$$d\left(i, Mi\right) = \log(\alpha) = \log(\alpha + \alpha^{-1}) = \log\left(\frac{a^2 + b^2 + c^2 + d^2}{\det(M)}\right).$$

Since $\text{Max}(a^2, b^2, c^2, d^2) \leq a^2 + b^2 + c^2 + d^2 \leq 4\,\text{Max}\left(a^2, b^2, c^2, d^2\right)$, we have $\log(a^2 + b^2 + c^2 + d^2) = \text{Max}\left(\log(a^2), \log(b^2), \log(c^2), \log(d^2)\right)$, and (22) follows.

One consequence of (22) is the following.

(23) If $P, Q, R \in HF^2$ then $d(P, Q) + d(Q, R) + d(R, P) \in 2\Lambda$.

Namely, by $GL_+(2,F)$ transitivity and invariance of d, we may assume $P = i, Q = Mi$, and $R = Ni$, where $M, N \in GL_+(2,F)$. Then $d(Mi, Ni) = d(i, M^{-1}Ni)$, hence by (22), modulo 2Λ

$$d(P,Q) + d(Q,R) + d(R,P) \equiv \log\det(M) + \log\det(M^{-1}N) + \log\det(N) ,$$

and the right side is clearly 0 modulo 2Λ. Alternatively, (23) is very easy to deduce from the first formula for d in (19), since the sum of the three distances becomes the log of a product whose numerator is a sum of squares and whose denominator is a square.

We conclude this section with a picture of the equivalence classes in HF^2, that is, points of $TF^2 = HF^2/\sim$. The picture is perhaps nicest in the Poincaré disk model B. First, suppose $A, A' \in B$ and let P, P' be the ideal endpoints in $\overline{S}^1 \subset \overline{F}^2$ of the hyperbolic line through A, A', with the points in the order $PAA'P'$. Let M be the midpoint of the arc PP'. Then, from (6), one deduces

$$d(A,A') = \begin{cases} \log|PA'| - \log|PA| & \text{if } AA'M \\ 2\log|PP'| - (\log|PA| + \log|P'A'|) & \text{if } AMA'. \end{cases} \tag{24}$$

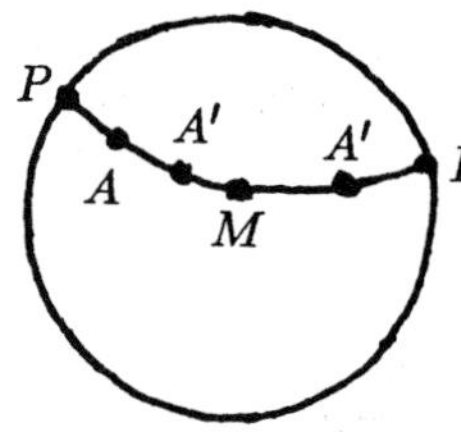

Here, as in (6), $|\ |$ denotes Euclidean segment lengths.

Next, the equivalence class in HF^2 of the origin $O \in B$ is certainly the union of all Euclidean circles centered at O whose radius r satisfies $\log(1-r) = 0$. Suppose $A \in B$ is a point with $[A] \neq [O] \in TF^2$ and let $s = 1 - |OA|$, the distance from A to the boundary circle $\overline{S}^1 \subset \overline{F}^2$. Then

(25) Proposition. *$d(A, A') = 0$ if and only if A' lies in a Euclidean circle in B, centered at A, with radius r, where r satisfies $\log(s-r) = \log(s)$, or, equivalently, $\log\left(1 - \frac{r}{s}\right) = 0$.*

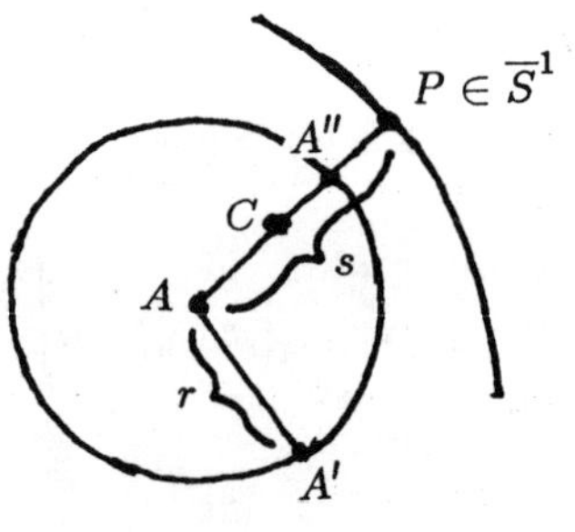

Proof. Let A'' be the point on the circle of radius r and center A, which is between A and P, and on the radius OP of B. By (24), $d(A, A'') = 0$ if $\log(s - r) = \log(s)$. Now, this *Euclidean* circle is also a *hyperbolic* circle in B, whose hyperbolic center C lies between A and A''. Thus, $0 = d(C, A'') = d(C, A')$, since $D(C, A'') \leq D(A, A'')$. Similarly, $d(C, A) = 0$, so $d(A, A') = 0$. ∎

(26) Corollary. *Let $\hat{F}$ be the completion of the microbial ordered field F. Then the inclusion $HF^2 \to H\hat{F}^2$ induces an isometry $TF^2 \xrightarrow{\sim} T\hat{F}^2$.*

Proof. Since $\log : F^+ \to \Lambda$ extends to $\hat{F}^+ \to \Lambda$, it is obvious that $TF^2 \to T\hat{F}^2$ exists and is a d-preserving inclusion. On the other hand, since F is dense in $\hat{F}$, every Euclidean circle in $\hat{F}^2$ contains points of F^2. Thus Proposition (25) implies that $TF^2 \to T\hat{F}^2$ is surjective. ∎

The Λ-Tree TF^2. If $\Lambda \subset \mathbf{R}$ is a subgroup, then a Λ-*metric space* T is a metric space such that all distances between points of T are numbers in Λ. Thus, the quotient space $TF^2 = HF^2/\sim$ defined in the previous section is a Λ-metric space. If $\lambda_1 \leq \lambda_2 \in \Lambda$, let $[\lambda_1, \lambda_2]_\Lambda = \left\{\lambda \in \Lambda \,\middle|\, \lambda_1 \leq \lambda \leq \lambda_2\right\}$. We call $[\lambda_1, \lambda_2]_\Lambda$ a *closed interval* in Λ. A subspace of a Λ-metric space T is called a *closed segment* in T if it is isometric to a closed interval in Λ. It is clear what is meant by the *endpoints* of a closed segment in T. A Λ-metric space T is called a Λ-*tree* if the following properties hold.

(27)(a) Every two points of T are endpoints of a unique closed segment in T.

(27)(b) If S and S' are closed segments in T with an endpoint in common, then $S \cap S'$ is a closed segment in T.

(27)(c) If S and S' are closed segments in T and if $S \cap S' = \{q\}$, where q is an endpoint of both S and S', then $S \cup S'$ is a closed segment in T.

We refer to Morgan and Shalen's paper [2] for a treatement of (even more general) Λ-trees. When $\Lambda \cong \mathbf{Z}$ then T is a $\mathbf{Z}$-tree if and only if T is a set of vertices of an ordinary simplicial tree (not necessarily locally finite) with distance given by the length of paths in T, where adjacent vertices are at distance 1. Here is our main result.

(28) Theorem. *TF^2 with metric d is a Λ-tree.*

The proof will occupy the rest of this section, although it is not especially difficult. The easiest case of Theorem (28) is the case when $F = \overline{F}$ is a Euclidean field. One advantage is that all hyperbolic lines in $H\overline{F}^2$ look alike, that is, $PGL_+(2, \overline{F})$ acts transitively on lines. Another advantage is that every hyperbolic triangle in $H\overline{F}^2$ has an incenter, that is, a point equidistant from all three sides. In general, passing from F to its Euclidean closure $\overline{F}$ replaces Λ by its 2-divisible hull $\overline{\Lambda}$, so $T\overline{F}^2$ has many more points that TF^2. On the other hand, passing from F to its completion $\hat{F}$ does not change Λ, and, by Corollary (26), $TF^2 \xrightarrow{\sim} T\hat{F}^2$ is an isometry of Λ-metric spaces. Thus it suffices to prove Theorem (28) when

$F = \hat{F}$ is complete. In our proof of Theorem (28) for general F we will make use of both $\hat{F}$ and $\overline{F}$.

(29) Lemma. *For a complete field $\hat{F}$, $PGL_+(2,\hat{F})$ acts transitively on hyperbolic lines in $H\hat{F}^2$ which project non-trivially to $T\hat{F}^2$, (that is, which do not project to a single point in $T\hat{F}^2$). In particular, every hyperbolic line in $H\hat{F}^2$ which projects non-trivially to $T\hat{F}^2$ is D-isometric to $\hat{F}^+$ hence its image in $T\hat{F}^2$ is d-isometric to Λ.*

Proof. Let $p, q \in T\hat{F}^2$ be two distinct points represented by $P, Q \in \hat{B}$, the unit disk model of $H\hat{F}^2$. Without loss of generality, we may assume $P = (0,0) \in \hat{B}$. Let $Q = (x,y)$ and write $r^2 = x^2 + y^2 < 1 \in \hat{F}$. We will prove $r \in \hat{F}$, that is, $x^2 + y^2$ is a square in $\hat{F}$.

Since $p \neq q \in T\hat{F}^2$, $d(p,q) = -\log(1-r^2) > 0 \in \Lambda$. Thus, $r^2 = 1 - s$, where $s \in \hat{\mathcal{M}} \subset \hat{\mathcal{O}} \subset \hat{F}$. Now, $1 - s = \left(1 - \frac{s}{2} - \frac{s^2}{8} - \frac{s^3}{16} \cdots\right)^2$, and this power series converges in the complete microbial field $\hat{F}$. (Any non-zero element of $\hat{\mathcal{M}}$ is a microbe.) thus $r = 1 - \frac{s}{2} - \frac{s^2}{8} - \frac{s^3}{16} \cdots \in \hat{F}$, as claimed. As a consequence, we have the unit vector $\left(\frac{x}{r}, \frac{y}{r}\right) \in \hat{S}^1 \subset \hat{F}^2$, which can be interpreted as a rotation of $\hat{B}$ which moves the x-axis in $\hat{B}$ to the line through $(0,0)$ and (x,y). Thus $SO(2,\hat{F})$ acts transitively on hyperbolic lines in $\hat{B}$ through $(0,0)$, which project non-trivially to $T\hat{F}^2$. ∎

(30) Remark. It is not generally true that all hyperbolic lines through $(0,0) \in \hat{B}$ look alike, even in the complete case. For example, if $\hat{F}$ is the completion of $F = \mathbf{Q}(X)$, with X infinitesimal and positive, then the entire line through $(0,0)$ and $(1/2, 1/2)$ in $\hat{B}$ collapses to a single point in $T\hat{F}^2$. Essentially this is because $\sqrt{2}$ has no close approximations in $\mathbf{Q}(X)$. On the other hand, the X-axis in $\hat{B}$ is D-isometric to $\hat{F}^+$, since $D((0,0),(x,0)) = \frac{1+x}{1-x}$. Such a hyperbolic line projects to an isometric copy of $\mathbf{Z} = \log(\hat{F}^+)$ in $T\hat{F}^2$. ∎

(31) Lemma. *Suppose $P, Q \in H\hat{F}^2$ and $d(P,Q) = 0 \in \Lambda$. Then for all $O \in H\hat{F}^2$, the hyperbolic segments OP and OQ in $H\hat{F}^2$ have the same image in $T\hat{F}^2$.*

Proof. First, since $d(O,P) = d(O,Q)$, the result is clear if OPQ are hyperbolically collinear in $H\hat{F}^2$. It also follows easily that if $d(O,P) = d(O,Q) = 0$ then all points in the hyperbolic triangle ΔOPQ have the same image in $T\hat{F}^2$.

Now assume $d(O,P) = d(O,Q) \neq 0$. We use the unit disk model $\hat{B}$ and assume O is the origin. By the proof of Lemma (29), there is a rotation $\rho \in SO(2,\hat{F})$ which takes the line OP to the line OQ in $\hat{B}$. Then $d(O,Q) = d(O,P) = d(O,\rho(P))$, so we have $d(Q,\rho(P)) = 0$ since $O, \rho(P)$ and Q are collinear. Thus also $d(P,\rho(P)) = 0$, and by the collinear case of Lemma (31), we may as well assume $\rho(P) = Q$. Now, if R is between O and P then $D(R,\rho(R)) \leq D(P,\rho(P))$, hence $d(R,\rho(R)) = 0$. Thus the rotation ρ takes each point of OP to a point of OQ with the same image in $T\hat{F}^2$. ∎

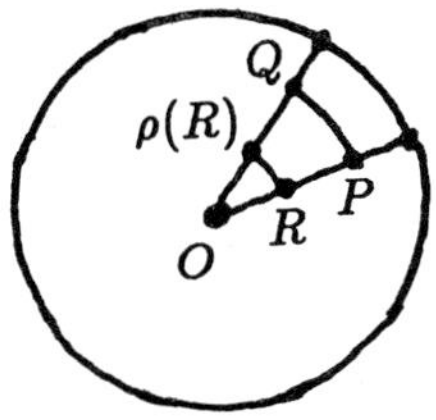

(32) Remark. Another proof of Lemma (31) can be given for $\overline{F}$, the Euclidean closure of F. We draw the picture in the Klein model $\overline{B}_0$. Let $P', Q' \in \overline{S}^1$ be the ideal endpoints of line PQ. If R is on segment OP then line $P'R$ intersects OQ at S and $D(R,S) \leq D(P,Q)$. Thus $d(R,S) = 0$, and projection from P' defines a bijection between the images of OP and OQ in $T\overline{F}^2$. ∎

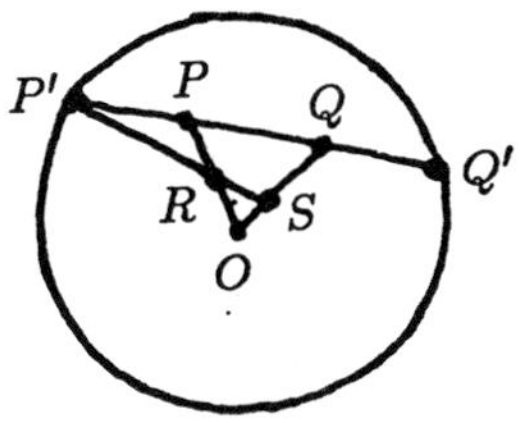

(33) Lemma. *Suppose $P, Q, R \in H\hat{F}^2$ have distinct images in $T\hat{F}^2$. Then the image of hyperbolic triangle ΔPQR in $T\hat{F}^2$ looks like one of the following:*

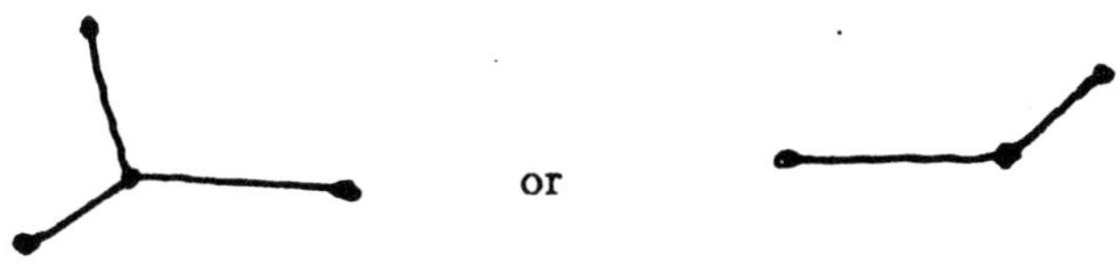

(Here, a segment represents a closed Λ-segment in $T\hat{F}^2$.)

Proof. We first prove Lemma (33) not for complete fields, but for Euclidean fields, that is, for $T\overline{F}^2$. In the Euclidean case, any hyperbolic triangle ΔPQR has an incenter O, that is, a point equidistant from all three sides. But for *any* interior point X of a hyperbolic triangle Δ, we have $D(X, \partial\Delta) < \sqrt{3}$, where $\partial\Delta$ is the boundary of the triangle.

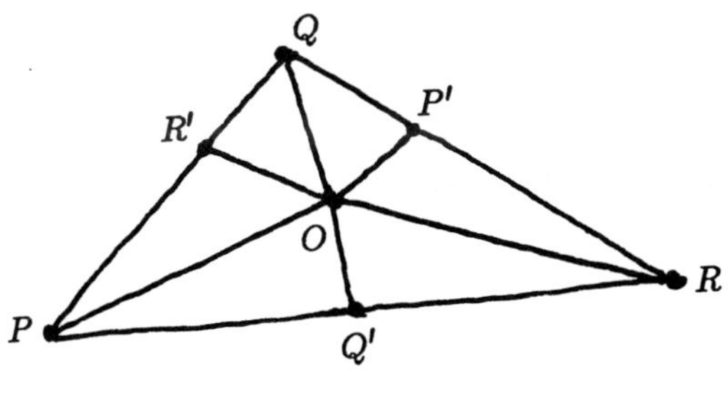

Since $\log(\sqrt{3}) = 0$, there are points R', P', Q' on the sides of ΔPQR with $d(O,P') = d(O,Q') = d(O,R') = 0$. Now, Lemma (31), as extended in Remark (32), implies that each of the six small triangles in the figure above has image a closed $\overline{\Lambda}$-segment in $T\overline{F}^2$. If the image of O in $T\overline{F}^2$ coincides with one of the images of P, Q, R, then ΔPQR projects to a single closed Λ-segment in $T\overline{F}^2$. Otherwise, ΔPQR projects to three closed $\overline{\Lambda}$-segments with common endpoint the image of O.

The assertion that $D(X, \partial\Delta) < \sqrt{3}$ for any interior point of a hyperbolic triangle, is proved just as in classical hyperbolic geometry over $\mathbf{R}$. The "worst case" is an ideal equilateral triangle, with X at the center.

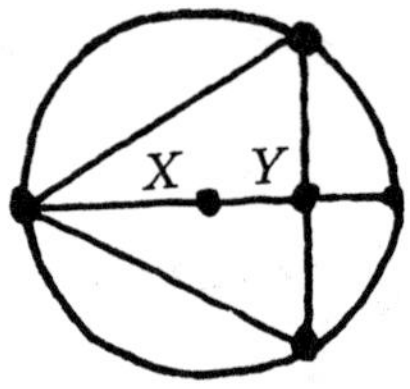

Direct computation, say in the Klein model picture above, gives $D(X, \partial\Delta) = D(X,Y) = \sqrt{3}$.

To extend Lemma (33) to complete fields $\hat{F}$, we use (23). The problem is, if $P, Q, R \in HF^2$ are rational points, then the incenter $O \in \Delta PQR$ lies in $H\overline{F}^2$, but probably not in HF^2 or even in $H\hat{F}^2$. However, if $d(P,Q) = r$, $d(P,R) = q$, and $d(Q,R) = p \in \Lambda$, then $p + q + r \in 2\Lambda$ by (23). Therefore, the equations $p' + q' = r$, $p' + r' = q$, $q' + r' = p$ can be solved in Λ.

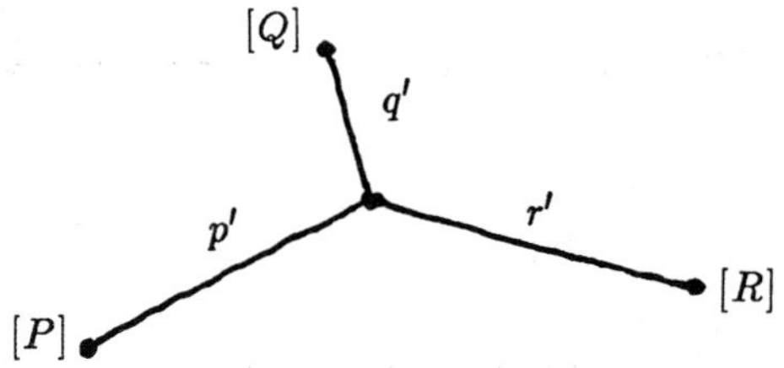

This means that the three closed Λ-segments in $T\hat{F}^2$ determined by PQ, PR, and QR (by Lemma (29)) do have a point in common in $T\hat{F}^2 = TF^2 \subset T\overline{F}^2$. ∎

Note that if the image of $\Delta PQR \subset H\hat{F}^2$ in $T\hat{F}^2$ is not a closed segment, then $d(P,R) + d(R,Q) > d(P,Q)$. Specifically, if $O \in H\overline{F}^2$ is the incenter of ΔPQR,

$$d(P,R) + d(R,Q) - d(P,Q) = 2d(O,Q) \tag{34}$$

We now give the proof of Theorem (28). We must establish conditions (27)(a), (27)(b), and (27)(c) for the Λ-metric space TF^2.

Proof of (27)(a). Let $P, Q \in HF^2$ project to $p, q \in TF^2$, $p \neq q$. Then $TF^2 = T\hat{F}^2$ by Corollary (26), and from Lemma (29), the hyperbolic segment $PQ \subset H\hat{F}^2$ projects to a closed Λ-segment in $T\hat{F}^2$. This gives existence of a closed segment in TF^2 with preassigned endpoints. (Note that we do *not* claim that the *rational* segment $PQ \subset HF^2$ itself projects to a closed segment in TF^2. It is necessary to pass to the completion $H\hat{F}^2$ first.)

To prove that there is a unique closed segment in TF^2 with endpoints p, q, reason as follows. Suppose $r \in TF^2$ with $d(p,r) + d(r,q) = d(p,q)$. Lift r to $R \in HF^2$. Then, by (34) and Lemma (33), the image in $T\hat{F}^2$ of $\Delta PQR \subset H\hat{F}^2$ degenerates to the image of segment PQ. Thus r lies on this closed segment in $T\hat{F}^2$, hence the closed segment pq is well-defined in $T\hat{F}^2 = TF^2$. ∎

Proof of (27)(b). Let $S = pq$ and $S' = pr$ be closed segments in $TF^2 = T\hat{F}^2$. Then immediately, from Lemma(33), $S \cap S' = po$, where $o \in TF^2$ is the image of the incenter $O \in H\overline{F}^2$ of ΔPQR. (Note that in the proof of Lemma (33), we showed that $o \in TF^2$ is rational even if $O \in H\overline{F}^2$ is not.) ∎

Proof of (27)(c). If $S = pq$ and $S' = qr$ then $S \cap S' = \{q\}$ implies $d(O,Q) = 0$, where $O \in H\overline{F}^2$ is the incenter of ΔPQR. Again from Lemma (33), the image in $T\hat{F}^2 = TF^2$ of $\Delta PQR \subset H\hat{F}^2$ degenerates to the image of segment PR, hence $S \cup S' = pr$. ∎

Comparison with the tree of a valued field. Suppose $v : F^* \to \Lambda$ is a valuation of a field F, with valuation ring $\mathcal{O} \subset F$. There is an associated Λ-tree T_v, the points of which are homothety classes of $\mathcal{O}$-lattices $L \subset F^2$. (That is, $L \simeq \mathcal{O}^2$ and L spans F^2; homothety classes means L is identified with aL, $a \in F^*$.) The distance $d_v(L, L') \in \Lambda$ is defined as follows. One can choose representatives such that $L' \subset L$, with bases $\{e_1', e_2'\}$, $\{e_1, e_2\}$, such that $e_1' = e_1$ and $e_2' = be_2$, $b \in \mathcal{O}$. Then $d_v(L, L') = v(b) \geq 0 \in \Lambda$. The case of discrete rank 1 valuations, that is, $\Lambda = \mathbf{Z}$, is studied in Serre's book [3]. A treatment of the more general case, including non-rank 1 valuations, can be found in Morgan and Shalen's paper [2].

If F is a microbial ordered field, we have the valuation $v : F^* \to \Lambda$, $v(a) = -\log(|a|)$, with $\Lambda \subset \mathbf{R}$. We want to compare the trees TF^2 and T_v. In fact, they are isometric. Let $I \subset F^2$ denote the standard lattice, with basis $\{(1,0),(0,1)\}$, and let $i = (0,1) \in HF^2 \subset C = F[i]$ be the usual basepoint of HF^2. Note that $GL_+(2,F)$, acting linearly on F^2, acts transitively on lattices, hence $PGL_+(2,F)$ acts transitively on the homothety classes of lattices, T_v.

(35) Proposition. *There is a $PGL_+(2,F)$-equivariant isometry $TF^2 \xrightarrow{\sim} T_v$ which takes the class of i in TF^2 to the class of the lattice I in T_v.*

Proof. Since $PGL_+(2,F)$ acts transitively on TF^2, we need to check that the only possible $PGL_+(2,F)$-equivariant map $TF^2 \to T_v$ defined by sending $[i]$ to $[I]$ is, in fact, well-defined and distance preserving. This map will be surjective since $PGL_+(2,F)$ acts transitively on T_v.

If $M = \begin{pmatrix} a & b \\ c & d \end{pmatrix} \in GL_+(2,F)$ then from (22),

$$d(i, Mi) = \operatorname{Max}(\log(a^2), \log(b^2), \log(c^2), \log(d^2)) - \log(\det(M)) \in \Lambda .$$

Now, Proposition II.3.13 of [2] asserts that if $M \in SL(2,F)$ then $d_v(I,MI) = -\mathrm{Min}(v(a^2), v(b^2), v(c^2), v(d^2))$. For arbitrary $M \in GL_+(2,F)$ apply that proposition to an extension of v to the field $F\left[\sqrt{\det(M)}\right]$ and the matrix $\left(1/\sqrt{\det(M)}\right)M \in SL\left(2, F\left[\sqrt{\det(M)}\right]\right)$. (Or, extend the argument of [2] to arbitrary $M \in GL_+(2,F)$.) In any event, one concludes

$$d_v(I,MI) = -\mathrm{Min}\left(v(a^2), v(b^2), v(c^2), v(d^2)\right) + v(\det(M)) \in \Lambda\ ,$$

and Proposition (35) is proved.

Remark. It is not surprising that the hyperbolic tree TF^2 is related to lattices. A point of the upper half plane HF^2 can be interpreted in the usual way as a $\mathbf{Z}$-lattice in $C = F[i]$, modulo C^*-homothety. Namely, if the $\mathbf{Z}$-lattice has basis $\{z_1 = x_1 + iy_1,\ z_2 = x_2 + iy_2\}$ ordered so that $\det\begin{pmatrix} x_1 & x_2 \\ y_1 & y_2 \end{pmatrix} > 0$, then the associated point in $HF^2 \subset C$ is the ratio z_2/z_1. The map of Proposition (35), $TF^2 \xrightarrow{\sim} T_v$, can be seen to be the map which assigns to the $\mathbf{Z}$-lattice represented by matrix $N = \begin{pmatrix} x_1 & x_2 \\ y_1 & y_2 \end{pmatrix}$, the $\mathcal{O}$-lattice represented by $N^t = \begin{pmatrix} x_1 & y_1 \\ x_2 & y_2 \end{pmatrix}$, that is, the $\mathcal{O}$-lattice with basis $\{(x_1, x_2), (y_1, y_2)\}$. ■

Directions and ends in the tree TF^2. We will compare in this section the interpretations of the directions at a point in TF^2 and T_v and the ends in TF^2 and T_v, as well as determine the isotropy groups of closed segments, directions, and ends. Of course, one could just use Proposition (35) and translate this information from T_v to TF^2. However, our point of view is that the results in TF^2 are geometrically motivated and should be interpreted directly.

A *direction* at $p \in TF^2$ is a germ of a closed segment, with p as endpoint. Two such closed segments S, S' determine the same direction if $S \cap S'$ is a nontrivial closed segment. An *end* in TF^2 is a germ of a path, where by a *path* we mean a subset of TF^2 isometric to $\Lambda^+ = \{\lambda \in \Lambda \mid \lambda \geq 0\}$. Two paths W, W' determine the same end if $W \cap W'$ is a path.

In Serre's book, [3], directions at a point in T_v are identified with points of the projective line $P^1(k)$, where $k = \mathcal{O}/\mathcal{M}$ is the residue field of the valuation $v:\ F^* \to \Lambda$. The ends in T_v are identified with points of the projective line $P^1(\hat{F})$, where $\hat{F}$ is the valuation completion of F. If $\lambda \in \Lambda$, $\lambda > 0$, let $\mathcal{M}_\lambda = \{a \in \mathcal{O} \mid v(a) \geq \lambda\}$. Then the points in T_v at distance λ from some fixed point can be identified with points of the projective line $P^1(\mathcal{O}/\mathcal{M}_\lambda)$. Of course, Serre's book only deals with the case $\Lambda = \mathbf{Z}$, but all these assertions generalize easily.

We will see that for a microbial ordered field F, the directions at a point, the ends, and the points at distance λ from a fixed point, identify naturally with the unit circles $S^1(k) \subset k^2$, $S^1(\hat{F}) \subset \hat{F}^2$, and $S^1\left(\mathcal{O}/\mathcal{M}_\lambda\right) \subset \left(\mathcal{O}/\mathcal{M}_\lambda\right)^2$, respectively. (If A is a ring, $S^1(A) = \{(a,b) \mid a^2 + b^2 = 1\} \subset A^2$.) The rational parametrization formula (3) can be used to identify the projective lines over k, $\hat{F}$, and $\mathcal{O}/\mathcal{M}_\lambda$ with the unit circles. In fact, this can be done $PGL_+(2,A)$-eqivariantly, for $A = k$, $\hat{F}$, and $\mathcal{O}/\mathcal{M}_\lambda$. One can trace through the assignments of points in $P^1(A)$ to certain lattices, and the assignments of points in $S^1(A)$ to certain points of HF^2, and see that everything is compatible with the Λ-tree isometry $TF^2 \xrightarrow{\simeq} T_v$ of Proposition (35). However, we do not do this here.

Our first step is to identify the isotropy group of a point and a closed segment in TF^2.

(37) Lemma. *The isotropy group of the point* $[i] \in TF^2$ *is* $PGL_+(2, \mathcal{O}) \subset PGL_+(2, F)$. *If* $M = \begin{pmatrix} a & b \\ c & d \end{pmatrix} \in GL_+(2, \mathcal{O})$. *then* M *fixes the closed segment* $[i][ri] \subset TF^2$, $r \in F^+$, *if and only if* $b/r \in \mathcal{O}$ *and* $rc \in \mathcal{O}$.

(38) Remark. If $r > 1$ then $rc \in \mathcal{O}$ is the only restriction, and if $r < 1$ then $b/r \in \mathcal{O}$ is the only restriction. If $\lambda = |\log(r)| > 0 \in \Lambda$ and if $\pi : \ GL_+(2, \mathcal{O}) \to GL_+(2, \mathcal{O}/\mathcal{M}_\lambda)$ is the obvious projection, then the lemma asserts that the isotropy group in $GL_+(2, \mathcal{O})$ of the segment $[i][ri]$ is $\pi^{-1}\left\{\begin{pmatrix} \bar{a} & \bar{b} \\ 0 & \bar{d} \end{pmatrix}\right\}$ when $r > 1$ and $\pi^{-1}\left\{\begin{pmatrix} \bar{a} & 0 \\ \bar{c} & \bar{d} \end{pmatrix}\right\}$ when $r < 1$. ∎

Proof of (37). From (22), if $M = \begin{pmatrix} a & b \\ c & d \end{pmatrix} \in GL_+(2, F)$ and if $d(i, Mi) = 0$ then $\log(\det(M)) \in 2\Lambda$. If $x \in F^+$ and $\log(x^2) = \log(\det(M))$ then $\log(\det((1/x)M)) = 0$, so $0 = d(i, (1/x)Mi) = \operatorname{Max}\left(\log\left(a^2/x^2\right), \log\left(b^2/x^2\right), \log\left(c^2/x^2\right), \log\left(d^2/x^2\right)\right)$, and we conclude $(1/x)M \in GL_+(2, \mathcal{O})$. Clearly every element of $GL_+(2, \mathcal{O})$ fixes $[i]$, hence we have shown that the isotropy group of $[i]$ is $PGL_+(2, \mathcal{O}) \subset PGL_+(2, F)$.

Next, suppose $M = \begin{pmatrix} a & b \\ c & d \end{pmatrix} \in GL_+(2, \mathcal{O})$ also fixes $[ri]$, that is, $d(ri, Mri) = 0$. This is the same as the assertion that $M' = \begin{pmatrix} \sqrt{r}^{-1} & 0 \\ 0 & \sqrt{r} \end{pmatrix} M \begin{pmatrix} \sqrt{r} & 0 \\ 0 & \sqrt{r}^{-1} \end{pmatrix}$ fixes $[i]$, in other words that $M' = \begin{pmatrix} a & b/r \\ rc & d \end{pmatrix} \in GL_+(2, \mathcal{O})$ also. ∎

Since $PGL_+(2, F)$ acts transitively on TF^2, the isotropy group of any point $p \in TF^2$ is a conjugate of the isotropy group $PGL_+(2, \mathcal{O})$ of $[i] \in TF^2$. It is not so obvious how $PGL_+(2, F)$ acts on closed segments of length λ in TF^2. However, we will see below that this action is also transitive, hence all isotropy groups of segments of length λ are conjugate.

Consider the completion $\hat{F}$ and the unit disk model $\hat{B}$ of $H\hat{F}^2$. In the proof of Lemma (29), we showed that the group of linear rotations $SO(2, \hat{F})$ acts transitively on the hyperbolic lines in the model $\hat{B}$ which pass through the origin $(0,0) = O$ and which project non-trivially to $T\hat{F}^2 = TF^2$. Thus $SO(2, \hat{F})$ also acts transitively on closed segments op of length λ in TF^2, $o = [O]$, $p \in TF^2$. To describe the set of such closed segments in TF^2 we must determine the subgroup of $SO(2, \hat{F})$ which fixes one segment, say ox, where $x = (x, 0) \in \hat{B}$, $\log\left(\frac{1+x}{1-x}\right) = \lambda$. Note ox in B corresponds to $[i][ri]$ in TF^2, $r = \frac{1+x}{1-x}$, under the identification of (2), $t : \ HF^2 \xrightarrow{\simeq} B$.

The linear rotation $\begin{pmatrix} \hat{c}, & -\hat{d} \\ \hat{d}, & \hat{c} \end{pmatrix} \in SO(2, \hat{F})$ of $\hat{B}$, $\hat{c}^2 + \hat{d}^2 = 1$, corresponds to the Möbius transformation $\begin{pmatrix} -\hat{d} & \hat{c} - 1 \\ 1 - \hat{c} & -\hat{d} \end{pmatrix} \in RO(2, \hat{F})$ of $H\hat{F}^2$, if $(\hat{c}, \hat{d}) \neq (1, 0)$. (See the section on hyperbolic planes over Euclidean fields for a discussion of the relation between $SO(2, \hat{F})$ acting on $\hat{B}$ and $RO(2,\hat{F}) \subset PGL_+(2,\hat{F})$ acting on $H\hat{F}^2$.) Now, $\det\begin{pmatrix} -\hat{d} & \hat{c}-1 \\ 1-\hat{c} & -\hat{d} \end{pmatrix} = 2(1-\hat{c})$, which may not be a unit in $\hat{\mathcal{O}}$. However, we can make computations over some Euclidean extension of $\hat{F}$, where $\begin{pmatrix} -\hat{d} & \hat{c}-1 \\ 1-\hat{c} & -\hat{d} \end{pmatrix} \equiv \begin{pmatrix} -\hat{d}/\sqrt{1-\hat{c}} & -\sqrt{1-\hat{c}} \\ \sqrt{1-\hat{c}} & -\hat{d}/\sqrt{1-\hat{c}} \end{pmatrix} \in PGL_+(2,\hat{\mathcal{O}})$. Then $\begin{pmatrix} \hat{c} & -\hat{d} \\ \hat{d} & \hat{c} \end{pmatrix}$ fixes ox if and only if $\begin{pmatrix} -\hat{d}/\sqrt{1-\hat{c}}, & -\sqrt{1-\hat{c}} \\ \sqrt{1-\hat{c}}, & -\hat{d}/\sqrt{1-\hat{c}} \end{pmatrix}$ fixes $[i][ri]$, and by Lemma (37) this holds precisely when $\sqrt{1-\hat{c}} \in \hat{\mathcal{M}}_\lambda \subset \hat{\mathcal{O}}$. Equivalently $\sqrt{(1-\hat{c})(1+\hat{c})} = \sqrt{1-\hat{c}^2} = |\hat{d}| \in \hat{\mathcal{M}}_\lambda$. Since $\hat{c}^2+\hat{d}^2 = 1$ implies $\hat{c},\hat{d} \in \hat{\mathcal{O}} \subset \hat{F}$, we have $SO(2,\hat{F}) = SO(2,\hat{\mathcal{O}})$ and we have shown that the subgroup which fixes the closed segment ox of length λ in $T\hat{F}^2$ is precisely, kernel $\left(SO(2,\hat{\mathcal{O}}) \to SO(2,\hat{\mathcal{O}}/\hat{\mathcal{M}}_\lambda)\right)$. Since $\hat{\mathcal{O}}/\hat{\mathcal{M}}_\lambda = \mathcal{O}\big/\mathcal{M}_\lambda$, we have proved the first statement of the following proposition.

(39) Proposition (i). *The closed segments of length λ in TF^2, $\lambda > 0$ in Λ, with one endpoint at p in TF^2 identify with points of the unit circle $SO(2,\mathcal{O}/\mathcal{M}_\lambda) \simeq S^1(\mathcal{O}/\mathcal{M}_\lambda) \subset (\mathcal{O}/\mathcal{M}_\lambda)^2$.*

(ii). *The group $PGL_+(2,F)$ acts transitively on segments of length λ in TF^2. More precisely, if $p = [i]$ then the action of the isotropy group of $[i]$, $PGL_+(2,\mathcal{O})$, on segments of length λ at $[i]$ is the usual action of $PGL_+(2,\mathcal{O}/\mathcal{M}_\lambda)$ on $S^1(\mathcal{O}/\mathcal{M}_\lambda)$, via the projection $PGL_+(2,\mathcal{O}) \to PGL_+(2,\mathcal{O}/\mathcal{M}_\lambda)$. In particular, $PGL_+(2,\mathcal{O})$ acts transitively on pairs of segments of length λ at $[i]$.*

Proof. Part (i) was proved in the discussion above. Part (ii) follows from the proof of part (i), which makes it clear that the action of $PGL_+(2,\mathcal{O})$ on the "circle" at distance λ from $[i]$ is determined by the action of $PGL_+(2,\hat{\mathcal{O}})$ on the boundary circle $\hat{S}^1$ of $\hat{B}$. ∎

If we let $\lambda \to 0$, and use the obvious fact that $\mathcal{M} = \bigcup_{\lambda>0} \mathcal{M}_\lambda \subset \mathcal{O}$, Proposition (39) has the following consequence concerning directions in TF^2.

(40) Corollary. *The directions at $p \in TF^2$ correspond bijectively with points on the circle $S^1(k) \subset k^2$, over the residue field $k = \mathcal{O}/\mathcal{M}$. The isotropy group of p in $PGL_+(2,F)$ acts transitively on pairs of directions at p. The isotropy group of one direction is conjugate to the subgroup $\pi^{-1}\left\{\begin{pmatrix} \bar{a} & \bar{b} \\ 0 & \bar{d} \end{pmatrix}\right\}$ of $PGL_+(2,\mathcal{O})$, where $\pi : PGL_+(2,\mathcal{O}) \to PGL_+(2,k)$ is the projection.* ∎

Now we consider ends in TF^2. Let $W \subset TF^2$ be a path, isometric to Λ^+. Choose $p_0 \in W$ and then choose $p_n \in W$ with $d(p_0p_n) = n \in \Lambda$, $n = 1,2,3,\cdots$. Since TF^2 is a Λ-tree, each closed segment p_0p_n is part of W. Lift the points $p_i \in TF^2$ to $P_i \in B$ the unit disk model of HF^2. We will prove that $\{P_i\}$ is a Cauchy sequence in B, which converges to a point $\hat{Q} \in \hat{S}^1$, the boundary of the unit disk $\hat{B}$ over the completion $\hat{F}$.

Without loss of generality, we may assume $P_0 = (0,0) \in B$. Let $\hat{Q}_n \in \hat{S}^1$ be the ideal endpoint of the hyperbolic ray $P_0P_n \subset \hat{B}$. The proof of Lemma (29) guarantees $\hat{Q}_n$ exists.

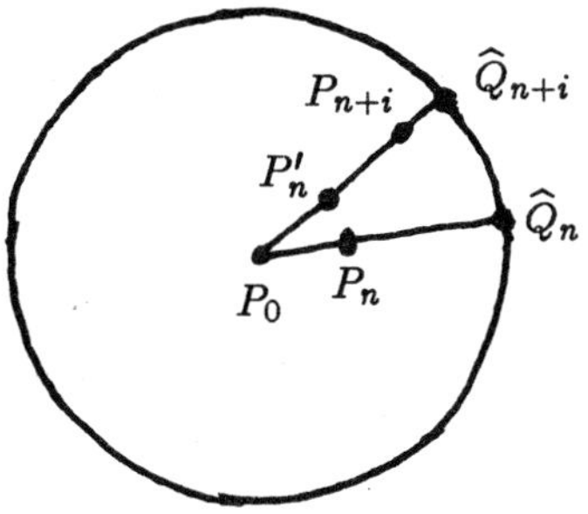

Now, in TF^2, $p_0p_n \subset p_0p_{n+i}$, since W is a path. Therefore there exists a point P'_n on hyperbolic segment P_0P_{n+i} with $d(P_n, P'_n) = 0$. From the proof of Lemma (39), we conclude that the points $\widehat{Q}_n$ and $\widehat{Q}_{n+i}$ in $S^1(\widehat{\mathcal{O}})$ represent the *same* point in $S^1\left(\widehat{\mathcal{O}}/\widehat{\mathcal{M}}_n\right)$, that is, the coordinates of $\widehat{Q}_n$ and $\widehat{Q}_{n+i}$ agree modulo $\widehat{\mathcal{M}}_n$. This bounds $\left\|\widehat{Q}_n - \widehat{Q}_{n+i}\right\|$ and proves $\{\widehat{Q}_n\}$ is a Cauchy sequence. (We use freely the fact that the valuation completion and the Cauchy order completion of F coincide.) Let $\widehat{Q}_n \to \widehat{Q} \in \widehat{S}^1$. If $s_n = \left\|\widehat{Q}_n - P_n\right\|$ then $n = d(P_0, P_n) = -\log(s_n)$, so s_n is a null sequence and we also have $P_n \to \widehat{Q}$.

(41) Proposition. The ends of the Λ-tree TF^2 correspond bijectively with points of the boundary circle $\hat{S}^1$ of $\widehat{B}$ over the completion $\widehat{F}$ of F. Each end is represented by a hyperbolic ray $O\widehat{Q}$ in $\widehat{B}$ from the origin O to $\widehat{Q} \in \widehat{S}^1$. The isotropy group in $PGL_+(2, F)$ of any end is conjugate in $PGL_+(2, \widehat{F})$ to the subgroup $\left\{\begin{pmatrix} a & b \\ o & d \end{pmatrix}\right\}$ which fixes $\infty = (1, 0) \in S^1$.

Proof. It is easy and well-known that ends in any tree are represented by paths from any basepoint. (If o, o' are two basepoints, tack the closed segment oo' onto the beginning of paths from o'.) The argument above shows that if $P_n \in B$ lift points p_n on a path W from $o \in TF^2 = T\widehat{F}^2$, then $P_n \to \widehat{Q} \in \widehat{S}^1$. Moreover, it is clear from that discussion that all p_n lie on the image of the ray $O\widehat{Q}$ in $T\widehat{F}^2$. Namely, if p_n is not on this path, then $\widehat{Q}_{n+i} \not\equiv \widehat{Q}$ modulo $\widehat{\mathcal{M}}_n$, for all $i > 0$, since p_n is on the closed segment op_{n+i}, which is contained in the image of ray $O\widehat{Q}_{n+i}$ in $T\widehat{F}^2$. But $\widehat{Q}_{n+i} \not\equiv \widehat{Q}$ modulo $\widehat{\mathcal{M}}_n$ contradicts $\widehat{Q}_n \to \widehat{Q}$. Thus all ends are represented by rays $O\widehat{Q}$.

If $\widehat{Q} \neq \widehat{Q}' \in \widehat{S}^1$ then for some n, $\widehat{Q} \not\equiv \widehat{Q}'$ modulo $\widehat{\mathcal{M}}_n$. This means the points in TF^2 at distance greater than n from O along $O\widehat{Q}$ and $O\widehat{Q}'$, respectively, are distinct. Therefore, $O\widehat{Q}$ and $O\widehat{Q}'$ represent distinct ends. The final statement about isotropy groups is an easy consequence of the transitivity of the action of $PGL_+(2, \widehat{F})$ on the boundary circle $\widehat{S}^1$. ∎

We point out that in general an end in $T\overline{F}^2$ is *not* represented by the image of any single hyperbolic ray in $\overline{B} \simeq H\overline{F}^2$, even for Euclidean fields $\overline{F}$. An end in $T\overline{F}^2$ is determined by a sequence of closed segments $p_0p_1 \subset p_0p_2 \subset p_0p_3 \subset \cdots$, which can be lifted to hyperbolic segments $P_0P_1, P_0P_2, P_0P_3, \cdots$ in $H\overline{F}^2$. But this sequence of hyperbolic segments need not converge to a hyperbolic ray in $H\overline{F}^2$. Thus, in general, $PGL_+(2, \overline{F})$ does not act transitively on ends in $T\overline{F}^2$, and to understand ends it is necessary to work with completions $\widehat{F}$.

Structure of $PGL_+(2,F)$ isometries of the tree TF^2. Our final results concern the structure of the isometries $\mu : TF^2 \xrightarrow{\sim} TF^2$ induced by matrices $M \in PGL_+(2,F)$. In [2] it is shown that μ has a fixed point if and only if $\log(\mathrm{tr}^2(M)/\det(M)) \le 0 \in \Lambda$. Note that if $M \in GL_+(2,F)$ then $\mathrm{tr}^2(M)/\det(M) \in F^+$ depends only on the class of M in $PGL_+(2,F)$, and also, of course, only on the conjugacy class of M. In [2] it is also shown that in the case $\log(\mathrm{tr}^2(M)/\det(M)) = \lambda > 0$, there is a unique invariant *axis* of $\mu : TF^2 \xrightarrow{\sim} TF^2$, that is, an invariant isometric copy of Λ. Moreover, points on this axis are translated distance λ by μ, and $\lambda = \operatorname{Min}_{p \in TF^2}(d(p, \mu p))$. We will establish the following refinement of these assertions.

(42) Proposition (i). If $M \in PGL_+(2,F)$, then the associated map $\mu : TF^2 \xrightarrow{\sim} TF^2$ has a fixed point if and only if $\log(\mathrm{tr}^2(M)/\det(M)) \le 0 \in \Lambda$.

(ii) If $\det(M)$ is a square in the completion $\widehat{F}$ of F and if $\log\left(\mathrm{tr}^2(M)/\det(M)\right) < 0$ then the map $M : H\widehat{F}^2 \xrightarrow{\sim} H\widehat{F}^2$ has a fixed point.

(iii) If $\log\left(\mathrm{tr}^2(M)/\det(M)\right) > 0$ then the map $M : H\widehat{F}^2 \xrightarrow{\sim} H\widehat{F}^2$ has a unique invariant hyperbolic line, and the D-translation distance along this line is precisely $\alpha > 1 \in F^+$, where $\mathrm{tr}^2(M)/\det(M) = 2 + \alpha + \alpha^{-1}$ and $\log\left(\mathrm{tr}^2(M)/\det(M)\right) = \log(\alpha)$.

Proof. Statement (i) is the result of Morgan-Shalen mentioned above and is proved as follows. If μ has a fixed point we may, after conjugating M, assume $[i] \in TF^2$ is fixed by μ. Then, by Lemma (37), we may assume $M \in GL_+(2,\mathcal{O})$. Then $\det(M) \in \mathcal{O}^*$ and $\mathrm{tr}^2(M) \in \mathcal{O}$, so $\mathrm{tr}^2(M)/\det(M) \in \mathcal{O}$, and $\log(\mathrm{tr}^2(M)/\det(M)) \le 0$. Conversely, given M with $\log(\mathrm{tr}^2(M)/\det(M)) \le 0$, first multiply M by a scalar so that $\log(\det(M)) \le 0$. Now conjugate M to rational canonical form, $M = \begin{pmatrix} 0 & -\det(M) \\ 1 & \mathrm{tr}(M) \end{pmatrix}$. From (22),

$$\begin{aligned} d(i, Mi) &= \operatorname{Max}\left(-\infty, 0, \log(\det{}^2(M)), \log(\mathrm{tr}^2(M))\right) - \log\left(\det(M)\right) \\ &= \operatorname{Max}\left(0, \log(\det(M)), \log(\mathrm{tr}^2(M)/\det(M))\right) = 0, \end{aligned}$$

hence $\mu([i]) = [i]$.

To prove Statement (ii), the equation for a fixed point $z \in H\widehat{F}^2 \subset \widehat{C} = \widehat{F}[i]$ of $M : H\widehat{F}^2 \xrightarrow{\sim} H\widehat{F}^2$ is $z = \frac{az+b}{cz+d}$, or $cz^2 - (a-d)z - b = 0$, where $M = \begin{pmatrix} a & b \\ c & d \end{pmatrix}$. Solving the quadratic equation,

$$\begin{aligned} z &= \frac{a - d + \left(\sqrt{4\det(M) - \mathrm{tr}^2(M)}\right) i}{2c} \\ &= \frac{a - d + \left(2\sqrt{\det(M)}\sqrt{1 - \left(\mathrm{tr}^2(M)/\det(M)\right)}\right) i}{2c}. \end{aligned}$$

Now $\log(\mathrm{tr}^2(M)/\det(M)) < 0$ implies that $1 - (\mathrm{tr}^2(M)/\det(M))$ is a square in $\widehat{F}$, by the proof of Lemma (29), hence a solution $z \in H\widehat{F}^2$ exists.

The proof of statement (iii) is quite similar. If $\log(\mathrm{tr}^2(M)\big/\det(M)) > 0$ then we find two boundary fixed points of $M : H\widehat{F}^2 \xrightarrow{\sim} H\widehat{F}^2$, namely,

$$r_1, r_2 = \frac{a - d \pm \sqrt{\mathrm{tr}^2(M) - 4\det(M)}}{2c} = \frac{a - d \pm |\mathrm{tr}(M)|\sqrt{1 - (4\det(M)\big/\mathrm{tr}^2(M))}}{2c} \in \widehat{F} \cup \infty .$$

(If $c = 0$, one of the boundary fixed points is ∞.) The two boundary fixed points determine an M-invariant hyperbolic line in $H\widehat{F}^2$. After a conjugation in $PGL_+(2, \widehat{F})$, we may assume the boundary fixed points are $0, \infty$. Then $M = \begin{pmatrix} \hat{a} & o \\ o & \hat{d} \end{pmatrix}$ and $\mathrm{tr}^2(M)\big/\det(M) = 2 + \hat{a}/\hat{d} + \hat{d}/\hat{a} = 2 + \alpha + \alpha^{-1}$, where $\alpha = \mathrm{Max}\left(\hat{a}/\hat{d}, \hat{d}/\hat{a}\right)$ is the D-translation distance of M along the imaginary axis. ∎

Remark. If $F = \overline{F}$ is a Euclidean field, much more can be said. The matrices $M \in SL(2, \overline{F})$, which represent non-identity elements of $PGL_+(2, \overline{F})$, are classified as elliptic, parabolic, or hyperbolic according to whether $\mathrm{tr}^2(M) < 4$, $\mathrm{tr}^2(M) = 4$, or $\mathrm{tr}^2(M) > 4$, respectively. In the elliptic case, $M : H\overline{F}^2 \xrightarrow{\sim} H\overline{F}^2$ has a fixed point. In the hyperbolic case, $M : H\overline{F}^2 \xrightarrow{\sim} H\overline{F}^2$ has two boundary fixed points, that is, an invariant hyperbolic line. The D-translation distance along this line is $\alpha > 1 \in \overline{F}^+$, where $\mathrm{tr}^2(M) = 2 + \alpha + \alpha^{-1}$. If $\log(\alpha) = 0 \in \overline{\Lambda}$, then the $\overline{\Lambda}$-axis in $T\overline{F}^2$ which is the image of this invariant hyperbolic line is fixed pointwise by the map $\mu : T\overline{F}^2 \xrightarrow{\sim} T\overline{F}^2$ induced by M. In the parabolic case, $M : H\overline{F}^2 \xrightarrow{\sim} H\overline{F}^2$ has exactly one boundary fixed point. After a conjugation, we may assume $M = \begin{pmatrix} 1 & b \\ 0 & 1 \end{pmatrix}$. Now, from formula (19), if $z \in H\overline{F}^2$, $z = x + iy$, then $d(z, Mz) = \mathrm{Max}\left(\log\left(\frac{b^2}{y^2}\right), 0\right)$. Thus $\mu([z]) = [z]$ in $T\overline{F}^2$ for all $z \in H\overline{F}^2$ with $b\big/\mathrm{Im}(z) \in \mathcal{O}$, in particular, if $\mathrm{Im}(z) \geq |b|$. ∎

In general, $F \subset \bar{F}$, $HF^2 \subset H\overline{F}^2$, $TF^2 \subset T\overline{F}^2$, and $PGL_+(2, F) \subset PGL_+(2, \overline{F})$. If $\Gamma \subset PGL_+(2, F)$ is a finitely generated subgroup then Γ acts on the $\overline{\Lambda}$-tree $T\overline{F}^2$, and this action may be adequate for the applications to Teichmüller theory, when $\Gamma = \pi_1(\Sigma)$ is the fundamental group of a surface Σ. As we have emphasized, the $\overline{\Lambda}$-tree $T\overline{F}^2$ is much easier to study than the Λ-tree TF^2, because hyperbolic geometry in $H\overline{F}^2$ is so well-behaved. Nonetheless, rationality results for $H\widehat{F}^2$, where $\widehat{F}$ is the completion of F, and for $T\widehat{F}^2 = TF^2$, such as those expressed in Corollary (26), Theorem (28), Lemma (29), Propositions (35) and (39), Corollary (40), and Propositions (41) and (42) seem interesting.

References.

1. *G. Brumfiel*, The real spectrum compactification of Teichmüller space, to appear.

2. *J. Morgan and P. Shalen*, Valuations, trees, and degenerations of hyperbolic structures, I, Ann. of Math., **120** (1984), p. 401–476.

3. *J. P. Serre*, Trees, Springer-Verlag, Berlin-Heidelberg-New York, 1980.

MATHEMATICS DEPARTMENT
STANFORD UNIVERSITY
STANFORD, CA 94305–2125

Contemporary Mathematics
Volume **74**, 1988

Gauge Theory and Representations of Kähler Groups

Kevin Corlette

§ 1 Introduction

The representation theory of fundamental groups of Kähler manifolds is a subject of great depth and beauty. It arises in the study of uniformization problems; in the guise of monodromy of families of algebraic varieties [13]; and, by means of the reinterpretation of unitary representations as flat holomorphic vector bundles, in the study of moduli spaces of semistable holomorphic vector bundles [22,7-9,29]. Recent developments have emphasized the mutual entanglements of these different aspects of the subject. This article is intended as a survey of some of these relationships.

The interaction between the gauge group of a vector bundle and geometry in the space of connections is the underlying theme in virtually every result I will discuss. This point of view was first made explicit by Atiyah and Bott in their study [1] of the Yang-Mills equations over Riemann surfaces. They observed that the space of Hermitian connections on a Hermitian vector bundle over a compact Riemann surface ought to be regarded as an infinite dimensional Kähler manifold, with the unitary gauge group acting by isometries, and the full group of vector bundle automorphisms acting holomorphically. They recognized that the relationship between semistability of holomorphic vector bundles and solutions of the Yang-Mills equations obtained by Narasimhan and Seshadri represented an infinite dimensional analogue of the relationship between moment maps and geometric invariant theory described in, e.g., [18], [14].

The idea of Atiyah and Bott has since been developed in several directions. First, the relationship between solutions of the Yang-Mills equations and stability, in particular the existence of compatible flat unitary structures, has been extended to higher dimensional Kähler manifolds by Don-

---

1980 Mathematics Subject Classification (1985 Revision). 53C05, 53C55, 58E20.

aldson (in the algebraic case) and Uhlenbeck and Yau (in the general case). Second, the idea extends naturally to more complex objects, called by Simpson [23] systems of Hodge bundles. In the case of curves, these have been studied, at least in forms related to the one I will describe, by Beilinson, Deligne, and Hitchin. The general case has been studied by Simpson. Among other things, this work gives characterizations of compact Kähler manifolds which are uniformized by bounded symmetric domains. Third, a different incarnation of this set of ideas has been used to show that, under suitable assumptions, a flat bundle over a Kähler manifold admits an underlying variation of Hodge structure. This leads to various rigidity results for representations of the fundamental group. Finally, Hitchin has used all of these ideas to study the moduli space of flat $Sl(2, \mathbf{C})$-bundles over a Riemann surface. He shows, for example, that the smooth part of this moduli space carries a natural complex symplectic structure, and that the natural metric is hyperkähler.

I would like to record here my indebtedness to the mathematicians who have contributed to my understanding of this subject. They include Raoul Bott, Dan Freed, Bill Goldman, John Millson, Carlos Simpson, Cliff Taubes, Karen Uhlenbeck, and Robert Zimmer.

§ 2 Geometry in the space of connections

Let M be a compact smooth manifold, and E a (real or complex) vector bundle over M. Take P to be the frame bundle of E. We may regard P as a right principal G-bundle, where G is the general linear group over the appropriate field. The tangent bundle of P contains as a subbundle the vertical tangent bundle, but there is no canonical choice of a complementary subbundle. A connection on E is a choice of such a complement, satisfying the condition that it be invariant under the action of G on P. A connection is therefore described by the projection of TP on the vertical tangent bundle with the chosen complement as kernel, hence by a one-form on P with values in the Lie algebra of G. It is easy to check that any two connections on E differ by a one-form which is the pullback of a one-form on M with values in the vector bundle $End(E)$ of endomorphisms of E, and furthermore that any such one-form may arise in this fashion. Thus, the space $\mathcal{C}$ of connections on E is an affine space modelled on $\mathcal{E}^1(M, End(E))$, the vector space of one-forms on M with values in $End(E)$.

There is a covariant derivative operator

$$D_A : \mathcal{E}^p(M,E) \longrightarrow \mathcal{E}^{p+1}(M,E)$$

associated to any connection $A \in \mathcal{C}$. The curvature F_A of A is the morphism

$$D_A^2 : \mathcal{E}^p(M,E) \longrightarrow \mathcal{E}^{p+2}(M,E),$$

which may be identified with an element of $\mathcal{E}^2(M, End(E))$.

If we choose geometric structures on M and E, then $\mathcal{C}$ inherits a geometric structure. Suppose M is a Kähler manifold, and E is a Hermitian vector bundle. Then $End(E)$ is also a Hermitian vector bundle, and $\mathcal{E}^1(M, End(E))$ inherits, to begin with, a nondegenerate symmetric real-valued inner product. Thus, at least heuristically, we may think of $\mathcal{C}$ as an infinite-dimensional Riemannian manifold. The complex structures on T^*M and $End(E)$ allow a number of choices of complex structure on $T^*M \otimes_{\mathbf{R}} End(E)$. Among the important ones are $1 \otimes i$, $i \otimes 1$, and $i \otimes \sigma$, where σ is the Cartan involution induced fiberwise on $End(E)$ by the Hermitian metric on E. The choice of any one of these gives $\mathcal{C}$ in addition a complex structure, and, in conjunction with the Riemannian structure, a Kähler metric.

The main thing to notice at this point is that all this structure is invariant under automorphisms of M and E which preserve their geometric structures. Here, I am thinking primarily of the unitary gauge group of E, which I shall denote by $\mathcal{U}$. This is the group of smooth sections of $End(E)$ which are unitary automorphisms at each fiber. This acts on $\mathcal{C}$ by pullback: if $u \in \mathcal{U}$ and $\omega \in \mathcal{C}$, then u acts on P, and $u^*\omega$ is again a connection on E. If we think in terms of the covariant derivative operator D associated to the connection, then u acts by $D \mapsto u \circ D \circ u^{-1} = D - (Du)u^{-1}$. By taking completions with respect to appropriate Sobolev norms, this may be made into an action of a Banach Lie group on a Banach manifold, but I will ignore this point for the time being.

What we now have before us is an action of the group $\mathcal{U}$ by isometries on the Kähler manifold $\mathcal{C}$. No matter which choice we have made for the complex structure on $\mathcal{C}$, the action of $\mathcal{U}$ extends to a holomorphic action of the complexified group $\mathcal{G}$ of automorphisms of E. This is an analogue of a finite-dimensional situation which has been studied by algebraic and symplectic geometers. I will briefly recall the main results. The reader may wish to consult [18] or [14].

Suppose V is a smooth quasiprojective variety over the complex numbers, with compatible Kähler metric, and K is a compact connected Lie group acting on V by isometries. The action extends to the complexification $K_{\mathbf{C}}$ of K. Geometric invariant theory is concerned with the problem of finding an appropriate quotient of V by $K_{\mathbf{C}}$ in the category of algebraic varieties. The usual topological quotient will not do, since it is generally not Hausdorff. The solution is to restrict the contributions given by pathological orbits in the quotient. Geometric invariant theory puts orbits in three categories: stable, semistable, and unstable. If V is a vector space, and $K_{\mathbf{C}}$ acts linearly, these are defined as follows:

1. $\mathcal{O}$ is stable if it is closed and any point in it has finite stabilizer,
2. $\mathcal{O}$ is semistable if its closure does not contain zero, and
3. $\mathcal{O}$ is unstable if it is not semistable.

The general case can be reduced to this by choosing a $K_{\mathbf{C}}$-linearization, i.e. and action of $K_{\mathbf{C}}$ on a projective space $\mathbf{P}^n$ and an equivariant embedding of V in $\mathbf{P}^n$.

Stability is an open condition, and an open set in the quotient is the usual quotient of the set of stable points in V by $K_{\mathbf{C}}$. The remaining points in the quotient correspond to certain equivalence classes of semistable points.

Now recall that V has a metric, and K acts by isometries. In particular, V has an invariant symplectic structure. Symplectic geometers have a parallel description for the process of equivariant linearization and the formation of the quotient by invariant theory. The construction of a $K_{\mathbf{C}}$-linearization roughly amounts to geometric quantization in this vocabulary. The algebro-geometric quotient construction is replaced in symplectic geometry by the Marsden-Weinstein reduction. The central object in all of this is the moment map, which may be understood as follows. Vector fields can be identified with one-forms, using the symplectic form. Vector fields which preserve the symplectic structure infinitesimally correspond to closed one-forms. Those vector fields which come from exact one-forms are called Hamiltonian, and form a Lie algebra $Ham(V)$. We have the following exact sequence of Lie algebras:

$$0 \longrightarrow \mathbf{R} \longrightarrow C^\infty(V) \longrightarrow Ham(V) \longrightarrow 0.$$

Here, we make $C^\infty(V)$ into a Lie algebra by means of the Poisson bracket, and the morphism into $Ham(V)$ is the composition of the exterior derivative with the identification of vector fields and one-forms. The action of K on V is Hamiltonian if the Lie algebra $\mathbf{k}$ acts by Hamiltonian vector fields. A moment map is said to exist if the morphism $\mathbf{k} \longrightarrow Ham(V)$ lifts to a Lie algebra morphism $\phi : \mathbf{k} \longrightarrow C^\infty(V)$. This is described by a map

$$\Phi : V \longrightarrow \mathbf{k}^*,$$

where, for $v \in V$, $\xi \in \mathbf{k}$, we have

$$[\Phi(v)](\xi) = [\phi(\xi)](v).$$

It will not be necessary to worry about the distinction between the Lie algebra and its dual, so the moment map will generally be taken to have its image in the Lie algebra itself.

Φ is an equivariant map, so if $\mathcal{N}$ is a coadjoint orbit, $\Phi^{-1}(\mathcal{N})/K$ is well-defined. This is the Marsden-Weinstein reduction of V with respect to $\mathcal{N}$. The fundamental result is the following.

Theorem 2.1 *The $K_{\mathbf{C}}$ orbit $\mathcal{O}$ is stable if and only if $\Phi|_{\mathcal{O}}$ has zero as a regular value.*

The Marsden-Weinstein reduction of V with respect to zero is the geometric invariant theory quotient of V, because the zeros of the moment map detect the appropriate equivalence classes of semistable orbits, as well as the stable orbits.

It is possible to hope for some analogue of this result when V is replaced by $\mathcal{C}$ with an appropriate choice of Kähler structure and K is replaced by $\mathcal{U}$. As an example, take $\mathcal{A} \subset \mathcal{C}$ to be the subspace of Hermitian connections in $\mathcal{C}$, and let the complex structure be that induced by $i \otimes 1$ on $T^*M \otimes End(E)$. $\mathcal{A}$ is then a complex subspace, and we have furthermore the $\mathcal{U}$-invariant complex subvariety $\mathcal{A}^{1,1} \subset \mathcal{A}$ of Hermitian connections with curvature of type $(1,1)$. Each such connection induces a holomorphic structure in E. Define E to be a stable holomorphic bundle if

$$\mu(F) < \mu(E)$$

for any torsion free coherent subsheaf F, where μ means the normalized degree, i.e.

$$\mu(F) = \frac{1}{(rkF)} \int_M c_1(F) \wedge \omega^{m-1},$$

and ω is the Kähler form on M. The moment map for the action of $\mathcal{U}$ is given by

$$A \mapsto \Lambda F_A,$$

where A is a Hermitian connection, F_A is the curvature of A, and the wedge means contraction with the Kähler form. $\mathcal{G}$ acts on $\mathcal{A}^{1,1}$ by permuting those connections which give rise to equivalent holomorphic structures on E. The correspondence we hoped for is true, as shown in increasing generality by Narasimhan and Seshadri [22], Donaldson [7-9], and Uhlenbeck and Yau [29].

Theorem 2.2 *If E is a holomorphic vector bundle, then it is stable if and only if $\wedge F_A$ is constant for some irreducible connection in the $\mathcal{G}$-orbit defined by E.*

By Chern-Weil theory, it is not possible to get the vanishing of $\wedge F_A$ unless the cohomology class represented by $c_1(E) \wedge \omega^{m-1}$ vanishes, where ω is the Kähler form of M.

This result has an equivalent formulation: E is stable if and only if it admits a Hermitian metric such that the canonical Hermitian connection is irreducible and $\wedge F$ is constant.

Now consider the implications of all this for unitary representations of the fundamental group. The character variety

$$X(\pi_1(M), U(n)) = Hom(\pi_1(M), U(n))/U(n)$$

is an affine variety over the real numbers, and it may be interpreted as the moduli space of flat unitary bundles over M. Flat connections satisfy the curvature condition in Theorem 2.2. If they are irreducible, then they define stable holomorphic structures on the underlying vector bundle. In general, they define semistable structures which are direct sums of stable bundles. Conversely, if a topological vector bundle with vanishing first and second Chern classes admits a stable holomorphic structure, the connection given by the theorem is automatically flat. This follows because $\wedge F_A = 0$, so F_A is a primitive $(1,1)$-form. This implies that the L^2 norm of the curvature is a constant times

$$\int_M Tr(F_A \wedge F_A) \wedge \omega^{m-2},$$

which can be expressed in terms of characteristic classes. $X(\pi_1(M), U(n))$ can therefore be identified with a moduli space of semistable holomorphic vector bundles over M. This sort of thing is expected to carry a complex structure. Indeed, when M is algebraic and there is an algebraic constuction of the moduli space of semistable vector bundles, this is the case. In general, the algebraic methods do not apply, but the construction of the moduli space from the action of $\mathcal{G}$ on $\mathcal{A}^{1,1}$ leads one to hope for the same result. The techniques from deformation theory which have been used in this problem by Kim [19] and Lübke and Okonek [20] seem difficult to apply to the semistable points, but they handle the stable points admirably, yielding the following result.

Theorem 2.3 *Suppose M is a compact Kähler manifold. The open subset of $X(\pi_1(M), U(n))$ consisting of irreducible connections carries a natural complex analytic structure.*

I shall use similar ideas in a somewhat more elaborate setting to describe a rigidity theorem for representations in certain noncompact Lie groups in §4. The difficulty with semistable points will assert itself in that context as well.

§ 3 Harmonic maps and strong rigidity in complex hyperbolic geometry

In this section, I will discuss the Kähler metric induced on $\mathcal{C}$ by the complex structure $1 \otimes i$ on $T^*M \otimes End(E)$. For part of this discussion, the complex structure on M will be irrelevant, so it will be assumed to have merely a Riemannian structure. For simplicity, we will fix a trivialization of $detE = \wedge^{rkE} E$, so that P is a principal $Sl(n, \mathbf{C})$-bundle, the gauge groups will contain only elements compatible with the structure on the determinant, and $\mathcal{C}$ contains only connections which annihilate the trivialization.

The complex gauge group $\mathcal{G}$ acts on $\mathcal{C}$ simply by conjugation:

$$D \mapsto g \circ D \circ g^{-1}.$$

Since we have the Hermitian metric on E at our disposal, the subspace $\mathcal{A} \subset \mathcal{C}$ of Hermitian connections is available. Any connection $D \in \mathcal{C}$ has a unique splitting

$$D = D^+ + \theta,$$

where $D^+ \in \mathcal{A}$, and θ is a one-form on M with values in the self-adjoint endomorphisms of E. A simple calculation shows that, up to a constant imaginary factor, the moment map for the action of $\mathcal{U}$ on $\mathcal{C}$ with the designated Kähler structure is

$$\Phi = D^{+,*}\theta,$$

where $D^{+,*}$ is the adjoint of D^+.

Consider the subvariety $\mathcal{C}_{flat} \subset \mathcal{C}$ of flat connections on E. $\mathcal{C}_{flat}$ is invariant under the action of $\mathcal{G}$. The vanishing of the moment map can be related to an appropriate notion of stability for $\mathcal{G}$-orbits in $\mathcal{C}_{flat}$. Before giving the result, let's give a reinterpretation of the moment map. Variations of Φ as D moves over a $\mathcal{G}$-orbit may be reinterpreted as variations in $D^{+,*}\theta$ when D is fixed, but the Hermitian metric is allowed to vary. A Hermitian metric, compatible with the trivialization of $detE$, is a section of the $Sl(n)/SU(n)$-bundle associated to P. In the flat case, this is simply an equivariant mapping from the universal cover $\widetilde{M}$ of M into $Sl(n)/SU(n)$. It turns out that the vanishing of the moment map is equivalent to the condition that this map should be harmonic; indeed, the one-form θ can be identified with the differential of the map, and D^+ with the pullback of the Levi-Civita connection on $Sl(n)/SU(n)$.

Nothing prevents us from doing the same construction with $Sl(n, \mathbf{C})$ replaced by any other semisimple real algebraic group G, although the symplectic interpretation is lacking if G is not complex. We obtain a theory of equivariant harmonic maps from $\widetilde{M}$ to G/K, where K is a maximal compact subgroup of G.

Under what circumstances can we expect to solve this problem? As motivation, consider the problem of finding a geodesic (i.e. a harmonic map from the circle) in a given homotopy class of loops on the locally symmetric space $\Gamma \backslash G/K$, where Γ is a torsion free discrete subgroup of G. It is a classical fact [2] that this is possible if and only if the corresponding element $\gamma \in \Gamma$ is a semisimple element of G. The general result is the following (see [4], [10]).

Theorem 3.1 *Suppose P is a flat bundle over the compact Riemannian manifold M. The associated G/K-bundle has a harmonic section if and only if the Zariski closure of the image of the holonomy homomorphism is a reductive subgroup of G.*

We recover the following fact, originally a special case of the existence result of Eells and Sampson [11]: if N is a compact nonpositively curved locally symmetric manifold and ν is a homotopy class of maps from M to N, then ν has a harmonic representative.

I will briefly describe the proof of the existence of the harmonic section when $dim_{\mathbf{R}}M = 2$, G is linear, and the holonomy is irreducible.

Fix a compatible metric on the associated vector bundle E, and let A be any flat connection with irreducible holonomy. We wish to deform it along the $\mathcal{G}$ orbit into a connection satisfying $D^{+,*}\theta = 0$. This condition is equivalent to minimizing the L^2 norm of θ along the orbit. Set up the nonlinear heat equation

$$\frac{\partial A}{\partial t} = -D_A D_A^{+,*} \theta_A.$$

Solutions of this equation give the formal gradient flow for $\| \theta_A \|_{L^2}^2$ along the orbit. We need to show that a solution with initial value A_0 exists, is defined for all time, and converges as $t \longrightarrow \infty$ to a connection in the same $\mathcal{G}$ orbit as the initial connection. The existence of a solution with given initial value for a short time is guaranteed by general theory [15], so I will concentrate on the convergence problem.

Let F^+ be the curvature of D^+. Since D_A remains flat as long as the solution is defined, we have

$$F^+ + \frac{1}{2}[\theta, \theta] = 0,$$

$$D^+\theta = 0.$$

A simple calculation shows that the L^2 norm of θ is bounded as A evolves. The maximum principle for the linear heat operator shows that the C^0 norm of Φ remains bounded.

The Weitzenböck formula [3] for the Laplacian

$$\Box^+ = D^+D^{+,*} + D^{+,*}D^+$$

on one-forms gives

$$\| \Phi \|_{L^2}^2 = (\Box^+\theta, \theta) = \| \nabla\theta \|_{L^2}^2 + (\theta \circ Ricci, \theta) + \frac{1}{4} \| [\theta, \theta] \|_{L^2}^2,$$

where $\nabla\theta$ is the full covariant derivative of θ. Hence,

$$\| \nabla\theta \|^2 \leq \| \Phi \|^2 + C \| \theta \|^2 \leq C'$$

is satisfied for all time.

In dimension 2, the Sobolev inequality implies that

$$\begin{aligned} \| \theta \|_{L^4}^2 &\leq C \textstyle\int_M | d(1 + | \theta |^2)^{\frac{1}{2}} |^2 \, dvol \\ &\leq C' \textstyle\int_M | \nabla\theta |^2 \, dvol \\ &\leq C''. \end{aligned}$$

Thus, F^+ is bounded in L^2 norm for all time. From the compactness result in [28], I can change $A(t)$ by a time dependent unitary gauge transformation so that the new Hermitian connections, which I shall continue to denote by D^+, remain in a subset of $\mathcal{C}$ which is bounded in the Sobolev $L^{2,1}$ norm. Now the L^2 bounds on Φ and θ and the vanishing of $D^+\theta$ give an $L^{2,1}$ bound on θ. Therefore, the connections converge weakly in $L^{2,1}$ to a limiting connection as $t \longrightarrow T$, where $[0,T)$ is the maximal time interval on which a solution to the nonlinear heat equation exists. If $T < \infty$, then we can use the convergence to extend the solution for a short time beyond T, so the solution exists for all time. Thus, we get a limiting connection D_∞ as $t \longrightarrow \infty$ satisfying $\Phi_\infty = 0$. By semicontinuity properties of the dimensions of the kernels of elliptic operators, there is a nonconstant homomorphism ϕ from E with initial connection D to E with the limit connection. If ϕ is an isomorphism, then we are done. If not, then $ker\ \phi$ is a D-invariant subbundle of E, contradicting the irreducibility of the holonomy. **QED**

Now return to the situation where M is Kähler. If G/K is a Hermitian symmetric space, then holomorphic sections of the G/K-bundle are special cases of harmonic sections. Under conditions on the rank of the mapping, ideas of Siu [24,25] can be used to show that any harmonic section is necessarily holomorphic or antiholomorphic. The necessary rank conditions can be guaranteed by nonvanishing in cohomology of characteristic forms pulled back to M from the symmetric space. For example, if $G = SU(n,1)$, then $K = U(n)$, and we have the Chern forms of the tangent bundle on the complex ball $B^n = SU(n,1)/U(n)$. This gives topological invariants of P which I shall denote (I hope not misleadingly) by $c_i(P)$. A typical result is the following, from [4].

Theorem 3.2 *Suppose M is a compact Kähler manifold and P is a flat $SU(n,1)$-bundle over M. If $c_1(P)^2 \neq 0$, then the associated B^n-bundle has a holomorphic or antiholomorphic section.*

This is proved by first showing the existence of a harmonic section. To do this, it must be shown that the holonomy has reductive Zariski closure under the assumptions on characteristic classes. Suppose it does not. There is a parabolic subgroup containing the image of the holonomy, and the flat connection may be continuously deformed until the holonomy is contained in the reductive piece of the Levi decomposition of the parabolic. But for $SU(n,1)$, this reductive piece is the product of a compact group with an abelian group. The characteristic classes then vanish, contradicting the assumption. Once the harmonic section is available, Siu's argument carries over to give the result. Note that we can always choose the complex structure on B^n so that the section is holomorphic.

This theorem was used in [4] to prove a strong rigidity result in complex hyperbolic geometry, originally conjectured by Goldman and Millson [12]. Let Γ be a cocompact torsion free discrete subgroup of $SU(m,1)$, and ρ a homomorphism from Γ to $SU(n,1)$, $n \geq m$. This homomorphism gives rise to a flat $SU(n,1)$-bundle over the compact complex hyperbolic manifold B^m/Γ. We want to know if ρ is a standard homomorphism, where standard means the following: there is a factorization

$$\rho : \Gamma \longrightarrow SU(m,1) \times H \longrightarrow SU(n,1),$$

where H is the centralizer of $SU(m,1)$ in $SU(n,1)$ under a standard embedding, and the homomorphism of Γ to $SU(m,1)$ is conjugate to the original inclusion. Such representations are characterized by the property that there is a totally geodesic section of the B^n-bundle associated to P.

Theorem 3.3 *Suppose $c_1(P)^m = c_1(M)^m$, $m \geq 2$. Then ρ is standard.*

The proof uses the existence of a holomorphic section and distance decreasing properties of holomorphic maps to produce the totally geodesic section. The appropriate analogue of this result for hyperbolic Riemann surfaces has been proved by Toledo [26,27].

This seems to be an optimal result, in the sense that the equality of characteristic classes cannot be weakened to nonvanishing of $c_1(P)^m$. Mostow [21] gave an example of a branched cover of a compact complex hyperbolic surface which again has a complex hyperbolic structure. The induced map on fundamental groups gives the required counterexample.

§ 4 Moduli spaces of representations

This section is devoted to a description of the moduli space of a certain class of representations of fundamental groups of Kähler manifolds.

Take M again to be a compact Kähler manifold, and let G be a simple real algebraic group such that G/K is a Hermitian symmetric space (and not the hyperbolic disk). I have pointed out that invariant differential forms on G/K give rise to topological invariants of flat G-bundles; in this section, I am interested in the volume form, which gives an invariant to be denoted by $vol(P)$, or $vol(\rho)$ for the corresponding holonomy homomorphism. The character variety $X(\pi_1(M), G)$ (i.e. the set of semisimple homomorphisms divided out by the action of G by conjugation) is a real algebraic variety. Let $\mathcal{M}$ be the component (in the classical topology) containing a representation ρ with $vol(\rho) \neq 0$. The main result is the following (see [5]).

Theorem 4.1 *$\mathcal{M}$ is compact and carries a natural complex structure.*

The compactness of $\mathcal{M}$ is a result of considerations from the preceding section. The condition on the volume invariant guarantees that the image of ρ is Zariski dense in G, and the existence of a holomorphic section of the associated flat G/K-bundle is obtained as before. The distance decreasing properties of holomorphic maps can again be applied, giving compactness of $\mathcal{M}$. Consult [5] for details. A similar argument is used by Deligne [6] to show that only finitely many representations of a given rank may arise as direct summands in the monodromy representation of a family of algebraic varieties over a fixed quasiprojective variety.

Why should $\mathcal{M}$ be expected to have a complex structure? In section two, we saw that the moduli space of unitary representations has a complex structure. On an infinitesimal level, this is a reflection of the fact that the cohomology group $H^1(M, End(E))$ (E is a flat unitary vector bundle) has a Hodge decomposition into classes of type (1,0) and (0,1). In the situation I am discussing, the naive Hodge decomposition does not work. However, Deligne observed (see [30]) that there is a more elaborate Hodge decomposition for the cohomology. Let P be the flat G-bundle associated to ρ, and $ad(P)$ the vector bundle associated to P by the adjoint representation of G. Because we have the holomorphic section of the G/K-bundle, $ad(P)$ becomes a real variation of Hodge structure [13]. As such, there is a natural bidegree in the fibers of the complexified bundle. Deligne's observation was that the bidegree in the fibers should be added to the bidegree of differential forms to give the appropriate bidegree in $\mathcal{E}^*(M, ad(P) \otimes \mathbf{C})$. This gives the Hodge decomposition of $H^1(M, ad(P)) \otimes \mathbf{C})$.

The Lie algebra $\mathbf{g}$ of G may be decomposed as $\mathbf{g} = \mathbf{k} \oplus \mathbf{p}$, where $\mathbf{k}$ is the Lie algebra of K. Since G/K is a Hermitian symmetric space, the complexification of $\mathbf{p}$ breaks up into (1,0) and (0,1) pieces, so we have $\mathbf{g} \otimes \mathbf{C} = (\mathbf{k} \otimes \mathbf{C}) \oplus \mathbf{p}^{1,0} \oplus \mathbf{p}^{0,1}$. There is a corresponding decomposition of the vector bundle:

$$ad(P) \otimes \mathbf{C} = F^{0,0} \oplus F^{1,-1} \oplus F^{-1,1}.$$

I have labelled the components with the appropriate bidegrees. A form in $\mathcal{E}^{p,q}(M, F^{p',q'})$ will be said to have total Hodge type $(p+p', q+q')$. It is easy to check that the flat connection D on $ad(P)$ splits into operators which shift the total Hodge type of a form of pure type either by (1,0) or (0,1), in complete analogy with the classical case. I shall denote this decomposition by

$$D = \mathcal{D}' + \mathcal{D}''.$$

These operators do not preserve the property of having a pure form type or that of having pure fiber type. However, there remain many analogies with the classical decomposition of a Hermitian connection into ∂ and $\overline{\partial}$ operators. For example, $\mathcal{D}'$ and $\mathcal{D}''$ both have square zero. If I am given a bundle with the complexification of G as structure group, a decomposition of the adjoint bundle as above, some choice of $\mathcal{D}''$, and a subbundle with structure group G, then this uniquely determines a $\mathcal{D}'$ operator and the connection $D = \mathcal{D}' + \mathcal{D}''$. Furthermore, if the given $\mathcal{D}''$ has square zero, then the resulting connection has curvature of total Hodge type (1,1). This is an analogue of the canonical connection associated to a holomorphic vector bundle with a Hermitian metric.

Let $\mathcal{C}$ be the set of all connections on the smooth G-bundle P associated to the representation ρ. Decompose the complexified adjoint bundle as above. This allows us to decompose any $D \in \mathcal{C}$ into operators which shift total Hodge type by (1,0), (0,1), (2,-1), and (-1,2). I would like to exclude the last two possibilities, so consider the affine subspace $\widehat{\mathcal{C}}$ of connections which only have components of the first two types. We decided above that all the flat connections we are interested in have underlying variations of Hodge structure, and thus lie in $\widehat{\mathcal{C}}$. The same ideas which were applied in §2 to $\mathcal{A}^{1,1}$ can now be applied to the subvariety $\widehat{\mathcal{C}}^{1,1}$ of connections with curvature of total Hodge type (1,1). By analogues of the arguments of Kim [19] and Lübke and Okonek [20], the existence of the complex structure follows. There is no trouble with semistable points, since the condition on $vol(\rho)$ implies that everything is stable.

There is an analogue of 2.2 for $\widehat{\mathcal{C}}^{1,1}$. Versions of it were proved for Riemann surfaces by Hitchin [16], and apparently also by Beilinson and Deligne. The general case has been proved by Simpson [23]. For simplicity, I will only formulate the result for symplectic bundles over Riemann surfaces. Let M be a compact Riemann surface, and P a right principal $Sp(n,\mathbf{R})$-bundle over M. Fix a reduction of the structure group of P to $U(n)$, the maximal compact subgroup of the symplectic group. Let E be the vector bundle associated to P by the complexification of the standard representation of $Sp(n,\mathbf{R})$. E is a Hermitian vector bundle. As usual $\mathcal{C}$ is the space of connections on P, and $\mathcal{A}$ the affine subspace of Hermitian connections. Recall the decomposition

$$D = D^+ + \theta$$

from the previous section. Using the complex structure on M, D^+ can be decomposed as

$$D^+ = \partial^+ + \overline{\partial}^+.$$

These operators shift the total Hodge type of a form in $\mathcal{E}^{p,q}(M, F^{p',q'})$ by (1,0) and (0,1), respectively. $\overline{\partial}^+$ defines a holomorphic structure on E. The one-form θ decomposes into four pieces:

$$\theta = \theta^{2,-1} + \theta^{1,0} + \theta^{0,1} + \theta^{-1,2},$$

where the superscripts represent the amount by which $ad(\theta^{p,q})$ shifts the total Hodge type. As above, $D \in \widehat{\mathcal{C}}$ means that

$$\theta^{2,-1} = \theta^{-1,2} = 0.$$

The condition $D \in \widehat{\mathcal{C}}^{1,1}$ is equivalent to the requirement that $\theta^{0,1}$ be a holomorphic form (note: $\theta^{0,1}$ is a (1,0)-form with values in $F^{-1,1}$) and $\theta^{1,0}$ is an antiholomorphic form. The holomorphic and antiholomorphic structures are given by $\overline{\partial}^+$ and ∂^+, respectively.

The moment map for the complex structure on $\widehat{\mathcal{C}}^{1,1}$ induced by $i \otimes 1$ on $T^*M \otimes ad(P)$ is, up to a constant,

$$\Phi = \Lambda(F^+ + [\theta^{1,0}, \theta^{0,1}]).$$

On $\widehat{\mathcal{C}}^{1,1}$ this is essentially just the curvature F of the connection D. This is to be related to the following notion of stability.

Definition 4.2 *Suppose $D \in \widehat{\mathcal{C}}^{1,1}$, and $\mathcal{D}''$ is the corresponding operator in the Deligne decomposition. D is stable if any $\mathcal{D}''$-invariant subbundle S of E satisfies*

$$\mu(S) < \mu(E).$$

Here, μ has the same meaning as in §2.

The gauge group I want to consider is $\mathcal{U}_{\mathbf{C}}$, the complexification of the group of unitary automorphisms of P. An element $g \in \mathcal{U}_{\mathbf{C}}$ acts on a $\mathcal{D}''$ operator by conjugation:

$$\mathcal{D}'' \mapsto g \circ \mathcal{D}'' \circ g^{-1}.$$

According to the discussion above, the $\mathcal{D}''$ operator determines the connection, once the metric has been fixed. Hence, this describes an action of $\mathcal{U}_{\mathbf{C}}$ on $\widehat{\mathcal{C}}$. The notion of stability is a property of $\mathcal{U}_{\mathbf{C}}$-orbits. Let $\mathcal{O}$ be such an orbit in $\widehat{\mathcal{C}}^{1,1}$.

Theorem 4.3 *If $\mathcal{O}$ is a stable orbit, there is some $\mathcal{U}$ orbit contained in $\mathcal{O}$ on which Φ is constant.*

This theorem contains the result of Narasimhan and Seshadri discussed in §2. In addition, it allows a rapid proof of the uniformization theorem for Riemann surfaces with negative Euler characteristic, as pointed out in [16] and [23]. Simpson gives very interesting uniformization results in higher dimensions.

§ 5 Hyperkählerian considerations

I will conclude with a brief discussion of some ideas of Hitchin. The proof of 4.1 relies on the possibility of relating the moduli space of flat bundles to two different moment maps. As observed by Hitchin [16], the general context in which to understand this is that of the hyperkähler moment map. A general discussion of this concept is given by Hitchin, Karlhede, Lindström and Rocek [17].

A hyperkähler manifold N is a Riemannian manifold together with three complex structures I, J and K, satisfying the following two conditions:

1. N, together with any one of I, J, K, is a Kähler manifold, and

2. I, J, K satisfy the relations of the quaternions ($\mathbf{H}$), i.e. $IJ = K$, $JK = I$ and $KI = J$.

Suppose G is a Lie group acting on N by isometries, preserving each of the complex structures. N has three distinct symplectic structures and, under favorable circumstances, there are moment maps Φ_1, Φ_2, Φ_3 corresponding to each of them. The hyperkähler moment map is the function $\mu : N \longrightarrow \mathbf{g}^* \otimes_{\mathbf{R}} \mathbf{H}$ defined by

$$\mu = \Phi_1 I + \Phi_2 J + \Phi_3 K.$$

The hyperkähler quotient of N is $\mu^{-1}(0)/G$. If it is smooth, then it also has a hyperkähler structure, hence has Ricci curvature zero.

Recall the discussion of the space $\mathcal{C}$ of all linear connections on a Hermitian vector bundle E over a compact Kähler manifold M. As in §2, let σ be the fiberwise Cartan involution on $End(E)$, and define $I = 1 \otimes i$, $J = i \otimes \sigma$, and $K = IJ$. It is easy to check that these choices define a hyperkähler structure on $\mathcal{C}$. If M is a Riemann surface, then the condition $\mu(A) = 0$ means that A is a flat connection which satisfies $D^{+,*}\theta = 0$. From this, Hitchin obtains the following.

Theorem 5.1 *Suppose M is a Riemann surface. The character variety $X(\pi_1(M), Sl(n, \mathbf{C}))$ carries a natural hyperkähler structure at smooth points.*

Previously, we have seen that the gradient flow for the norm of the moment map is a very special object. The corresponding flow for the norm of the hyperkähler moment map does not seem to have received any published consideration, although this would probably be of some interest, even in very simple cases. One cannot expect to relate the flow to the orbit structure of a quaternionic form of the group, since the latter is generally ill-defined. Hence, one of the main problems would be to characterize the set of points for which the gradient flow of $\| \mu \|^2$ leads down to a fixed G-orbit in $\mu^{-1}(0)$, and another would be to give a characterization of stationary points for the flow.

This problem is of special interest in the case of $\mathcal{C}$. In two different situations (3.2 and 4.3), the analytic properties of this flow allowed us to conclude the existence of a nonconstant holomorphic map from the universal cover of M to a bounded symmetric domain. Clearly, one would like to weaken the hypotheses to obtain both results at a single stroke. As a first step, it might be useful to understand the heat flow

$$\frac{\partial A}{\partial t} = -D_A^* F_A - D_A D_A^{+,*} \theta_A$$

for a Riemann surface. Since $\mathcal{C}$ is roughly four times the dimension of the unitary gauge group, it seems fair to expect that this should converge as t approaches ∞ for any initial choice of connection. If this could be proved, it might give useful clues about the situation in higher dimensions.

References

1. M.F. Atiyah and R. Bott, The Yang-Mills equations over Riemann surfaces, Phil. Trans. R. Soc. Lond. A 308 (1982), 523-615.

2. W. Ballman, M. Gromov and V. Schroeder, *Manifolds of Nonpositive Curvature*, Birkhäuser, Boston, 1985.

3. J.-P. Bourguignon and H.B. Lawson, Jr., Stability and isolation phenomena for Yang-Mills fields, Comm. Math. Phys. 79 (1981), 189-230.

4. K. Corlette, Flat G-bundles with canonical metrics, to appear in J. Diff. Geo.

5. K. Corlette, in preparation.

6. P. Deligne, Un théorème de finitude pour la monodromie, *Discrete Groups in Geometry and Analysis: papers in honor of G.D. Mostow on his sixtieth birthday*, R. Howe, ed., Birkhäuser, Boston, 1987.

7. S.K. Donaldson, A new proof of a theorem of Narasimhan and Seshadri, J. Diff. Geo. 18 (1983), 269-277.

8. S.K. Donaldson, Anti-self-dual Yang-Mills connections over complex algebraic surfaces and stable vector bundles, Proc. Lond. Math. Soc. 50 (1985), 1-26.

9. S.K. Donaldson, Infinite determinants, stable bundles and curvature, Duke Math. J. 54 (1987), 231-247.

10. S.K. Donaldson, Twisted harmonic maps and the self-duality equations, Proc. Lond. Math. Soc. 55 (1987), 127-131.

11. J. Eells and J.H. Sampson, Harmonic mappings of Riemannian manifolds, Amer. J. Math. 86 (1964), 109-160.

12. W. Goldman and J. Millson, Local rigidity of discrete groups acting on complex hyperbolic space, Inv. Math. 88 (1987), 495-520.

13. P. Griffiths, ed., *Topics in Trancendental Algebraic Geometry*, Ann. of Math. Studies 106, Princeton University Press, Princeton, 1984.

14. V. Guillemin and S. Sternberg, Geometric quantization and multiplicities of group representations, Inv. Math. 67 (1982), 515-538.

15. R. Hamilton, *Harmonic Mappings of Manifolds with Boundary*, Lecture Notes in Mathematics 471, Springer Verlag, New York, 1975.

16. N.J. Hitchin, The self-duality equations on a Riemann surface, Proc. Lond. Math. Soc. 55 (1987), 59-126.

17. N.J. Hitchin, A. Karlhede, U. Lindström, and M. Rocek, Hyperkähler metrics and supersymmetry, Comm. Math. Phys. 108 (1987), 535-589.

18. G. Kempf and L. Ness, The length of vectors in representation spaces, *Algebraic Geometry, Proceedings, Copenhagen 1978,* K. Lonsted, ed., Lecture Notes in Mathematics 732, Springer Verlag, New York, 1982.

19. H.-J. Kim, Moduli of Hermite-Einstein vector bundles, Math. Z. 195 (1987), 143-150.

20. M. Lübke and C. Okonek, Moduli spaces of simple bundles and Hermitian-Einstein connections, Math. Ann. 276 (1987), 663-674.

21. G.D. Mostow, On a remarkable class of polyhedra in complex hyperbolic space, Pac. J. Math. 86 (1980), 171-276.

22. M.S. Narasimhan and C.S. Seshadri, Stable and unitary vector bundles on a compact Riemann surface, Ann. Math. 82 (1965), 540-567.

23. C. Simpson, Systems of Hodge bundles and uniformization, Doctoral dissertation, Harvard University (1987).

24. Y.-T.Siu, The complex-analyticity of harmonic maps and the strong rigidity of compact Kähler manifolds, Ann. Math. 112 (1980), 73-112.

25. Y.-T.Siu, Strong rigidity of compact quotients of exceptional bounded symmetric domains, Duke J. Math. 48 (1981), 857-871.

26. D. Toledo, Harmonic maps from surfaces to certain Kaehler manifolds, Math. Scand. 45 (1979), 13-26.

27. D. Toledo, Representations of surface groups in $PSU(1,n)$ with maximum characteristic number, to appear.

28. K. Uhlenbeck, Connections with L^p bounds on curvature, Comm. Math. Phys. 83 (1982), 31-42.

29. K. Uhlenbeck and S.T. Yau, On the existence of Hermitian-Yang-Mills connections in stable vector bundles, Comm. Pure and Appl. Math. 34 (1986), S257-293.

30. S. Zucker, Hodge theory with degenerating coefficients: L_2 cohomology in the Poincaré metric, Ann. Math. 109 (1979), 415-476.

Department of Mathematics
University of Chicago
5734 South University Avenue
Chicago, Illinois 60637

Contemporary Mathematics
Volume **74**, 1988

THE DIOPHANTINE NATURE OF SOME CONSTRUCTIONS AT INFINITY

Daniel R. Farkas

What follows is a footnote to a footnote. In 1971, George Bergman published a landmark paper ([2]) that led to breakthroughs in infinite group theory ([9]) and fundamental work in group rings ([10]). It was also related to new invariants discovered for solvable groups ([12]). Bergman's paper included the following footnote: "D. Mumford has pointed out to me that this point of view is similar to that of the theory of Néron models..." To the best of my knowledge, this remark has not been pursued.

In a similar vein, several mathematicians at this conference wondered whether it would be helpful to learn Diophantine techniques as a way of extending the same philosophy of logarithmic behavior at infinity in the work of Shalen, Culler, and Morgan ([3], [7]).

My belief is that the Diophantine connection is fundamental rather than coincidental. Not only is there an overlap of methods, but it is possible that arithmetic end-products of Diophantine geometry can be directly applied in these other areas. In the short illustration that follows, I will examine Roseblade's generalization ([10], [1]) of Bergman's theorem. It turns out to be an immediate consequence of a recent theorem which has appeared in connection with a question of Lang ([5]).

I wish to thank Darrell and Felice for providing sage hospitality while these ideas were being written up.

§1. A Diophantine connection.

The Diophantine theorem we shall refer to is a relative of the Thue-Siegel-Roth Theorem. It is essentially a result about number fields.

Theorem ([4], [11]; cf. [6]): Assume K is a field of characteristic zero and W is a finitely generated multiplicative subgroup of K. Choose an integer $d \geq 2$. Then the collection of all $(w_1, ..., w_d) \in W^d$ such that $w_1 + ... + w_d = 0$ but no subsum is zero comprises a finite set up to homothety in K.

We compare this result to Roseblade's Intersection Theorem. In order to state the proposition, we review the notion of the "controller" of an ideal in a group algebra. Let I be a two-sided ideal of the group algebra $k[\Gamma]$. Then there is a unique minimal subgroup $\Lambda \subset \Gamma$ such that $I = (I \cap k[\Lambda])k[\Gamma]$. The subgroup Λ is called the controller of T (see,e.g., Chapter 8 of [8]).

Roseblade's Intersection Theorem: Let k be a field and let A be a finitely generated abelian group. Assume that G is a group of automorphisms of A and extend it to a group of k-automorphisms of $k[A]$. If P is a G-invariant prime ideal of $k[A]$ and D is its controller then G acts like a finite group on the image of D modulo P.

We establish a link by re-examining the controller Λ for an ideal I of $k[\Gamma]$. Consider $t \in I$. By a fission of t under I we mean a collection $s_1, ..., s_n \in k[\Gamma]$ such that

$$(i) t = s_1 + ... + s_n$$

$$(ii) s_j \in I \text{ for } j = 1, ..., n$$

$$(iii) supp(s_i) \cap supp(s_j) = \emptyset \text{ for } i \neq j.$$

Since the support of t is finite, it has only finitely many fissions under I. The subgroup afforded by the fission of t is $< ab^{-1} | a, b \in supp(s_j)$ for $1 \leq j \leq n >$. The key observation is that the subgroup afforded by a maximal fission of t is contained in Λ.

If $M = \{t_1, ..., t_m\}$ is a finite set of elements in I then by a fission of M under I we mean a union of fissions for each t_i; the subgroup afforded by M is the group generated by the subgroups afforded by t_i for $i = 1, ..., m$. Notice that if M is a finite set of generators for the ideal I then Λ is precisely the subgroup afforded by M.

A Diophantine proof of Roseblade's Theorem (characteristic zero): We might as well assume $D = A$. Let $\theta : k[A] \to k[A]/P$ be the canonical k-algebra map and let K be the field of fractions of $k[A]/P$. Choose a finite set of generators for P and consider a maximal fission under P of these generators, denoted T.

Given $t \in T$, write $t = \sum_{i=1}^{d} \lambda_i a_i$ with $\lambda_i \in k^*$ and $a_i \in A$. Suppose W is the finitely generated multiplicative subgroup of K generated by the image of A and all of the λ_i as t ranges over T. Set $w_i = \lambda_i \theta({}^g a_i)$ for a fixed $g \in G$. Then $w_i \in W$ and

$$\sum_{i=1}^{d} w_i = \theta({}^g t) = 0.$$

Similarly, any proper subsum of the w_i which equals zero gives rise to a further fission of T. By the Diophantine theorem, there are only finitely many such w_i up to homothety in K.

What does it mean that $(\lambda_1\theta({}^g a_1), ..., \lambda_d\theta({}^g a_d))$ is a multiple of $(\lambda_1\theta({}^h a_1), ..., \lambda_d\theta({}^h a_d))$? For all i, j

$$\theta({}^g(a_i a_j^{-1})) = \theta({}^h(a_i a_j^{-1})).$$

Thus if A_t is the subgroup of A afforded by t then $\theta({}^g b) = \theta({}^h b)$ for all $g \in A_t$.

Since we are assuming A controls P, we have $A = \Pi_{t \in T} A_t$. Thus we have shown that for each $a \in A$ there are only finitely many values for $\theta({}^g a)$ as g ranges over G. ∎

Notice that we never used the full force of the Diophantine theorem. We only required the following property:

(D) Let W be a finitely generated multiplicative subgroup of a field K and let G be a group of automorphisms of K which stabilizes W. If $w_1, ..., w_d \in W$ are such that $w_1 + ... + w_d = 0$ but no subsum is zero then there is a subgroup of finite index in G which fixes $(w_1, ..., w_d)$ up to homothety (under the diagonal action).

In fact, (D) is equivalent to Roseblade's Intersection Theorem. To derive (D), let k be the prime subfield of K. The action of G on W extends to an action on the group algebra, so that the obvious map $k[W] \to K$ is G-equivariant. Let the prime ideal P be its kernel. If E is the controller of P, there is, by assumption, a subgroup H with finite index in G such that H fixes each element of E. If we consider $w_1, ..., w_d$ in $k[W]$ then $w_1 + ... + w_d \in P$ but no subsum lies in P. Since $P = (P \cap k[E])k[W]$, it must be that $w_1, ..., w_d$ all lie in the same coset of E. Property (D) is immediate when this is interpreted in K.

The full Diophantine theorem undoubtedly has other implications for group algebras. For instance, let P be a faithful prime ideal of $k[A]$ with A a finitely generated abelian group. ("Fidelity" here means that there is no nontrivial kernel when sending A to its image modulo P.) Then P contains only finitely many elements of minimum support up to multiplication by a unit.

REFERENCES

1. G.M. Bergman, The logarithmic limit-set of an algebraic variety, Trans. Amer. Math. Soc. 157 (1971), 459-469.
2. R. Bieri and J.R.J. Groves, A rigidity property for the set of all characters induced by valuations, Trans. Amer. Math. Soc. 294(1986), 425-434.
3. M. Culler and P.B. Shalen, Varieties of group representations and splittings of 3-manifolds, Ann. of Math. (2) 117 (1983), 109-146.
4. J.H. Evertse, On sums of S-units and linear recurrences, Compositio Math. 53 (1984), 225-244.
5. S. Lang, Fundamentals of Diophantine Geometry, Springer-Verlag, Berlin, 1983.
6. M. Laurent, Equations diophantiennes exponentielles, Invent. Math. 78 (1984), 299-327.
7. J.W. Morgan and P.B. Shalen, Valuations, trees, and degenerations of hyperbolic structures I, Ann. of Math (2) 120(1984), 401-476.
8. D.S. Passman, The Algebraic Structure of Group Rings, Wiley-Interscience, New York, 1977.
9. J. E. Roseblade, Group rings of polycyclic groups, J. Pure Appl Algebra 3(1973), 307-328.
10. J.E. Roseblade, Prime ideals in group rings of polycyclic groups, Proc. London Math. Soc (3) 36 (1978), 385-447.
11. H.P. Schlickewei and A.J. Van der Poorten, The growth conditions for recurrence sequences,

Rep. 82.0041, Macquarie Univ. NSW Australia (1982).

12. R. Strebel, Finitely presented soluble groups, in Group Theory: Essays for Philip Hall, Academic Press, London, 1984.

Department of Mathematics

Virginia Polytechnic Institute and State University

Blacksburg, VA 24061

Contemporary Mathematics
Volume 74, 1988

COMPLEX REPRESENTATIONS AND ONE-RELATOR PRODUCTS OF CYCLICS

Benjamin Fine and Gerhard Rosenberger

ABSTRACT. A one-relator product of cyclics is a group of the form $(*A_i)/N(S)$ where each A_i is cyclic and S is a non-trivial cyclically reduced word in the free product $*Ai$. If $S = R^m$ where $m\geq 2$ and R is also a cyclically reduced word in the free product then a Freiheitssatz exists. The proof of this Freiheitssatz depends on the existence of essential representations of such groups in $PSL_2(C)$. In the present paper we discuss the Freiheitssatz and then survey a series of results which depend on both the Freiheitssatz and the existence of complex representations. In particular we examine three areas: the Tits alternative, the property of being virtually torsion-free and amalgam decompositions.

Table Of Contents

1. Introduction. A one-relator product of a family of groups $\{A_i\}$, $i\in I$ is a group $G=(*A_i)/N(S)$ where S is a cyclically reduced word in the free product $*A_i$ and N(S) is its normal closure. S is called the relator. A one-relator group is just a one-relator product of free groups. In general one-relator products share many properties with one-relator groups [13]. A good general discussion of one-relator products is in the Groups St. Andrews survey by J.Howie. [13].

In this paper we concentrate on the case where each A_i is cyclic and the relator is a proper power - that is $S = R^m$ where R is cyclically reduced in $(*A_i)$ and $m \geq 2$. Thus the groups we

1980 Mathematics Subject Classification: Primary 20F05,20E06, Secondary 20E07

consider have presentations of the form

$$(1)\quad G = \langle a_1, a_2, \ldots, a_n;\ a_i^{e_i} = 1,\ i = 1, \ldots, n,\ R^m(a_1, \ldots a_n) = 1 \rangle$$

where $e_i=0$ or $e_i \geq 2$ for each i and $R(a_1,\ldots,a_n)$ is cyclically reduced in the free product on $a_1,\ldots,a_n$.

Of independent interest are the 2-generator case of the above. These groups then have the form

$$(2)\quad G = \langle a,b;\ a^p = b^t = R^m(a,b) = 1 \rangle$$

and are called the generalized triangle groups. Notice that if $R(a,b) = ab$ then we have an ordinary triangle group.

In [6] Fine, Howie and Rosenberger proved a Freiheitssatz for groups of form (1). The proof of this Freiheitssatz involved a refinement of a technique in Baumslag. Morgan and Shalen [3] used to show that the generalized triangle groups (2) are always non-trivial. The technique involved showing that groups (1) have non-trivial representations in $PSL_2(C)$ which are "almost" faithful in the sense that the image of each generator has exactly the order given in the presentation and the image of R has order m.

In this paper we survey and announce a series of results which are consequences of both the Freiheitssatz and its proof. These results are geometric in that they depend not only on the existence of representations in $PSL_2(C)$ but also on the geometric nature of the image groups - in particular their fixed point structure. The general sense of these results is that one-relator products of cyclics (1) share many properties with linear groups - although they may or may not be linear.

In section 2 we discuss the Freiheitssatz and an extension of it to more general one-relator products. In section 3 we examine the Tits Alternative for groups of form (1) - that is when do they have either free subgroups of rank 2 or solvable subgroups of finite index. In section 4 we show that many one-relator products of cyclics possess torsion-free normal subgroups of finite index a property held by many linear groups in particular finitely generated Fuchsian Groups. In section 5 we use the Helling and Culler-Shalen theory of the variety of group representations to show that if groups of form (1) have 3 or greater generators they must decompose as non-trivial free products with amalgamation. Finally in section 6 we consider the case where the relator is not a proper power.

2. The Freiheitssatz. The cornerstone of the theory of one-relator groups is the Freiheitssatz, originally proved by Magnus [16]. In [6], a direct analog of the classical Freiheitssatz was given by Fine, Howie and Rosenberger for one-relator products of cyclics with proper power relators. This and its proof is the starting off point for all further results.

Theorem 1. [6] Suppose the group G is given by

$$G = < a_1,a_2,...,a_n;\ a_1^{e_1} = a_2^{e_2} =....= a_n^{e_n} = R^m(a_1,a_2,...a_n) = 1 >$$

with $n \geq 2$, $m \geq 2$, $e_i = 0$ or $e_i \geq 2$ for $i = 1,...,n$ and $R(a_1,...,a_n)$ a cyclically reduced word in the free product on $a_1,a_2,...,a_n$ which involves all $a_1,...,a_n$. Then

$$<a_1,a_2,...,a_{n-1}> \ = \ <a_1;a_1^{e_1}> * <a_2;a_2^{e_2}>*.......*<a_{n-1};a_{n-1}^{e_{n-1}}>$$

i.e. $<a_1,...,a_{n-1}>$ is the free product of cylics of the obvious orders.

The proof of this is a refinement of a technique in Baumslag, Morgan and Shalen [3] and we briefly outline it. The key idea is in showing that G has a non-trivial representation $\rho:G\to PSL_2(C)$ such that ρ maps the subgroup generated by $a_1,...,a_{n-1}$ epimorphically on the free product of cyclics $<A_1,...,A_{n-1};A_1^{e_1} =....=A_{n-1}^{e_{n-1}}=1>$.

It is well-known that projective matrices $A_1,...,A_{n-1}$ in $PSL_2(C)$ can be chosen so that;

$$<A_1,A_2,...,A_{n-1}> \ = \ <A_1;A_1^{e_1}>*<A_2;A_2^{e_2}>*.......*<A_{n-1};A_{n-1}^{e_{n-1}}>$$

Consider $A_n = \pm\begin{pmatrix} w & wt-w^2-1 \\ 1 & t-w \end{pmatrix}$ where $t = 2\cos(\pi/e_n)$ and w is a variable to be determined.

Since $tr(A_n) = 2\cos(\pi/e_n)$ we have $A_n^{e_n} = 1$.

Now consider $R(A_1,...,A_n)$ formed by substituting $A_1,...,A_n$ into the expression for R. Then

$$R(A_1,....A_n) = \pm \begin{pmatrix} f_1 & f_2 \\ f_3 & f_4 \end{pmatrix}$$

where the f_i are polynomials in the coefficients of $A_1,...,A_n$ and thus polynomials in w. If

$$f(w) = tr\, R(A_1,...A_n) = f_1(w) + f_4(w)$$

is non-constant in w then by the fundamental theorem of algebra there exists a solution w_o to

$$f(w) = 2\cos(\pi/m).$$

For this w_o, $tr\, R = 2\cos(\pi/m)$ and then $R^m = 1$. Thus $<A_1,...,A_n)$ provides a representation of

G in $PSL_2(C)$ with $<A_1,...,A_{n-1}>$ the appropriate free product of cyclics. Since the image then of $<a_1,...,a_{n-1}>$ is a free product of cyclics (each of maximal possible order) $<a_1,...,a_{n-1}>$ will also be the appropriate free product of cyclics.

The proof is completed by a technical argument [6] showing that there does exist a choice of $A_1,...,A_{n-1}$ such that f(w) is a non-constant polynomial in w.

The proof actually handled a more general Freiheitssatz. Since all that was necessary was that free products of cyclics can be faithfully represented in $PSL_2(C)$ we obtain;

Theorem 2. [6] Let A and B be groups which admit faithful representations in $PSL_2(C)$ and suppose R is a cyclically reduced word in the free product of A and B of length ≥ 2. Suppose $G = (A*B)/N(R^m)$ where $m \geq 2$. Then G admits a representation $\rho: G \to PSL_2(C)$ such that

$B \to G \xrightarrow{\rho} PSL_2(C)$ and $A \to G \xrightarrow{\rho} PSL_2(C)$ are faithful and (R) has order m. In particular the maps A G and B G are injective - that is the Freiheitssatz holds.

An instructive way to motivate the subsequent results is to think of the representation constructed above as being "almost" faithful. This leads to questions related to exactly which properties of linear groups are shared by one-relator products of cyclics.

3. The Tits Alternative. As a straightforward consequence of the Freiheitssatz {Theorem 1} it follows that if $n \geq 4$, one-relator products of cyclics of form (1) possess non-abelian free subgroups. Further Baumslag, Morgan and Shalen [3] using a deficiency argument were able to show that in the generalized triangle groups (2) if $S(G) = (1/m)+(1/t)+(1/p) < 1$ then G contains a non-abelian free subgroup. This is a direct generalization of the same statement for ordinary triangle groups.{For the ordinary triangle groups S(G) represents the Euler characteristic.} These results raised the question of the existence and nature of non-abelian free subgroups in one-relator products of cyclics.

Recall that for a linear group G the Tits Alternative says that G either contains a free subgroup of rank 2 or a solvable subgroup of finite index. Our investigations on the free subgroups of one-relator products led us to ask.

Question. Do one-relator products of cyclics (with proper power relator) satisfy the Tits Alternative?

The answer is yes if $n \geq 3$ and probably yes if $n = 2$. In this latter case we were able to show that a large subclass of the generalized triangle groups satisfy the Tits Alternative. Whether the remaining ones do is still open.

The next two theorems are summaries of our results in this area. These actually represent a collection of results done by Fine, Howie, Levin and Rosenberger. [7],[8],[9],[10].

Theorem 3. Let G be a one-relator product of cyclics

$$G = \langle a_1,\dots,a_n;\ a_i^{e_i} = 1, i = 1,\dots n,\ R^m(a_1,\dots,a_n)=1 \rangle$$

where $e_i = 0$ or $e_i \geq 2$ for $i = 1,\dots n$, $m \geq 2$ and R is cyclically reduced in the free product on $a_1,\dots,a_n$ which involves all $a_1,\dots,a_n$. Then

(1) If $n \geq 4$, G contains a free subgroup of rank 2

(2) If $n = 3$ and $(e_1,e_2,e_3) \neq (2,2,2)$ then G contains a free subgroup of rank 2

(3) If $n = 3$ and $(e_1,e_2,e_3) = (2,2,2)$ then either G contains a free subgroup of rank 2 or a free abelian subgroup of rank 2 and of index 2.

The case $n = 2$, that of the generalized triangle groups is the most complicated. The next result says that a generalized triangle group satisfies the Tits Alternative if either $n \geq 3$ or if one generator has infinite order.

Theorem 4. Let $G = < a,b; a^p = b^t = R^m(a,b) = 1 >$ with $t \leq p, t=0$ or $t \geq 2$ and $R(a,b)$ cyclically reduced in the free product on a and b involving both a and b. If $m \geq 3$ or $t = 0$ and $m \geq 2$ then either G contains a free subgroup of rank 2 or a solvable subgroup of finite index.

As remarked this is a combination of several results [7],[8],[9],[10] in which we actually show more than just the existence of free subgroups. In most cases we prove that G contains a generating pair $\{u,v\}$ and a sufficiently large integer q such that $\{u^q,v^q\}$ is a basis for a free subgroup of rank 2. These results generalize a well-known theorem of Ree and Mendelsohn on one-relator groups with torsion. [18].

If G is a 2-generator group we call a generating pair $\{u,v\}$ a Ree-Mendelsohn pair or RM-pair if there exists an integer q such that $<u^q,v^q>$ is free of rank 2. In particular theorem 4 includes the following results.

(1) If $m \geq 3, t = 0$ and $p \geq 2$ then G contains an RM-pair

(2) If $m = 2, t = 0$ and $p \geq 3$ then G contains an RM-pair

(3) If $m = 2, t = 0$ and $p = 2$ and suppose $R = ab^{n_1}ab^{n_2}....ab^{n_k}$ with $k \geq 1$. Then

(a) If $k \geq 2$ or $|n_i| \geq 3$ for some i then G has a free subgroup of rank 2. Further if k is even or k is odd with $k \geq 3$ and $n_1+...+n_k \neq 0$ then G has an RM-pair.

(b) If $k = 1$ and $|n_1| \geq 3$ then G has an RM-pair

(c) If $k = 1$ and $|n_1| \leq 2$ then G is infinite and solvable.

Notice that $t = 0$ implies that b has infinite order. {we could just as well have $p = 0$.} The most complicated case to handle is when both generators have finite order.

(4) If $p \geq 2, t \geq 2$, and $m \geq 3$ then G satisfies the Tits Alternative.

(5) If $p \geq 2, t \geq 2$, $m=2$ and $R= a^{q_1} b^{h_1}.....a^{q_k}b^{h_k}$ then if $1 \leq k \leq 2$ the Tits Alternative holds.

Thus the only open situation relative to the Tits alternative is when both generators have finite order, R has order two and the syllable length of the relator is greater than 2.

The reason that there are two types of results above - ones that give the existence of RM-pairs and ones that only say G satisfies the Tits Alternative comes from the method of proof. To obtain an RM-pair we represent G in $PSL_2(C)$ by the technique described in part 2. Let G = <A,B> be the image group in $PSL_2(C)$. If we can show that we can find a representation such that <A,B> is non-elementary and non-elliptic then we can apply results of Majeed [18] and Rosenberger [20], to guarantee the existence of RM pairs. If we cannot show that the image group has this property we can still {in the remaining cases listed above } use results of Majeed and Mason [17] or combinatorial arguments to get that G either has a non-abelian free subgroup or a solvable subgroup of finite index.

We outline the proof of theorem 3 to give an idea of the method. With 3 or greater generators it becomes somewhat less technical than the generalized triangle group case.

First of all if $n \geq 4$, then from the Freiheitssatz the group G must contain a non-trivial free product of cyclics of rank 3 and thus a free subgroup of rank 2. Similarly if n = 3 and $(e_1,e_2,e_3) \neq (2,2,2)$ then G contains a free product of cyclics which is not infinite dihedral ($\mathbf{Z}_2 * \mathbf{Z}_2$) and therefore a free subgroup of rank 2.

Finally we suppose that n = 3 and $(e_1,e_2,e_3) = (2,2,2)$. Let $a = a_1a_2$, $b = a_1a_3$ and let H = <a,b>. Then $| G:H | \leq 2$. We show that H contains a free subgroup of rank 2 or H is free abelian of rank 2 and | G:H | = 2.

Let $R(a_1,a_2,a_3) = a_1{}^eS(a,b)$ with e = 0 or e = 1.

<u>Case 1</u>. $m \geq 3$ or m = 2 and e = 0. Then choose

$$A_1 = \pm\begin{pmatrix} 0 & 1 \\ -1 & 0 \end{pmatrix}, \quad A_2 = \pm\begin{pmatrix} -1 & -2 \\ 1 & 1 \end{pmatrix}, \quad A_3 = \pm\begin{pmatrix} w & -w^2-1 \\ 1 & -w \end{pmatrix}$$

and let $\qquad A = A_1A_2, \quad B = A_1A_3$

By a technical argument we show that there exists a choice of w_o for w in A_3 such that $<A_1,A_2,A_3>$ represents G in $PSL_2(C)$ and further the linear transformations A and $B = B(w_o)$ have no common fixed points { as linear fractional transformations} and the group <A,B> is not infinite dihedral.

We now argue on the classification of A and B.

If $\operatorname{tr} B \in \mathbf{R}$ and $\operatorname{tr}(AB) \in \mathbf{R}$ we may assume after a suitable conjugation that $\langle A,B\rangle \subset PSL_2(\mathbf{R})$. From the work of Rosenberger [20] it follows that $\langle A,B\rangle$ contains a generating pair $\{u,v\}$ such that $\langle u^q,v^q\rangle$ is free of rank 2 for a sufficiently large integer q.

If both B and AB are loxodromic $\{ \operatorname{tr}(B) \notin \mathbf{R}$ and $\operatorname{tr}(AB) \notin \mathbf{R}\}$ then there exists a positive integer q such that $|\operatorname{tr}(B^q)| > 2$ and $|\operatorname{tr}(AB)^q| > 2$. From results of Majeed [18] the subgroup $\langle B^q,(AB)^q\rangle$ is free of rank 2 for a sufficiently large integer q. . But $\langle A,B\rangle = \langle B,AB\rangle$ and thus G contains a free subgroup of rank 2.

Finally if $\operatorname{tr}(B) \in \mathbf{R}$ but $\operatorname{tr}(AB) \notin \mathbf{R}$ then $\operatorname{tr}(BAB) \notin \mathbf{R}$ since $\operatorname{tr}(BAB) = \operatorname{tr}(B)\operatorname{tr}(AB) - \operatorname{tr}(A)$. Then again from Majeed $\langle BA,BAB\rangle$contains an RM-pair. An analagous argument works if $\operatorname{tr}(B) \notin \mathbf{R}$ but $\operatorname{tr}(AB) \in \mathbf{R}$. This completes case 1.

Case 2: $m = 2$ and $e = 1$. The relator now has the form $R(a_1,a_2,a_3) = a_1S(a,b)$. The subgroup $H = \langle a,b\rangle$ now has index 2 in G and by applying the Reidemeister-Schreier process we find that H has the presentation

$$H = \langle a,b \,;\; S(a^{-1},b^{-1})S(a,b) = 1\rangle .$$

We can assume without loss of generality that S(a,b) is cyclically reduced in the free product on a and b and that there is nor free cancellation between $S(a^{-1},b^{-1})$ and S(a,b). The exponent sum on b is zero so we can express H as an HNN group with stable letter {b}. For each i $\mathbf{Z}$ let $x_i = b^{-i}ab^i$. Then $S(a^{-1},b^{-1})S(a,b)$ can be expressed as a freely reduced word T on the $\{x_i\}$.

Let M be the greatest integer such that x_M appears in T and m the least integer such that x_m appears in T. Clearly $|M-m| \geq 1$. H can then be expressed as;

$$H = \langle x_m,\ldots\ldots,x_M,b \,;\; T(x_m,\ldots\ldots,x_M)=1,\ b^{-1}x_ib = x_{i+1},\ i = m,\ldots.M-1\rangle.$$

Now let

$$K = \langle x_m,\ldots\ldots x_M;\ T(x_m,\ldots x_M) = 1\rangle$$

$$K_1 = \langle x_m,\ldots\ldots x_{M-1};\ \rangle$$

$$K_2 = \langle x_{m+1},\ldots\ldots x_M;\rangle$$

Then H is an HNN group with base K, free part generated by {b} and associated subgroups K_1 and K_2. If $|M-m| \geq 2$ then K_1 is free of rank at least 2 and is contained in H so H contains a free subgroup of rank 2. If $|M-m| = 1$ then $M = m+1$. It follows that either $x_m = x_{m+1}$ or

$x_m \neq x_{m+1}$ in H. If $x_m = x_{m+1}$ then $b^{-1}ab = a$ or $ab = ba$. Then $H = <a,b>$ is free abelian of rank 2 and is of index 2 in G. Finally if $x_m \neq x_{m+1}$ then there exists no p,q $\mathbf{Z}/\{0\}$ such that

$(x_m x_{m+1}^{-1})^p = x_m^q$ or $(x_m x_{m+1}^{-1})^p = x_{m+1}^q$ since the abelianization of H is free abelian of rank 2. Then from Britton's lemma on HNN groups $<x_m, x_{m+1}^{-1}, b>$ is a free subgroup of rank 2.

This completes case 2 and the outline of the proof.

We close this section with two additional results prompted by the Baumslag, Morgan and Shalen theorem on free subgroups of generalized trianglke groups. We note that in their proof they actually show more than just the existence of a free subgroup but rather the existence of a subgroup of finite index which maps epimorphically onto a free group of rank 2. We give this below.

Theorem B. (Baumslag,Morgan,Shalen) Let G be a generalized triangle group of form (2). If $S(G) = (1/m)+(1/t)+(1/p) < 1$ then G contains a subgroup of finite index mapping onto a free group of rank 2.

Recall that a group G is SQ-universal if every countable group can be embedded isomorphically as a subgroup of a quotient of G. Free groups of rank 2 and groups which map epimorphically onto such free goups are SQ-universal. Thus as an immediate consequence of theorem B we get.

Corollary. Let G be as in theorem B above. Then G is SQ-universal.

Using the general Freiheitssatz (theorem 1) for one-relator products of cyclics we can generalize theorem B and its corollary.

Theorem 5. Let G be a one-relator product of cyclics as in theorem 3. If $n \geq 4$ or $n = 3$, $(e_1,e_2,e_3) \neq (2,2,2)$ or $n = 3, (e_1,e_2,e_3)=(2,2,2)$ and $m \geq 3$, then G contains a subgroup of finite index mapping epimorphically onto a free group of rank 2. In particular the group G is SQ-universal.

4. Torsion-Free Normal Subgroups. The well-known Fenchel-Fox Theorem [16] asserts that a finitely generated Fuchsian group must contain a normal torsion-free subgroup of finite index. This

has been extended by Selberg to much more general finitely generated linear groups [23]. A similar result for general one-relator groups with torsion was given by Fischer, Karrass and Solitar [11].

Using the representation of theorem 1 we were able to give a relatively simple proof of the Fischer-Karrass-Solitar theorem. This raised the question of when do general one-relator products of cyclics with proper power relators possess normal torsion-free subgroups of finite index.

Our main result along these lines is that if the relator word $R(a_1,...,a_n)$ can be written in the form $R=UV$ with $U=U(a_1,...,a_p)$, $V=V(a_{p+1},...,a_n)$ and U,V cyclically reduced in the free products of the respective generators which they involve then G contains a normal torsion-free subgroup of finite index. In proving this we actually can handle a more general result concerning one-relator products of groups which admit faithful representations in $PSL_2(C)$. We first present the Fischer-Karrass-Solitar result.

Theorem 5. [11]{Fischer-Karrass-Solitar} Let G be a 1-relator group with torsion. Then G contains a torsion-free normal subgroup of finite index.

Proof. [6]. Suppose $G = < a_1,...a_n; R^m(a_1,...a_n) >$ with $m \geq 2$, R cyclically reduced involving all a_i and R is not a proper power. Let $\rho:G \to PSL_2(C)$ be the representation constructed as in theorem 1 so that $\rho \mid_{<a_1,...,a_{n-1}>}$ and $\rho\mid_{<a_n>}$ are faithful and $\rho(R(a_1,...a_n))$ has order m.

Since G is finitely generated, $\rho(G)$ is a finitely generated subgroup of $PSL_2(C)$. From a theorem of Selberg [23] $\rho(G)$ contains a normal torsion-free subgroup of finite index. This subgroup $\overline{H}$ lifts to a normal subgroup H of finite index in G. We claim H is torsion-free. If $g \in H$ and g has finite order then $g \in Ker\rho$ since $\overline{H}$ is torsion-free. But the elements of G of finite order are all conjugates of powers of the relator R [16]. Since $\rho(R)$ has order m it follows that $Ker \rho$ is torsion-free. Therefore $g = 1$ and H is torsion-free.

Results of J. Howie allow us to extend this result directly to one-relator products of locally indicable groups which admit faithful representations in $PSL_2(C)$. Recall that a locally indicable group is a group where every finitely generated non-trivial subgroup has an infinite cyclic homomorphic image. Specifically we have.

Theorem 6. Let $G = (A*B)/N(R^m)$ where A and B are finitely generated locally indicable groups which admit faithful representations in $PSL_2(C)$. Suppose R is cyclically reduced in the free product A*B of length ≥ 2 and R is not a proper power. Then if $m \geq 2$, G contains a normal torsion-free subgroup of finite index.

The proof depends as in the proof of the Fischer-Karrass-Solitar result on the fact that for groups as in theorem 6 the elements of finite order are conjugates to powers of R. [14].

Now for general one-relator products of cyclics we have.

Theorem 7.[10] Let $G = < a_1,...,a_n;\ a_1^{e_1}=...=a_n^{e_n} = (UV)^m = 1 >$ where $n \geq 2, m \geq 2$ and m maximal, $e_i = 0$ or $e_i \geq 2$ for $i = 1,...n$, $U = U(a_1,...,a_p)$, $V = V(a_{p+1},...,a_n)$, $1 \leq p < n$ and U,V are non-trivial words in the free product on $a_1,...,a_p$ and $a_{p+1},...,a_n$ respectively. Then G has a normal torsion-free subgroup of finite index.

The proof of this depends on being able to handle conjugacy classes of elements of finite order. This is handled by the following lemma which is of independent interest.

Lemma.[10] Let G be as in theorem 6. Then any element of finite order is conjugate to a power of UV or a power of some a_i.

The proof [10] is combinatorial and depends on the fact that since R = UV the group G can be decomposed as a free product with amalgamation if U and V are cyclically reduced, UV involves all generators and $n \geq 3$..

Once this lemma is established the proof of theorem 7 follows exactly as in the proof of the Fischer-Karrass -Solitar theorem. Since this depends on being able to classify elements of finite order as well as the linear representation results of Collins and Peraud [4] and J. Howie [14] allow us to extend theorem 6 to further one-relator products of cyclics.

Theorem 8. [10] Let $G = < a_1,..a_n;\ a_i^{e_i}=1, i=1,...n,\ R^m(a_1,..a_n) = 1 >$ with $n \geq 2, e_i=0$ or $e_i \geq 2$

for i=1,..n,and R cyclically reduced in the free product on $a_1,...a_n$ involving all a_i. Suppose m $\geq$4 and R is not conjugate in the free product on $\{a_1,...a_n\}$ to a word of the form $XUYU^{-1}$ for some word U and 2 elements X and Y of order p $\geq$ 2, q $\geq$2 respectively where (1/m) +(1/p) +(1/q) > 1. Then G has a torsion-free normal subgroup of finite index.

The proof of the above depends on results in [4],[14] which allow for the classification of elements of finite order - if R is not a proper power then any such element must be conjugate to either a power of R or a power of some a_i.

This last statement cannot be reduced to $2 \leq m \leq 3$. Suppose $G = \langle a,b;a^2 = b^3 = R^3=1 \rangle$ with R the word $R = abababab^2$. R is not conjugate in the free product on a and b to a word of the form $XUYU^{-1}$ for some word U and two elements X and Y or order p$\geq$2,q$\geq$2 respectively with (1/3)+(1/p)+(1/q) > 1 and further G is infinite solvable. This was proven independently by Marston Condor (private communication) From his proof the following can be deduced.

Let $x = ababab^2ab^2$ and $y = ab^2ab^2abab$. Then the commuator [x,y] has order 2 but is not conjugate to a power of a, thus giving an example of an element not conjugate to either a power of R or a power of a generator.

Using analagous proofs, theorems 7 and 8 can be generalized to one-relator products of groups which admit faithful representations in $PSL_2(C)$. In particular.

Theorem 9. [10] Let $G = (A*B)/N(R^m)$ where A and B are finitely generated groups which admit faithful representations in $PSL_2(C)$ and R is a cyclically reduced word in the free product A*B of length $\geq$ 2. Then:

(1) If m$\geq$ 2 and R = UV where U is a non-trivial element of A and V is a non-trivial element of B then G contains a torsion-free normal subgroup of finite index.

(2) If m $\geq$ 4 and R is not conjugate in A*B to a word of the from $XUYU^{-1}$ for some word U and elements X,Y with respective orders p $\geq$ 2, q$\geq$ 2 with (1/m) +(1/p) +(1/q) > 1 then G contains a normal torsion-free subgroup of finite index.

5. Amalgam Decompositions. The powerful variety of representations techniques of Culler and Shalen [5] and Helling [12] allow us to determine if a group decomposes as a non-trivial amalgam. Their results are based on the dimension of the space of characters of representations of a group G in $SL_2(C)$ considered as an affine variety. Combining this with the existence of the

representation in part 2 we obtain that if the relator is a proper power and there are at least 3 generators then a one-relator product of cyclics decomposes as a non-trivial free product with amalgamation. Similar work was done by Baumslag and Shalen on one-relator groups with torsion. Specifically.

Theorem 10.[10] Suppose $G = \langle a_1,...,a_n;\ a_i^{e_i} = 1, i = 1,...,n, R^m(a_1,...,a_n) = 1 \rangle$. If $n \geq 3$, $m \geq 2$ and $e_i = 0$ or $e_i \geq 2$ for $i = 1,...,n$ then G is a non-trivial free product with amalgamation.

Proof. We may assume that each $e_i \geq 2$. If not we introduce the relations $a_i^{f_i} = 1$ with $f_i \geq 2$ for each a_i with $e_i = 0$ and write $f_i = e_i$ if $e_i \geq 2$. The group $G^* = \langle a_1,,,a_n;\ a_i^{f_i} = 1, R^m = 1\rangle$ is then an epimorphic image of G and it is known that if $\omega: G_1 \to G_2$ is an epimorhism and G_2 is a non-trivial free product with amalgamation then G_1 is also via ω.

So let $e_i \geq 2$ for $i = 1,...n$. This implies that G is generated elements of finite order. We may assume that R is cyclically reduced in the free product on $a_1,...,a_n$ and involves all the generators. From theorem 1 we know that there exists an irreducible representation of G in $PSL_2(C)$ and therefore the space of representations of G in $(PSL_2(C))^n$ is non-empty. From a result of Culler and Shalen [5] if the dimension of the space of characters of representations of a group in $PSL_2(C)$ is positive {as an affine variety} then the group admits a splitting as a non-trivial free product with amalgamation or an HNN group. From the method of proof in theorem 1 each of (n-1) matrices have two degrees of freedom with the trace and determinant being specified while the final matrix has one degree of freedom - the trace,determinant and relator condition specified. Therefore from Culler and Shalen {or Helling [12] and Rosenberger [21] in a different setting } the dimension of the character space is $2(n-1) + 1 - 3 = 2n-4$. This is positive if $n \geq 3$. Thus if $n \geq 3$ G splits as a non-trivial free product with amalgamation or is an HNN group. However G is generated by elements of finite order so its abelianization G^{ab} is finite and therefore G cannot be an HNN group.

The case of 2 generators - the generalized triangle groups is quite different. These groups sometimes admit splittings as non-trivial free products with amalgamation and sometimes don't. The ordinary triangle groups $T(m,n,p) = \langle a, b;\ a^m = b^n = (ab)^p = 1 \rangle$ with $m,n,p \geq 2$ are never non-trivail free products with amalgamation. However the Fuchsian group of the form $\langle a,b;\ [a,b]^n = 1\rangle$ are free products with amalgamation if $n \geq 3$ and n is not a power of 2 [22].

Using a theorem of Bass [1] we can give a decomposition result for 2 generator one-relator groups with torsion - if the relator is not in the commutator group. In [1] Bass proved that if G is a finitely generated subgroup of $GL_2(C)$ then one of the following cases must occur:

(1) There is an epimorphism $f:G \to \mathbf{Z}$ such that $f(U)) = 0$ for all unipotent elements U in G

(2) G is an amalgamated free product $G = G_o *_H G_1$ with $G_o \neq H \neq G_1$, such that every finitely generated unipotent subgroup of G is contained in a conjugate of G_o or G_1.

(3) G is conjugate to a group of triangular matrices $\begin{pmatrix} a & b \\ 0 & d \end{pmatrix}$ with a and d roots of unity

(4) G is conjugate to a subgroup of $GL_2(A)$ where A is a ring of algebraic integers

Using Bass' result we obtain.

Theorem 11. [10] Let $G = < a,b; R^m(a,b) = 1 >$ with $m \geq 2$ and R(a,b) cyclically reduced involving both a and b and not contained in the commutator subgroup of the free group on a and b. Then G is decomposable as a non-trivial free product with amalgamation.

The proof consists of showing that the group G of theorem 10 has an epimorphic image in $PSL_2(C)$ which cannot satisfy the correponding conditions (1),(3) or (4) of Bass' theorem. A slight extension of this argument yields.

Theorem 12.[10] Let $G = <a,b; a^n = R^m(a,b)$ with $m,n \geq 2$ and R cyclically reduced in the free product on a and b involving both a and b. If $R(a,b) = a^p b^q c$ with $0 \leq p < n$ and $q = 0$ and c in the commutator subgroup in the free product on a and b then G admits a splitting as a non-trivial free product with amalgamation.

We conjecture that any two-generator one-relator group with torsion decomposes as a non-trivial free product with amalgamation.

Although our results mentioned in the previous sections indicate that one-relator product products of cyclics share many properties with linear groups there are situations which are emphatically non-linear. Consider the group $G = <a,b;a^3 = b^5 = (aba^{-1}b^2)^2 = 1>$. Then for any representation of this group in $PSL_2(C)$ the resulting image group is either trivial, cyclic or isomorphic to the alternating group A_5. However G does contain a free subgroup of rank 2. Specifically let N be the normal closure in G of the generator b. Then N has index 3 and has a presentation

$$N = <x,y,z;\ x^5 = y^5 = z^5 = (xy^2)^2 = (yz^2)^2 = (zx^2)^2 = 1>.$$

For N there exists a representation $\rho : N \to PSL_2(C)$ such that

$$<x,y;x^5=y^5=(xy^2)^2=1> \to N \overset{\rho}{\to} PSL_2(C) \qquad \text{is faithful.}$$

The group $(x,y;x^5 = y^5 = (xy^2)^2=1>$ has a rank 2 free subgroup from our previous results.

6. Non-Proper Power Relators. Our major technique for handling one-relator products of cyclics has been the linear representation developed in the proof of theorem 1. This in turn depended on the relator $S = R^m$ being a proper power. The situation is more complicated if S is not a proper power. In this case it is not clear even if there is an appropriate Freiheitssatz.

Straightforward examples show that if R is not a proper power then even (n-2) or less of the generators need not generate a free product of cyclics. For example consider the group $G = <a_1,a_2,a_3,a_4;\ a_1{}^2=a_2{}^2=a_3{}^2=a_4{}^2 = a_3a_1a_2a_3a_4=1>$ Then in this group a_1a_2 is conjugate to a_4 so a_1a_2 has order 2. Therefore $<a_1,a_2> = \mathbf{Z}_2 \times \mathbf{Z}_2$.

However for certain special forms of the relator - in particular when G is a finitely generated Fuchsian group - we do obtain as an application of thoerem 1 a straightforward analog of the Freiheitssatz. Specifically.

Theorem 13. [10] Let F be a finitely generated Fuchsian group with standard presentation

$$F = < a_1,\ldots,a_p,b_1,\ldots,b_t,c_1,d_1,\ldots c_g,d_g;\ a_i{}^{e_i} = 1, i=1,\ldots p,\ a_1\ldots a_pb_1\ldots b_t[c_1,d_1]..[c_g,d_g]=1>$$

If $n = p+t+2g$ and $n \geq 3$ then any subset of (n-2) of the given generators generates a free product of cyclics.

The proof involves an argument on the signature of F and uses a result of Hoare,Karrass and Solitar [15]. The same arguments also apply in the non-orientable case. That is.

Theorem 14. Let the group F have the presentation

$$F = <a_1,\ldots a_p,b_1,\ldots b_t,c_1,\ldots c_g;\ a_i{}^{e_i} = 1,\ i = 1,\ldots,p\ ,\ a_1\ldots a_pb_1\ldots b_tc_1{}^2\ldots c_g{}^2 = 1 >$$

with $e_i \geq 2$ for i = 1,..p. Then if n = p + t + g and $n \geq 3$ any subset of (n-2) of the given generators generates a free product of cyclics.

Finally cancellation arguments in free products with amalgamation allow us to give the following additional results.

Theorem 15. Let $G = \langle a_1,...,a_n;\ a_i^{e_i} = 1,\ U(a_1,..,a_p)=V(a_{p+1},...,a_n)\rangle$ with $n \geq 4$, $1 \leq p \leq n-2$, $e_i=0$ or $e_i \geq 2$ for $i=1,...,n$, $U=U(a_1,...,a_p)$ and $V=V(a_{p+1},...,a_n)$ cyclically reduced words in the free product of the respective generators they involve. Then any subset of $(n-2)$ of the given generators generates a free product of cyclics of the obvious orders.

By similar arguments we obtain a two-generator subgroup theorem.

Theorem 16. Let G be as in theorem 15. Suppose further that in the free product on $a_1,...,a_p$ and $a_{p+1},...,a_n$ respectively neither U nor V is a proper power or is conjugate to a word of the from $XWYW^{-1}$ for some word W and elements X,Y of order 2. Then any two generator subgroup of G is a free product of cyclics.

References

[1] H. Bass "Finitely generated subgroups of $GL_2(C)$" in The Smith Conjecture edited by John Morgan, Wiley, New York (1984)

[2] G. Baumslag and P. Shalen - preprint

[3] G. Baumlsag, J. Morgan and P.Shalen - "On the generalized triangle groups " - to appear. Presented at Groups St. Andrews 1985.

[4] D.Collins and H. Perraud "Cohomology and finite subgroups of small cancellation quotients of free products " Math. Proc. Camb. Phil. Soc. 37 (1985) 243-259

[5] M. Culler and P.Shalen "Varieties of group representations and splittings of 3 manifolds" Ann. of Math. Vol. 117, No. 1 (1983) 109-147

[6] B. Fine, J. Howie and G. Rosenberger "One-relator quotients and free products of cyclics" - to appear Pro. of the Amer. Math. Soc.

[7] B. Fine, J. Howie and G. Rosenberger "Ree-Mendelsohn Pairs in generalized triangle groups" - to appear

[8] B. Fine and G.Rosenberger "A note on generalized triangle groups " - Abh. Math. Sem. Univ. Hamburg 56 (1986) 233-244

[9] B. Fine,F. Levin and G. Rosenberger "Free subgroups and decompositions of one-relator products of cyclics; Part 1: The Tits Alternative" - to appear Arch. der Math

[10] B.Fine,F. Levin and G. Rosenberger "Free subgroups and decompositions of one-relator products of cyclics; Part 2: Normal Torsion-Free Subgroups and FPA Decompositions" to appear J. of the Indian Math. Soc.

[11] J. Fischer,A.Karrass and D.Solitar "On one-relator groups having elements of finite order "- Proc. Amer. Math Soc. 33 (1972) 297-301

[12] H. Helling "Diskrete Untergruppen von $SL_2(\mathbf{R})$" - Inventiones Math. 17 (1972) 217-229

[13] J. Howie "One-relator products of groups" - Proceedings of Groups St. Andrews 1985, Cambridge University Press, 216-220

[14] J. Howie "Cohomology of one-relator products of locally indicable groups"- J. London Math. Soc. (2) 30 (1984) 419-430

[15] A.H.M. Hoare,A.Karrass and D.Solitar " Subgroups of infinite index in Fuchsian groups" - Math. Z. 125,(1972) 56-7

[16] R.Lyndon and P. Schupp Combinatorial Group Theory - Springer-Verlag 1977

[17] A. Majeed and A.W. Mason "Solvable-by-finite subgroups of GL(2,F)" - Glasgow Math J. 19 (1978) 45-48

[18] A. Majeed " Freeness of the group $<a^n,b^n>$ for some integer n; a,b in $SL_2(\mathbf{C})$" - J. Math Soc. Japan 29 (1977) 30-33

[19] R. Ree and N.S. Mendelsohn "Free subgroups of groups with a single defining relation" - Arch. Math. 19 (1968) 577-580

[20] G. Rosenberger " On discrete free subgroups of linear groups " - J. London Math. Soc.(2) 17 (1978) 79-85

[21] G. Rosenberger "Some remarks on a paper of A.Beardon and P.L. Waterman on strongly discrete subgroups of $SL_2(\mathbf{C})$" J. London Math Soc. (2) 27 (1983) 39-42

[22] G. Rosenberger "Bemerkungen zu einer Arbeit von H. Zieschang " Arch. der Math. 29 (1977) 623-627

[23] A. Selberg " On discontinuous groups in higher dimensional symmetric spaces" Intern. Colloq. Function Theory, Tata Institute Bombay 1960 , 147-164

[24] H. Zieschang " On decompositons of discontinuous groups of the plane " Math Z. 151 (1976) 165-168

[25] H. Zieschang, E. Vogt and H.D. Coldeway Surfaces and Planar Discontinuous Groups Lecture Notes in Math 835 - Springer-Verlag 1980

Benjamin Fine
Department of Mathematics
Fairifeld University
Fairfield, Connecticut 06430
United States

Gerhard Rosenberger
Abteilung Mathematik
Universitat Dortmund
Postfach 500
D-4600 Dortmund 50
Federal Republic of Germany

Contemporary Mathematics
Volume **74**, 1988

Sometimes H^1 is H^2 and Discrete Groups Deform

Murray Gerstenhaber and Samuel D. Schack

In studying deformations of algebraic objects, what is sometimes called a first cohomology group might better be called a second—not because numerology matters, but for clarity. The prime examples, which involve deformations of an algebra morphism $\varphi: B \longrightarrow A$, will follow a brief recapitulation of deformation theory for a single algebra since we shall need the formulas.

For simplicity we will assume that all our algebras are unital and over a field K, to start. A **deformation** A_t of an algebra A is an algebra over the power series ring $K[\![t]\!]$ given by a $K[\![t]\!]$-bilinear multiplication α_t on the $K[\![t]\!]$-module $A[\![t]\!]$ of the form

$$\alpha_t = \alpha + t\alpha_1 + t^2\alpha_2 + \dots,$$

where

i) α is the original multiplication in A (extended to be $K[\![t]\!]$-bilinear—something which, in what follows, will always be tacitly understood, when necessary),

ii) the α_i are K-bilinear maps $A \times A \longrightarrow A$, and

iii) α_t is associative.

It is an easy exercise to show that A_t is again unital. A second deformation A'_t with multiplication α'_t is **equivalent** to A_t if

---

Both authors gratefully acknowledge partial support from the National Science Foundation.

there is an isomorphism $F_t: A_t \longrightarrow A'_t$ of the form

$$F_t = \mathrm{Id}_A + tF_1 + t^2F_2 + \cdots,$$

where the F_i are linear maps $A \longrightarrow A$. Equivalently,

$$\alpha'_t(a,b) = F_t\alpha_t(F_t^{-1}a, F_t^{-1}b),$$

so if $F_t: A[\![t]\!] \longrightarrow A[\![t]\!]$ is given then this formula *defines* an α'_t equivalent to α_t. In this circumstance it is easy to check that α_1 is an element of the group $Z^2(A,A)$ of Hochschild 2-cocycles of A with coefficients in itself (considered as an A-bimodule), and that $\alpha'_1 = \alpha_1 - \delta F_1$. Therefore we may view the class $[\alpha_1] \in H^2(A,A)$ as the infinitesimal of (the equivalence class of) the deformation α_t. (Actually, one may have $\alpha_1 = \ldots = \alpha_{r-1} = 0$ and $\alpha_r \neq 0$, in which case $\alpha_r \in Z^2$ and $[\alpha_r]$ is the infinitesimal—at least in characteristic 0—but that is a more delicate matter which we can here safely overlook.) A **trivial deformation** is one equivalent to the original multiplication, *i.e.* the usual power series *algebra* $A[\![t]\!]$. For an algebra A to have only trivial deformations it suffices, but is not necessary, that $H^2(A,A) = 0$.

Now if A and $\alpha_1 \in Z^2(A,A)$ are given then we can define a new multiplication $\alpha + t\alpha_1$ on $A[\![t]\!]/t^2$ and ask if there is an $\alpha_2 \in C^2(A,A)$—the group of Hochschild 2-cochains (K-bilinear maps $A \times A \longrightarrow A$)—such that $\alpha + t\alpha_1 + t^2\alpha_2$ defines an associative multiplication on $A[\![t]\!]/t^3$. This is possible if and only if the trilinear map $\alpha_1 \bar{\circ} \alpha_1$ given by

$$\alpha_1 \bar{\circ} \alpha_1(a,a',a'') = \alpha_1(\alpha_1(a,a'),a'') - \alpha_1(a,\alpha_1(a',a'')),$$

which is always guaranteed to lie in $Z^3(A,A)$, is a coboundary, *i.e.* is $\delta\alpha_2$ for some $\alpha_2 \in C^2(A,A)$ in which case this α_2 will serve. Since $[\alpha_1 \bar{\circ} \alpha_1]$ in fact depends only on $[\alpha_1]$, we have a quadratic "obstruction map" $Sq\colon H^2(A,A) \longrightarrow H^3(A,A)$ such that α_2 exists if and only if $Sq[\alpha_1] = 0$. The **primary obstruction** to "integrating" α_1 to a deformation is $Sq[\alpha_1]$.

Hochschild observed that $H^\bullet(A,A)$, the direct sum of all $H^n(A,A)$, carries an associative **cup product**, where if $\eta^r \in H^r$ and $\zeta^s \in H^s$ then $\eta^r \smile \zeta^s \in H^{r+s}$; this is induced simply by multiplication of the values of cocycles representing η^r and ζ^s. However, the preceding simple observation led to the discovery that $H^\bullet(A,A)$ actually has a much richer structure. First, the cup product is what is now called **supercommutative**, *i.e.* it respects the gradation $(H^r \smile H^s \subseteq H^{r+s})$ and $\eta^r \smile \zeta^s = (-1)^{rs}\zeta^s \smile \eta^r$. Moreover, there is a **superLie** product with $[\eta^r,\zeta^s] \in H^{r+s-1}$, (so the "reduced degrees" $r-1$ and $s-1$ of η and ζ now add); the defining laws are

$$[\eta^r,\zeta^s] = -(-1)^{(r-1)(s-1)}[\zeta^s,\eta^r], \quad \text{and}$$
$$(-1)^{(q-1)(s-1)}[\omega^q,[\eta^r,\zeta^s]] + (-1)^{(r-1)(q-1)}[\eta^r,[\zeta^s,\omega^q]] +$$
$$(-1)^{(s-1)(r-1)}[\zeta^s,[\omega^q,\eta^r]] = 0.$$

One has

$$[\omega^q, \eta^r \smile \zeta^s] = [\omega^q, \eta^r] + (-1)^{(q-1)r} \eta^r \smile [\omega^q, \zeta^s].$$

Finally, for each m there is a quadratic map

$$\mathrm{Sq}\colon H^{2m}(A,A) \longrightarrow H^{4m-1}(A,A)$$

with $[\eta,\eta] = 2\mathrm{Sq}\eta$; if the characteristic of K is 2 then there is a quadratic map $H^n(A,A) \longrightarrow H^{2n-1}(A,A)$ for all n.

The formulas for the superLie multiplication and Sq are not difficult: If $\alpha^r \in C^r(A,A)$ and $\beta^s \in C^s(A,A)$, define $\alpha^r \mathbin{\bar{\circ}_i} \beta^s \in C^{r+s-1}(A,A)$ for $i = 1,\ldots,r$ by

$$\begin{aligned}\alpha^r \mathbin{\bar{\circ}_i} \beta^s(a_1,\ldots,a_{i-1},b_1,\ldots,b_s,a_{i+1},\ldots,a_r)\\ = \alpha^r(a_1,\ldots,a_{i-1},\beta^s(b_1,\ldots,b_s),a_{i+1},\ldots,a_r)\end{aligned}$$

and set

$$\alpha^r \mathbin{\bar{\circ}} \beta^s = \Sigma(-1)^{(i-1)(s-1)} \alpha^r \mathbin{\bar{\circ}_i} \beta^s.$$

Then

$$\begin{aligned}[\alpha^r, \beta^s] &= \alpha^r \mathbin{\bar{\circ}} \beta^s - (-1)^{(r-1)(s-1)} \beta^s \mathbin{\bar{\circ}} \alpha^r, \quad \text{and}\\ \mathrm{Sq}\alpha^r &= \alpha^r \mathbin{\bar{\circ}} \alpha^r.\end{aligned}$$

If α and β are cocycles then so is $[\alpha^r,\beta^s]$ and its class depends only on those of α and β, so this multiplication is defined on $H^\bullet(A,A)$; if r is even or the characteristic of K is 2 then the same is true of Sq.

If $\alpha + t\alpha_1 + \dots + t^{r-1}\alpha_{r-1}$ is an associative multiplication on $A[\![t]\!]/t^r$ then $\sum_{i+j=r} \alpha_i \bar{\circ} \alpha_j$ is automatically a 3-cocycle and is a coboundary if and only if the multiplication lifts to one on $A[\![t]\!]/t^{r+1}$; any such lifting has the form $\alpha + t\alpha_1 + \dots + t^{r-1}\alpha_{r-1} + t^r\alpha_r$ where $\delta\alpha_r = \sum \alpha_i \bar{\circ} \alpha_j$. The class of $\sum \alpha_i \bar{\circ} \alpha_j$ in $H^3(A,A)$—which depends on the *actual* cochains $\alpha_2,\dots,\alpha_{r-1}$—is **an rth order obstruction** to integrating α_1 to a deformation.

We turn now to the examples.

Example 1. (Nijenhuis). Suppose that we wish to vary a subalgebra B within a fixed algebra A in the sense that we seek vector subspaces of the same dimension which are "close" to B and which are closed under the multiplication in A. Since these subalgebras need not be isomorphic to B, one poses the following formal problem: Find a deformation B_t of B and a morphism $\varphi_t\colon B_t \longrightarrow A[\![t]\!]$ of the form $\varphi_t = \varphi_0 + t\varphi_1 + \dots$ where φ_0 (restricted to B) is the inclusion of B into A (followed by that of A into $A[\![t]\!]$). The infinitesimals of this deformation problem are identified by Nijenhuis as the elements of $H^1(B,A/B)$ where A/B is the quotient of the B-bimodule A by the sub-bimodule B.

Example 2. Suppose that $\varphi: \Gamma \longrightarrow \mathrm{End}_K V$ is a linear representation of an abstract group Γ as endomorphisms of a vector space V over K. We seek to vary φ. Letting $K\Gamma$ be the group algebra of Γ over K, we pose the formal problem: Construct morphisms $\varphi_t: K\Gamma \longrightarrow (\mathrm{End}_K V)[[t]]$ having the form $\varphi_t = \varphi_0 + t\varphi_1 + \dots$ where the φ_i are linear maps from $K\Gamma$ to $\mathrm{End}_K V$ and $\varphi_0 = \varphi$. Here the infinitesimals are generally described as the elements of $H^1(K\Gamma, \mathrm{End}_K V)$, where $\mathrm{End}_K V$ is viewed as a $K\Gamma$-bimodule by virtue of φ.

The common feature in these examples is that one is deforming a unital morphism of K-algebras $\varphi: B \longrightarrow A$. This is a special case of the theory introduced in our earlier paper "On the Deformation of Algebra Morphisms and Diagrams" (*Trans. Amer. Math. Soc.* **279**), but we recapitulate it here, along with the cohomology theory associated with such a morphism. First, we define a **bimodule** T over φ to be a morphism $T: N \longrightarrow M$ of (additive) abelian groups, where

i) N is a B-bimodule;

ii) M is an A-bimodule which, by virtue of φ, we can also view as a B-bimodule; and

iii) T is a morphism of B-bimodules.

With this, the Hochschild cochain complexes $C^\bullet(B,N)$, $C^\bullet(A,M)$, and $C^\bullet(B,M)$ are well-defined. We can now define a new complex, denoted $C^\bullet(\varphi,T)$, an n-cochain Γ of which is a triple $(\Gamma^B, \Gamma^A; \Gamma^{AB})$ with $\Gamma^B \in C^n(B,N)$, $\Gamma^A \in C^n(A,M)$, and $\Gamma^{AB} \in C^{n-1}(B,M)$. Define $\Gamma^A\varphi$ and $T\Gamma^B$ in $C^n(B,M)$ by

$$(\Gamma^A\varphi)(b_1,\ldots,b_n) = \Gamma^A(\varphi b_1,\ldots,\varphi b_n) \quad \text{and}$$
$$T\Gamma^B(b_1,\ldots,b_n) = T(\Gamma^B(b_1,\ldots,b_n)).$$

Then $\delta\colon C^n(\varphi,T) \longrightarrow C^{n+1}(\varphi,T)$ is given by

$$\delta(\Gamma^B,\ \Gamma^A;\ \Gamma^{AB}) = (\delta\Gamma^B,\ \delta\Gamma^A;\ T\Gamma^B - \Gamma^A\varphi - \delta\Gamma^{AB}).$$

Since $\delta(T\Gamma^B) = T(\delta\Gamma^B)$ and $\delta(\Gamma^A\varphi) = (\delta\Gamma^A)\varphi$, one has indeed $\delta\delta = 0$. We denote cocycles, coboundaries, and cohomology by $Z^n(\varphi,T)$, $B^n(\varphi,T)$, and $H^n(\varphi,T) = Z^n(\varphi,T)/B^n(\varphi,T)$, respectively. We can, of course, in particular take $T = \varphi$.

(It is not necessary for what we do here but worth remarking that above we really have a contravariant functor from the poset $\{0 < 1\}$ to the category of unital K-algebras, *i.e.* a presheaf of algebras, together with the natural concept of a bimodule over this presheaf. This construction can be extended to the case where we begin with an arbitrary small category $\mathcal{C}$, *e.g.* the poset of opens in a topological space.)

By a **deformation** of φ we mean a morphism $\varphi_t\colon B_t \longrightarrow A_t$ in which

i) B_t is a deformation of B;

ii) A_t is a deformation of A; and

iii) φ_t is an algebra morphism of the form

$\varphi_t = \varphi + t\varphi_1 + \ldots$, where each φ_i is a K-linear map from B to A.

Writing the multiplication in A_t as $\alpha_t = \alpha + t\alpha_1 + \dots$ and that in B_t as $\beta_t = \beta + t\beta_1 + \dots$, it is easy to check that $(\alpha_1, \beta_1; \varphi_1) \in Z^2(\varphi,\varphi)$. A second deformation $\varphi'_t: B'_t \longrightarrow A'_t$ is **equivalent** to the first if there exist morphisms $G_t = \mathrm{Id}_B + tG_1 + \dots$ from B_t to B'_t and $F_t = \mathrm{Id}_A + tF_1 + \dots$ from A_t to A'_t such that $\varphi'_t G_t = F_t \varphi_t$, *i.e.* the obvious square commutes. (As before, if only the original deformation and $F_t: A[\![t]\!] \longrightarrow A[\![t]\!]$ and $G_t: B[\![t]\!] \longrightarrow B[\![t]\!]$ are given then one can immediately construct the second deformation to which the pair G_t, F_t provides the equivalence.) An easy calculation shows that we now must have $\beta'_1 = \beta_1 - \delta G_1$, $\alpha'_1 = \alpha_1 - \delta F_1$, and $\varphi'_1 = \varphi_1 - (\varphi G_1 - F_1\varphi)$. Thus, $(\beta'_1, \alpha'_1; \varphi'_1) = (\beta_1, \alpha_1; \varphi_1) - \delta(G_1, F_1; 0)$. So passing to an equivalent deformation, we can replace the 2-cocycle $\Gamma = (\beta_1, \alpha_1; \varphi_1)$ with $\Gamma - \delta(G_1, F_1; 0)$. This really exhausts the class of Γ, since although the general 1-cochain has the form $(G_1, F_1; a)$ with $a \in A = C^0(B,A)$, one has $\delta(G_1, F_1; a) = \delta(G_1, F_1 + \delta a; 0)$. We may therefore properly identify the **infinitesimal deformations** of $\varphi: B \longrightarrow A$ with elements of $H^2(A,A)$. (Be warned, however, that one must modify this somewhat when the nerve of the base category is not homologically trivial.)

Now, when an infinitesimal deformation of an algebra morphism (or, more generally, an infinitesimal deformation of a presheaf of algebras over an arbitrary small category) is given, we should like, as in the case of a single algebra, to know when it can be extended to a deformation modulo t^3. In this we are aided by the following remarkable fact, a consequence of the *Cohomology Comparison*

Theorem (which would be too lengthy to state in full, but holds for presheaves of algebras over arbitrary small categories):

> To the morphism $\varphi\colon B \longrightarrow A$ and φ-bimodule $T\colon N \longrightarrow M$ we can assign in a functorial way a single K-algebra $\varphi!$ and a single $\varphi!$-bimodule $T!$ such that there is a natural isomorphism $H^{\bullet}(\varphi,T) \longrightarrow H^{\bullet}(\varphi!,T!)$.

(In general, the proof relies upon an isomorphism of Yoneda cohomologies.) In fact, for $\varphi!$ we can take the 2×2 matrices of the form $\left(\begin{smallmatrix} b & 0 \\ a & a' \end{smallmatrix}\right)$ with $b \in B$ and $a, a' \in A$. The addition in $\varphi!$ is obvious while the multiplication is given by

$$\begin{bmatrix} b & 0 \\ a_0 & a_1 \end{bmatrix} \cdot \begin{bmatrix} b' & 0 \\ a_0' & a_1' \end{bmatrix} = \begin{bmatrix} bb' & 0 \\ a_0\varphi b' + a_1 a_0' & a_1 a_1' \end{bmatrix}.$$

The bimodule $T!$ consists of matrices $\left(\begin{smallmatrix} n & 0 \\ m & m' \end{smallmatrix}\right)$ with the obvious meanings and operations. Taking $T = \varphi$, we can now transport to $H^{\bullet}(\varphi,\varphi)$ all of the structure of $H^{\bullet}(\varphi!,\varphi!)$. Thus, $H^{\bullet}(\varphi,\varphi)$ has, in particular, a supercommutative cup product, a superLie product, and—most important for the present problem—a quadratic map $Sq\colon H^2(\varphi,\varphi) \longrightarrow H^3(\varphi,\varphi)$ which indeed gives the primary obstruction to an infinitesimal deformation. The formulas, which are not difficult, are given in the aforementioned paper. We reproduce first that for $Sq\colon H^2 \longrightarrow H^3$. This one, of course, can be deduced directly by observing what relations must hold between the first and second order terms of a deformation of φ. Writing the infinitesimal $\Gamma \in Z^2(\varphi,\varphi)$ in the form $(\beta_1, \alpha_1; \varphi_1)$ we have

$$\mathrm{Sq}\Gamma = (\beta_1 \bar{\circ} \beta_1, \alpha_1 \bar{\circ} \alpha_1; \alpha_1 \bar{\circ} \varphi_1 - \varphi_1 \bar{\circ} \beta_1 + \varphi_1 \smile \varphi_1).$$

The third term perhaps requires comment since two algebras are involved, but it is clear that the definition of $\varphi_1 \bar{\circ} \beta_1$ is formally meaningful, while $\varphi_1 \smile \varphi_1$ is defined since the values of φ_1 can be multiplied in A. To interpret $\alpha_1 \bar{\circ} \varphi_1$ extend the definition of $\bar{\circ}$ to $f \in C^m(A,A)$ and $g \in C^n(B,A)$ by first setting

$$\begin{aligned} f \bar{\circ}_i g(b_1,\ldots,b_{m+n-1}) \\ = f(\varphi b_1,\ldots,\varphi b_{i-1}, g(b_i,\ldots,b_{i+n-1}), \varphi b_{i+n},\ldots,\varphi b_{m+n-1}) \end{aligned}$$

and then

$$f \bar{\circ} g = \sum (-1)^{(i-1)(n-1)} f \bar{\circ}_i g,$$

as before. In general, if $\Gamma = (\Gamma^B, \Gamma^A; \Gamma^{AB}) \in C^m(\varphi,\varphi)$ and $\Delta = (\Delta^B, \Delta^A; \Delta^{AB}) \in C^n(\varphi,\varphi)$ we define their cup product by

$$\Gamma \smile \Delta = (\Gamma^B \smile \Delta^B, \Gamma^A \smile \Delta^A; \Gamma^{AB} \smile \varphi\Delta^B + (-1)^m \Gamma^A \varphi \smile \Delta^{AB})$$

and we set

$$\Gamma \bar{\circ} \Delta = (\Gamma^B \bar{\circ} \Delta^B, \Gamma^A \bar{\circ} \Delta^A; \Gamma^A \bar{\circ} \Delta^{AB} + (-1)^{n-1} \Gamma^{AB} \bar{\circ} \Delta^B + \Delta^{AB} \smile \Gamma^{AB}).$$

When Γ and Δ are cocycles, set

$$\mathrm{Sq}\Gamma = \Gamma \bar{\circ} \Gamma$$

and

$$[\Gamma,\Delta] = \Gamma \bar{\circ} \Delta - (-1)^{(m-1)(n-1)} \Delta \bar{\circ} \Gamma.$$

Part of the power of the Cohomology Comparison Theorem is in providing such formulas. A direct proof that $[-,-]$ defines a superLie product in $H^{\bullet}(\varphi,\varphi)$ is at best exceedingly involved. (Try it!)

As in the case of a single algebra, there are higher order obstructions to integrating a 2-cocycle $(\beta_1, \alpha_1; \varphi_1)$ to a deformation of $\varphi: B \longrightarrow A$. Once $(\beta_i, \alpha_i; \varphi_i)$ for $1 < i < r$ have been constructed, $\sum_{i+j=r} (\beta_i, \alpha_i; \varphi_i) \bar{\circ} (\beta_j, \alpha_j; \varphi_j)$ is automatically a 3-cocycle whose cohomology class in $H^3(\varphi,\varphi)$ is an rth order obstruction. If that class vanishes then $(\beta_r, \alpha_r; \varphi_r)$ can be constructed as a cochain cobounding $\sum (\beta_i, \alpha_i; \varphi_i) \bar{\circ} (\beta_j, \alpha_j; \varphi_j)$.

We return now to the examples to see how the "classical" infinitesimals in a certain H^1 can be reconciled with those given by the general theory in H^2.

Example 1. As stated the problem is ambiguous because equivalence of deformations has not been defined! We want $\varphi_t: B_t \longrightarrow A[\![t]\!]$ with the multiplication in A in effect remaining fixed. (Note that φ_t is necessarily a monomorphism.) The general theory would therefore identify this φ_t with $\varphi'_t: B'_t \longrightarrow A'_t$ if there is an automorphism $f_t = f_0 + tf_1 + t^2f_2 + \dots$ of $A[\![t]\!]$ with $f_0 = \mathrm{Id}_A$ such that $\varphi'_t = f_t\varphi_t$. We get some of these f_t by choosing an $a_t = 1 + ta_1 + t^2a_2 + \dots$ in $A[\![t]\!]$ and conjugating by it. This finer equivalence is all that Nijenhuis allows; his theory cannot cope with the first. With the problem so defined, there is a simple argument identifying the infinitesimals with $H^1(B,A/B)$. For $\varphi_t(B[\![t]\!])$ will be closed under multiplication in A_t if and only if for every $b',b'' \in B$ there is a $b_t = b_0 + tb_1 + t^2b_2 + \dots$ in $B[\![t]\!]$ such that $(\varphi_t b')(\varphi_t b'') = \varphi_t b_t$. Comparing the first order terms shows that $b'(\varphi_1 b'') + (\varphi_1 b')b'' = \varphi_1(b_0) + b_1$ while setting $t = 0$ reveals that $b_0 = b'b''$. If we now reduce modulo B then this implies that the induced map $\bar{\varphi}_1: B \longrightarrow A/B$ is a derivation, *i.e.* $\bar{\varphi}_1 \in Z^1(B,A/B)$. Following φ_t with conjugation by a_t replaces φ_1 by $\varphi_1 - \delta a_1$. In Nijenhuis' theory one therefore views $[\bar{\varphi}_1] \in H^1(B,A/B)$ as the infinitesimal of the class of φ_t.

To compare this with our general theory for a deformation $\varphi_t: B_t \longrightarrow A_t$ of a morphism, observe that since A is fixed we have $\alpha_t = \alpha$. The associated $\Gamma \in Z^2(\varphi,\varphi)$ is therefore $(\beta_1, 0; \varphi_1)$ and the cocycle condition becomes $\delta\beta_1 = 0$ and $\varphi\beta_1 = \delta\varphi_1$. Consider now the coefficient sequence (of B-bimodules) $0 \longrightarrow B \longrightarrow A \longrightarrow A/B \longrightarrow 0$.

This induces a long exact sequence, a part of which is

$$\cdots \longrightarrow H^1(B,A) \longrightarrow H^1(B,A/B) \longrightarrow H^2(B,B) \longrightarrow H^2(B,A) \longrightarrow \cdots.$$

The equation $\varphi\beta_1 = \delta\varphi_1$ says both that $[\beta_1] \in H^2(B,B)$ is sent to 0 in $H^2(B,A)$ and, because φ_1 is clearly a lifting of $\bar{\varphi}_1$ to A, that $[\beta_1]$ is the image of $[\bar{\varphi}_1] \in H^1(B,A/B)$. But this does *not* say that $[\bar{\varphi}_1]$ determines the class of $(\beta_1, 0; \varphi_1)$ because of the ambiguity in the meaning of "the class." Nijenhuis has tacitly introduced the subcomplex $C_0^{\bullet}(\varphi,\varphi)$ of $C^{\bullet}(\varphi,\varphi)$ consisting of those $(\Gamma^B, \Gamma^A; \Gamma^{AB})$ with $\Gamma^A = 0$. It is easy to prove that because φ is a monomorphism there is an isomorphism $H_0^n(\varphi,\varphi) \longrightarrow H^{n-1}(B,A/B)$ given on cocycles by $(\Gamma^B, 0; \Gamma^{AB}) \longmapsto \bar{\Gamma}^{AB}$ where $\bar{\Gamma}^{AB}$ is the reduction of Γ^{AB} modulo B and the cohomology class of the cocycle on the left is, of course, computed within $C_0^n(\varphi,\varphi)$. However, the morphism $H_0^{\bullet}(\varphi,\varphi) \longrightarrow H^{\bullet}(\varphi,\varphi)$ induced by the inclusion of the Nijenhuis subcomplex $C_0^{\bullet}(\varphi,\varphi)$ into $C^{\bullet}(\varphi,\varphi)$ is generally not a monomorphism.

Two things are lost by the "efficient" replacement of $H_0^2(\varphi,\varphi)$ by $H^1(B,A/B)$. First, one cannot see the relation between the two problems into which the original ambiguously stated one splits. Second, while Nijenhuis computes the obstruction morphism $Sq\colon H^1(B,A/B) \longrightarrow H^2(B,A/B)$, its derivation is hard to understand and the fact that it works seems, at first, bewildering. Working in the more general context, it is clear that the subcomplex $C_0^{\bullet}(\varphi,\varphi)$ is closed under the operations $\smile$, Sq, and $[-,-]$ previously introduced in all of $C^*(\varphi,\varphi)$, from which one can, in particular, directly obtain Nijenhuis' formulation of the obstruction morphism. For the obstruction to the class of $(\beta_1, 0; \varphi_1)$ is the class of $(\beta_1 \bar{\circ} \beta_1, 0; -\varphi_1 \bar{\circ} \beta_1 + \varphi_1 \smile \varphi_1)$, where $\beta_1 = \delta\varphi_1$. Thus, we see immediately that if $[\bar{\varphi}_1] \in H^1(B,A/B)$ then to get its obstruction in $H^2(B,A/B)$ we must first lift $\bar{\varphi}_1$ to $\varphi_1 \in C^1(B,A)$ and then take the class of the reduction of $-\varphi_1 \bar{\circ} \delta\varphi_1 + \varphi_1 \smile \varphi_1$ modulo B. This is, of course, what Nijenhuis says, but we can understand now why it works. However, in fairness to Nijenhuis, who clearly had a deep insight into the problem, the general formulas for $\smile$, Sq, and $[-,-]$ would be equally mysterious without the Cohomology Comparison Theorem. (As an exercise the reader may compute the cup product in $H^*(B,A/B)$, being wary of the dimensions, which now do not add.)

Example 2. Here we consider a deformation

$$\varphi_t : K\Gamma[\![t]\!] \longrightarrow (\mathrm{End}_K V)[\![t]\!]$$

of a linear representation $\varphi: K\Gamma \longrightarrow \mathrm{End}_K V$ of a group Γ on a (tacitly finite-dimensional) vector space V. We wish to keep the group Γ, and therefore the group algebra $K\Gamma$, fixed. Of course, we also want to keep $\mathrm{End}_K V$ fixed but need not separately assume it since $\mathrm{End}_K V$ is rigid, *i.e.* all deformations are equivalent to the trivial deformation.

Although we use Γ for the group, there will be no conflict in our notation for cocycles if we read $K\Gamma$ as a single symbol. The ambiguity of the previous problem disappears completely because $\mathrm{End}_K V$ has only inner automorphisms and has vanishing cohomology in positive dimensions. Because both $K\Gamma$ and $\mathrm{End}_K V$ are held fixed the element of $Z^2(\varphi,\varphi)$ associated with φ_t now simply has the form $\Gamma = (0, 0; \varphi_1)$ where $\varphi_1 \in Z^1(K\Gamma,\mathrm{End}_K V)$. If φ_t is equivalent to another such deformation φ_t' then $(0, 0; \varphi_1 - \varphi_1') = \delta(\delta\Delta^B, \delta\Delta^A; 0)$ for some $(\delta\Delta^B, \delta\Delta^A; 0) \in C^1(\varphi,\varphi)$. Then $\varphi_1 - \varphi_1' = \delta(\varphi\Delta^B - \Delta^A\varphi)$ and the infinitesimal of the equivalence class of φ_t may therefore be regarded as $[\varphi_1] \in H^1(K\Gamma,\mathrm{End}_K V)$.

The subcomplex $C^\bullet_{00}(\varphi,\varphi)$ of $C^\bullet(\varphi,\varphi)$ which is appropriate to the present problem is that whose cocycles have the form $(0, 0; \Gamma^{AB})$. The inclusion of $C^\bullet_{00}(\varphi,\varphi)$ into $C^\bullet(\varphi,\varphi)$ induces a morphism $H^\bullet_{00}(\varphi,\varphi) \longrightarrow H^\bullet(\varphi,\varphi)$ which—because $\mathrm{End}_K V$ has vanishing cohomology—will be a monomorphism whenever $K\Gamma$ likewise has vanishing cohomolgy, *e.g.* when Γ is finite and K has characteristic zero. Such a situation rarely obtains; however, $H^n_{00}(\varphi,\varphi) \longrightarrow H^n(\varphi,\varphi)$ is a monomorphism for $n \geqslant 2$ if, for example, Γ is free (and K is arbitrary)—as occurs when Γ is the fundamental group of a punctured Riemann surface. Now, clearly, $(0, 0; \Gamma^{AB}) \longmapsto \Gamma^{AB}$ induces an isomorphism $H^n_{00}(\varphi,\varphi) \longrightarrow H^{n-1}(K\Gamma,\mathrm{End}_K V)$ for $n \geqslant 1$. Using this isomorphism certainly seems efficient, since there is no difficulty in transferring the formulas for $\smile$, Sq, and $[-,-]$ to $H^\bullet(K\Gamma,\mathrm{End}_K V)$.

If $f^m \in C^m(K\Gamma, \mathrm{End}_K V)$ and $g^n \in C^n(K\Gamma, \mathrm{End}_K V)$ then we may form $f^m \smile g^n \in C^{m+n}(K\Gamma, \mathrm{End}_K V)$ by multiplying the values of f and g in $\mathrm{End}_K V$, *i.e.* by composing them. However, this cup product of cocycles is generally no longer a cocycle. In fact, since the cup product generally defined on $C^\bullet(\varphi,\varphi)$ vanishes identically on $C^\bullet_{00}(\varphi,\varphi)$, there seems to be no cup product in the usual sense on $H^\bullet(K\Gamma, \mathrm{End}_K V)$. On the other hand, Sq and $[-,-]$ survive. In particular, $\mathrm{Sq}[(0, 0; \varphi_1)] = [(0, 0; \varphi_1 \smile \varphi_1)]$, so the primary obstruction to the infinitesimal deformation $[\varphi_1] \in H^1(K\Gamma, \mathrm{End}_K V)$ is $[\varphi_1 \smile \varphi_1]$. If it vanishes and $\delta\varphi_2 = -\varphi_1 \smile \varphi_1$ then a secondary obstruction is $[\varphi_1 \smile \varphi_2 + \varphi_2 \smile \varphi_1]$, which is the graded (cup-) commutator of φ_1 and φ_2. In general, if $[f] \in H^{2n+1}(K\Gamma, \mathrm{End}_K V)$ then $\mathrm{Sq}[f] = [f \smile f]$ is well-defined in $H^{4n+2}(K\Gamma, \mathrm{End}_K V)$, and we can extend this also to even dimensions if the characteristic of K is 2. Finally, for $(0, 0; f^{m-1}) \in Z^m_{00}(\varphi,\varphi)$ and $(0, 0; g^{n-1}) \in Z^n_{00}(\varphi,\varphi)$ the general formula gives

$$[(0, 0; f^{m-1}),(0, 0; g^{n-1})] = (0, 0; g \smile f - (-1)^{(m-1)(n-1)} f \smile g).$$

Therefore, increasing m and n by 1, if $[f] \in H^m(K\Gamma, \mathrm{End}_K V)$ and $[g] \in H^n(K\Gamma, \mathrm{End}_K V)$ then $[[f],[g]] = [g \smile f - (-1)^{mn} f \smile g]$. One now has a superLie product with the usual grading on $H^\bullet(K\Gamma, \mathrm{End}_K V)$.

Since everything works so well in this example, it is difficult to complain of the use of the isomorphism $H^\bullet_{00}(\varphi,\varphi) \longrightarrow H^{\bullet-1}(K\Gamma, \mathrm{End}_K V)$, which in particular makes an H^2 look like an H^1. On the other hand, if one set out to discover the structure of $H^{\bullet}(K\Gamma, \mathrm{End}_K V)$ directly, without the guidance of the Cohomology Comparison Theorem, the results just presented might seem accidental and mysterious.

What has been lost, aside from clarity, is a whole theory, for with the reduced complex one is not led to ask whether a deformation problem of the form $\varphi_t\colon (K\Gamma)_t \longrightarrow (\mathrm{End}_K V)[[t]]$ is meaningful with $(K\Gamma)_t$ again a group algebra. Indeed, one group algebra can deform to another! For a finite group Γ such a deformation must be a "jump deformation" and can happen at all only if the characteristic of K divides the order of Γ, but for infinite groups it is possible even over $\mathbb{R}$ or $\mathbb{C}$. One is led to the following extraordinary observation: there is a natural partial order on all groups of a fixed finite order; it is defined by setting $\Gamma \leqslant \Gamma'$ if $K\Gamma$ deforms to $K\Gamma'$ for some field K. (One can make the same definition for all groups, but conceivably the ordering may contain loops.)

The theory is easiest to understand for p-groups since there one must fix the characteristic of K to be p (else $K\Gamma$ will have no cohomology in positive dimensions and so will be rigid). To show that deformation is possible, let $\Gamma = \mathbf{Z}/2\mathbf{Z} \times \mathbf{Z}/2\mathbf{Z}$; if the characteristic of K is 2 then $K\Gamma \cong K[x,y]/(x^2,y^2)$. Now consider

$$(K\Gamma)_t \;=\; K[[t]][x,y]/(x^2 + ty,\, y^2) \;\cong\; K[[t]](\mathbf{Z}/4\mathbf{Z}).$$

We have deformed $\mathbf{Z}/2\mathbf{Z} \times \mathbf{Z}/2\mathbf{Z}$ to $\mathbf{Z}/4\mathbf{Z}$.

For abelian p-groups Γ, the partial order is easy to intuit and to prove. Let the group order be p^n. Then $\Gamma \cong \mathbb{Z}/p^{n_1}\mathbb{Z} \times \dots \times \mathbb{Z}/p^{n_r}\mathbb{Z}$ where $n_1 \geqslant \dots \geqslant n_r$ and $n_1 + \dots + n_r = n$, *i.e.* $(n_1,\dots,n_r)$ is a partition of n. There is a natural partial order on these partitions: $(n_1,\dots,n_r) \leqslant (n_1',\dots,n_s')$ if $n_1 \leqslant n_1'$, $n_1 + n_2 \leqslant n_1' + n_2'$, and so forth for all partial sums. If Γ and Γ' are groups of order p^n, then $\Gamma \leqslant \Gamma'$ if and only if their corresponding partitions of n, which we denote $\nu\Gamma$ and $\nu\Gamma'$, have $\nu\Gamma \leqslant \nu\Gamma'$. What the partial order may be for all p-groups of order p^n is (at the time of writing) unknown, but it seems reasonable to conjecture that every p-group is a deformation of an abelian one, specifically, of the direct product of the quotient groups of the upper central series. However, other nonabelian groups, *e.g.* S_3 (see below), need not be deformations of abelian ones.

Some relations are evident: every finite cyclic group is maximal in the partial order, since deformation will not decrease the orders of elements. Abelian groups may deform to nonabelian ones, but not conversely. For $n = 6$ the poset therefore has two disconnected points, $\mathbb{Z}/2\mathbb{Z} \times \mathbb{Z}/3\mathbb{Z} = \mathbb{Z}/6\mathbb{Z}$ and S_3. One suspects that all S_n are rigid. (This would be trivial if $H^r(KS_n,KS_n) = 0$ for $r > 0$ or even just $r = 2$ and all n, K. Of course, this fails for $n = 2$ and $K = \mathbb{F}_2$, but may be true for all sufficiently large n.) Perhaps even all finite simple groups are maximal.

The more interesting part of the theory for geometry will

probably be that of the deformation of infinite groups over $\mathbb{C}$. That such deformation is possible is evidenced by the group $\Gamma = \mathbb{Z} \times \mathbb{Z} \times \mathbb{Z}$ which deforms over $\mathbb{R}$ to the group Γ' of all upper triangular 3×3 matrices over $\mathbb{Z}$ of the form $I + qE^{12} + rE^{13} + sE^{23}$ where I is the 3×3 identity matrix and E^{ij} is the 3×3 matrix with a 1 in row i, column j and 0's elsewhere: set $E^{12}E^{23} = tE^{13}$. Lubotzky and Magid have given examples of rigid group representations, not withstanding the fact that there exist nontrivial infinitesimal deformations. (The same has been done for finite-dimensional associative algebras in prime characteristic, but the existence of such is unknown in characteristic zero.)

We have no conjectures in characteristic zero (except that the theory will be interesting), and close with a description of the partial order in the smallest nontrivial case for p-groups, groups of order 8. There are five these: the abelian groups $\mathbb{Z}/8\mathbb{Z}$, $\mathbb{Z}/4\mathbb{Z} \times \mathbb{Z}/2\mathbb{Z}$, and $\mathbb{Z}/2\mathbb{Z} \times \mathbb{Z}/2\mathbb{Z} \times \mathbb{Z}/2\mathbb{Z}$ —which we will denote by their corresponding partitions of 3, namely (3), (2,1), and (1,1,1) —the quaternion group $Q = \{\pm 1, \pm i, \pm j, \pm k\}$, and the dihedral group D_4 of symmetries of the square. The latter has generators σ and τ with $\sigma^2 = \tau^4 = 1$ and $\sigma\tau\sigma = \tau^3$. Note that Q and D_4 have the same group algebras over any algebraically closed field K whose characteristic is not 2 (although, for example, over $\mathbb{R}$ they are distinct). However, if the characteristic of K is 2 then $\{x \in KQ \mid x^2 = 0\}$ is a linear space (contained in the center) of dimension 4, while $\{x \in KD_4 \mid x^2 = 0\}$ contains several five-dimensional linear subspaces. It follows that KQ

cannot deform to KD_4 since the variety of elements having square 0 cannot grow larger under deformation. Also D_4 cannot deform to Q since KQ has an automorphism of order 3 with a four-dimensional fixed space and eigenspaces for the eigenvalues ω (= a primitive cube root of 1) and ω^2 both of dimension 2. Since the order is prime to the characteristic, such an automorphism must descend to any algebra of which KQ is a deformation, but there is none such in KD_4. For the same reason Q cannot be a deformation of (2,1). It remains only to show explicitly that D_4 is a deformation of (2,1) and that Q is a deformation of (1,1,1). For the first part, let (2,1) have generators a and b with $a^2 = b^4 = 1$, and let the deformation parameter be t. Setting $ba = (1 + t)ab + tab^3$ defines a deformation of K(2,1); we may now take $\sigma = a$ and $\tau = ab$. For the second, let the generators of (1,1,1) be a, b, and c and define a deformation by setting $a^2 = b^2 = (1 + t) + tc$ and $ba = (1 + t + tc)ab$. Since $1 + t + tc$ is a central element $\neq 1$ with square equal to 1, it together with a (which serves as i) and b (serving as j) generates a group isomorphic to Q. The poset of groups of order 8 is therefore

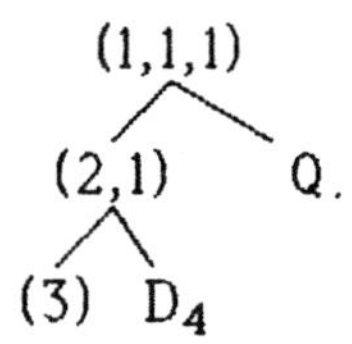

University of Pennsylvania, Philadelphia, PA 19104-6395
State University of New York at Buffalo, Buffalo, NY 14214-3093

Contemporary Mathematics
Volume **74**, 1988

Geometric structures on manifolds and varieties of representations

WILLIAM M. GOLDMAN

Department of Mathematics
University of Maryland
College Park, Maryland 20742

16 November 1987

Abstract. Many interesting geometric structures on manifolds can be interpreted as structures locally modelled on homogeneous spaces. Given a homogeneous space (X, G) and a manifold M, there is a *deformation space* of structures on M locally modelled on the geometry of X invariant under G. Such a geometric structure on a manifold M determines a representation (unique up to inner automorphism) of the fundamental group π of M in G. The deformation space for such structures is "locally modelled" on the space $\mathrm{Hom}(\pi, G)/G$ of equivalence classes of representations of $\pi \to G$. A strong interplay exists between the local and global structure of the variety of representations and the corresponding geometric structures. The lecture in Boulder surveyed some aspects of this correspondence, focusing on: (1) the "Deformation Theorem" relating deformation spaces of geometric structures to the space of representations; (2) representations of surface groups in $\mathrm{SL}(2; \mathbf{R})$, hyperbolic structures on surfaces (with singularities), Fenchel-Nielsen coordinates on Teichmüller space; (3) convex real projective structures on surfaces; (4) representations of Schwarz triangle groups in $\mathrm{SL}(3; \mathbf{C})$. This paper represents an expanded version of the lecture.

According to Felix Klein's Erlanger program of 1872, geometry is the study of the properties of a space which are invariant under a group of transformations. A geometry in Klein's sense is thus a pair (X, G) where X is a manifold and G is a Lie group acting (transitively) on X. If (X, G) is Euclidean geometry (so $X = \mathbf{R}^n$ and G is its group of isometries), then one might "infinitesimally model" a manifold M on Euclidean space by giving each tangent space a Euclidean geometry, which varies from point to point; the resulting *infinitesimally Euclidean* geometry is a Riemannian metric. Infinitesimally modelling a space on other homogeneous spaces gives rise to other G-structures and connections. But to model a space *locally* on a geometry (and not just infinitesimally) requires certain integrability conditions to be satisfied; these can be expressed in terms of vanishing curvature. For example, a flat Riemannian metric is a *locally* Euclidean structure since every point has a neighborhood isometric to an open subset of Euclidean space. The study of such geometric structures was initiated in general by Charles Ehresmann [13] under the name of "locally homogeneous structures." By assuming a standard local structure, the existence and classification of such structures becomes a global topological problem, involving the fundamental group of the underlying space and in particular its representations in the structure group of the geometry. One constructs a "space of geometric structures"

The author gratefully acknowledges partial support by National Science Foundation grant DMS-86-13576 and an Alfred P. Sloan Foundation fellowship.

on a given manifold, the points of which correspond to equivalence classes of geometric structures; this *deformation space* is itself "locally modelled" on the space of equivalence classes of representations of the fundamental group. This paper surveys several aspects of the deformation theory of geometric structures on a manifold and its relation to the variety of representations of the fundamental group in a Lie group.

Let G be a Lie group acting transitively on a manifold X and let M be a manifold of the same dimension as X. An *(X,G)-atlas* on M is a pair $(\mathcal{U},\Phi)$ where $\mathcal{U}$ is an open covering of M and $\Phi = \{\phi_\alpha : U_\alpha \longrightarrow X\}_{U_\alpha \in \mathcal{U}}$ is a collection of coordinate charts such that for each pair $(U_\alpha, U_\beta) \in \mathcal{U}\times\mathcal{U}$ and connected component C of $U_\alpha \cap U_\beta$ there exists $g_{C,\alpha,\beta} \in G$ such that $g_{C,\alpha,\beta}\circ\phi_\alpha = \phi_\beta$. An *$(X,G)$-structure* on M is a maximal (X,G)-atlas and an *(X,G)-manifold* is a manifold together with an (X,G)-structure on it. It is clear that an (X,G)-manifold has an underlying real analytic structure, since the action of G on X is real analytic.

Suppose that M and N are two (X,G)-manifolds and $f : M \longrightarrow N$ is a map. Then f is an *(X,G)-map* if for each pair of charts $\phi_\alpha : U_\alpha \longrightarrow X$ and $\psi_\beta : V_\beta \longrightarrow X$ (for M and N respectively) and a component C of $U_\alpha \cap f^{-1}(V_\beta))$ there exists $g = g(C,\alpha,\beta) \in G$ such that the restriction of f to C equals $\psi_\beta^{-1}\circ g\circ\phi_\alpha$. In particular we only consider (X,G)-maps which are local diffeomorphisms. It is easy to see that if $f : M \longrightarrow N$ is a local diffeomorphism where M and N are smooth manifolds, then for every (X,G)-structure on N, there is a unique (X,G)-structure for which f is an (X,G)-map. In particular every covering space of an (X,G)-manifold has a canonical (X,G)-structure. In the converse direction, if $\Gamma \subset G$ is a discrete subgroup which acts properly and freely on X then X/Γ is an (X,G)-manifold. Such a structure is said to be *complete*.

Examples of Geometric Structures

Many familiar geometric structures are (X,G)-structures, for various choices of X and G. Here is a list of a few of them:

(1) Euclidean structures.

Here $X = \mathbf{R}^n$ and G is the group of all isometries. An (X,G)-structure is then identical to a flat Riemannian metric. Completeness of the Euclidean structure is equivalent to geodesic completeness of the metric which is in turn equivalent to completeness as a metric space. Thus a closed Euclidean manifold is covered by $\mathbf{R}^n$ and by Bieberbach's theorem is finitely covered by a flat torus. Furthermore every homotopy equivalence between closed Euclidean manifolds is homotopic to an affine diffeomorphism.

(2) Locally homogeneous Riemannian manifolds.

A natural generalization of Euclidean structures are constant (nonzero) curvature manifolds — a manifold with a Riemannian metric of constant sectional curvature is locally isometric either to Euclidean space, the sphere, or hyperbolic space, depending on whether the curvature is zero, positive, or negative respectively. If X is one of these model spaces and G is its isometry group, then a constant curvature metric on a manifold M is exactly an (X,G)-structure on M.

This is part of the following more general construction. Suppose that (X, g_X) is a Riemannian manifold upon which a Lie group G acts transitively by isometries. (This is basically equivalent to requiring that the homogeneous space $X = G/H$ have the property that the isotropy group H is compact.) Then every (X, G)-manifold M inherits a Riemannian metric g_M locally isometric to g_X. Each coordinate chart $\phi_\alpha : U_\alpha \longrightarrow X$ determines a Riemannian metric $\phi_\alpha^*(g_X)$ on U_α; since G acts isometrically on X with respect to g_X, any two of these local metrics agree on the intersection of two coordinate patches. Thus there exists a metric g_M on M whose restriction to U_α equals $\phi_\alpha^*(g_X)$. It is easily seen that any Riemannian metric on M which is everywhere locally isometric to X arises from an (X, G)-structure as above. This collection of structures includes all the geometric structures used to uniformize 2- and 3-dimensional manifolds in the sense of the geometrization conjecture of Thurston [67] (see also Scott [61]). Completeness of the (X, G)-structure as above is equivalent to geodesic completeness of the Riemannian metric and hence metric completeness; see Thurston [66] for discussion. Since every such (X, G)-structure on a closed manifold M is complete and hence determines a subgroup $\Gamma \subset G$ which will be discrete and cocompact, the subject of (X, G)-structures on closed manifolds is essentially equivalent to the study of discrete cocompact subgroups of G which act freely on X.

A basic example is the following. If S is a closed orientable surface of genus $g > 1$, then the deformation space of hyperbolic structures on S is a manifold, the *Teichmüller space* $\mathfrak{T}(S)$ of S, and is homeomorphic to a cell of dimension $6g - 6$. One can alternatively describe $\mathfrak{T}(S)$ as the quotient of the space of discrete embeddings $\pi_1(S) \hookrightarrow \mathrm{PGL}(2; \mathbf{R})$ by the action of $\mathrm{PGL}(2; \mathbf{R})$ by conjugation.

(3) Locally homogeneous indefinite metrics.

Other interesting geometric structures are modelled on homogeneous pseudo-Riemannian manifolds, where the invariant metric tensor is allowed to be indefinite. The arguments that compact implies complete completely break down; however no example of a closed manifold with an incomplete locally homogeneous indefinite metric seems to be known. Complete flat pseudo-Riemannian metrics on closed manifolds have been classified in dimension 3 (Fried-Goldman [18]), dimension 4 and signature (1,3) (Fried [17]), dimension 4 and signature (2,2) (Wang [71]), and a general structure theorem has been obtained for flat Lorentz metrics (Goldman-Kamishima [28]).

As observed and analyzed by Kulkarni and Raymond [49], a large class of Seifert-fibered 3-manifolds admit *constant negative curvature* Lorentz metrics. In particular every Seifert 3-manifold over hyperbolic base having nonzero (rational) Euler class has such a structure. Such structures are modelled on the Lie group $X = \mathrm{PSL}(2; \mathbf{R})$ with automorphism group generated by left-multiplications and automorphisms of the group $\mathrm{PSL}(2; \mathbf{R})$. The most tractable class of such structures are those which Kulkarni and Raymond call "standard" — these are characterized by having a timelike Killing vector field on the time-orientable double covering. Equivalently, such structures have a Riemannian metric locally isometric to a left-invariant Riemannian metric on $\mathrm{PSL}(2; \mathbf{R})$. Clearly every standard structure is complete and Kulkarni-Raymond [49] prove that a closed 3-manifold which admits a standard structure must be a Seifert manifold of the above type (the deformation theory of standard structures is worked out in Kulkarni-Lee-Raymond [47]). However nonstandard complete structures exist, although every closed 3-manifold which admits a complete structure also admits a standard structure and therefore is a Seifert 3-manifold over hyperbolic base with nonzero Euler class ([22]).

(4) Flat affine structures.

Here $X = \mathbf{R}^n$ and $G = \mathrm{Aff}(n; \mathbf{R})$ is the group of affine transformations. An (X, G)-manifold here is called an *affinely flat* or just an *affine* manifold and an *affine structure* on manifold M is just a flat torsionfree affine connection on M. In the absence of a compatible metric, an affine manifold does not possess a canonical sense of distance; yet there is a well-defined notion of *parallel*. Thus while it makes no sense to speak of a particle moving at "unit speed" on an affine manifold (as there is on a Riemannian manifold) there is nonetheless a notion of moving at "constant speed" along a straight line. Geodesic completeness can then be expressed as the property that a particle moving at constant speed in a straight line will continue forever. Geodesic completeness in this sense is equivalent to completeness of the (X, G)-structure above. Complete affine manifolds are discussed in Milnor [55] and the classification of complete affine structures on closed 3-manifolds is given in Fried-Goldman [18]. It is conjectured that a *compact* complete affine manifold is finitely covered by quotients of solvable Lie groups with left-invariant complete affine structures; this conjecture is equivalent to the assertion that the fundamental group of a compact complete affine manifold contains a solvable subgroup of finite index. Such structures can be classified in terms of representations of solvable Lie algebras. A simple case occurs for complete affine structures on the two-torus, where there are two types of left-invariant complete affine structures; the deformation space of complete affine structures on the two-torus may be identified with the real plane $\mathbf{R}^2$ with the topology whose only proper open sets are the open subsets of $\mathbf{R}^2 - \{0\}$; in particular in even such a simple example as this the deformation space is non-Hausdorff. A homotopy equivalence between two compact complete affine manifolds with virtually solvable fundamental group is generally not necessary homotopic to an affine diffeomorphism, but is homotopic to a *polynomial* diffeomorphism (reminiscent of the rigidity in Bieberbach's theorem); see [18] for details. Boyom [3] has recently shown that every nilmanifold admits a homogeneous complete affine structure. Recently Margulis [52,53] has constructed a very surprising example of a *noncompact* complete flat affine 3-manifold (in fact a flat Lorentz manifold) whose fundamental group is a nonabelian free group.

Kostant and Sullivan [44] proved that the Euler characteristic of a compact complete affine manifold must vanish, affirming a conjecture first stated by Chern. Benzécri [2] proved that a closed surface which admits a (possibly incomplete) affine structure must have zero Euler characteristic. It is unknown in higher dimension whether the Euler characteristic of a compact affine manifold in higher dimensions is zero; for more information on affine structures on compact manifolds, see Goldman-Hirsch [27], Kobayashi [42], Shima [62] and the references cited there.

(5) Flat projective structures.

Let X be a projective space (either real or complex) and G its group of projective automorphisms. In traditional terminology, a *(flat) projective structure* modelled on this projective geometry is a flat normal projective connection. Projective structures on Riemann surfaces (called in [24] "$\mathbf{CP}^1$-structures") arose classically in the study of second order differential equations on Riemann surfaces and in connection with the uniformization problem. For a survey of the analytic theory of $\mathbf{CP}^1$-structures, the reader is referred to Gunning [32] (see also Earle [11], Hejhal [34], Hubbard [37], Kra-Maskit [45], and the references cited there). The deformation space $\mathbf{CP}^1(S)$ of $\mathbf{CP}^1$-structures on a closed orientable surface S of genus $g > 1$ can be naturally identified with the cotangent bundle of the Teichmüller space of S. An alternative description of this space is due to Thurston (unpublished), who identifies $\mathbf{CP}^1(S)$ with the (trivial) bundle over Teichmüller space $\mathfrak{T}(S)$ having for fiber the the space $\mathfrak{ML}(S)$ of measured geodesic laminations on S. This identification

$$\Theta : \mathbf{CP}^1(S) \longrightarrow \mathfrak{T}(S) \times \mathfrak{ML}(S)$$

is based on the fundamental insight that a $\mathbf{CP}^1$-structure can be identified with a locally convex pleated map of the universal covering surface into hyperbolic 3-space (compare Epstein-Marden [14] and Canary-Epstein-Green [7]); the parameters in $\mathfrak{T}(S) \times \mathfrak{ML}(S)$ describe the intrinsic geometry and the extrinsic geometry (the bending parameters) of the pleated surface respectively.

Projective structures modelled on the real projective plane (i.e. $\mathbf{RP}^2$-structures) are not nearly as well understood. In 1976, J.Smillie [63] and W. Thurston [65] independently discovered an $\mathbf{RP}^2$-structure on the 2-torus whose developing map (see below) fails to be a covering onto its image (the complement of 3 points in $\mathbf{RP}^2$). Moreover, every closed surface of genus $g > 1$ enjoys an $\mathbf{RP}^2$-structure whose developing map is a surjection onto $\mathbf{RP}^2$. Even so, the deformation space of $\mathbf{RP}^2$-structures on a compact surface S with $\chi(S) < 0$ is a Hausdorff real analytic manifold of dimension $-8\chi(S)$ (Goldman [26]). This is a consequence of the fact that the holonomy group of an $\mathbf{RP}^2$-structure on a closed surface with $\chi(S) < 0$ cannot fix a point in $\mathbf{RP}^2$; however it seems to be unknown whether the holonomy group can leave invariant a projective line in $\mathbf{RP}^2$ (such a line would necessarily intersect the developing image, by the result of Benzècri above). It seems that a tractable class of $\mathbf{RP}^2$-structures are the *convex* $\mathbf{RP}^2$*-manifolds,* i.e. $\mathbf{RP}^2$-manifolds of the form Ω/Γ, where $\Omega \subset \mathbf{RP}^2$ is a convex domain and $\Gamma \subset \mathrm{PGL}(3;\mathbf{R})$ is a discrete group acting properly and freely on Ω. As a special case, the hyperbolic structures on S determine convex $\mathbf{RP}^2$-structures on S and thus the Teichmüller space of S lies inside the deformation space $\mathcal{P}(S)$ of convex $\mathbf{RP}^2$-structures on S. Projective structures modelled on $\mathbf{CP}^n$ for $n \geq 2$ are discussed in Kobayashi-Ochiai [41] as well as Gunning [31] and Yoshida [74].

(6) Flat conformal structures.

Here $X = \mathrm{S}^n$ and $G = \mathrm{SO}(n+1,1)$ is the group of conformal transformations of X where $n > 2$. An (X,G)-structure on M is the same as a conformal class of Riemannian metrics, each locally conformal to a flat metric; hence such structures are called *flat conformal structures.* A large class of closed 3-manifolds admit flat conformal structures, although relatively simple ones might not — in particular an nontrivial oriented S^1-bundle over a 2-torus admits no such structure [21] (an analytic proof of this is contained in the theory developed in the recent paper of Schoen-Yau [60]). However, Gromov, Lawson and Thurston [30] have produced startling examples of flat conformal structures on certain nontrivial S^1-bundles over closed hyperbolic surfaces. Hyperbolic structures determine flat conformal structures; although hyperbolic structures on closed manifolds of dimension ≥ 3 satisfy Mostow rigidity (every homotopy equivalence is homotopic to a unique isometry), there is often a rich supply of deformations of the flat conformal structure; see Apanosov [1], Gusevskii-Kapovich [76], Johnson-Millson [38], Kamishima [75],

Kapovich [40,77], Korouniotis [43], Kulkarni [46], Kulkarni-Pinkall [48], Millson [56] and Schoen-Yau [60] for further discussion.

(7) Spherical CR structures.
Here $X = S^{2n-1}$ the boundary of the unit ball in $\mathbf{C}^n$ and $G = \mathrm{PU}(n,1)$ is the group of biholomorphisms of the unit ball acting on its boundary by CR-automorphisms. An (X,G)-structure on a real manifold M^{2n-1} is then a CR-structure all of whose local invariants (the analogue of curvature) vanish. Every Seifert 3-manifold with nonzero Euler class admits such a structure; however, the 3-torus is not sufficiently complicated to admit such a structure [21]. No example seems to be currently known of a closed hyperbolic 3-manifold which admits such a structure, although one expects them to exist in abundance. For a general discussion of these geometric structures, see Burns-Shnider [5] and Ehlers-Neumann-Scherk [12].

The Graph of a Geometric Structure

We seek a concrete geometric object which so to speak is the "graph" of a geometric structure. We begin by graphing the coordinate charts. For each coordinate chart (U_α, ϕ_α) we consider the following information:

(1) The product $E_\alpha = U_\alpha \times X$ is an X-bundle over the coordinate patch U_α with projection $\pi_\alpha : E_\alpha \longrightarrow U_\alpha$;
(2) The sets $U_\alpha \times \{x\}$, parametrized by $x \in X$, are the leaves of a foliation $\mathcal{F}_\alpha$ of U_α transverse to the fibers of $\pi_\alpha : E_\alpha \longrightarrow U_\alpha$;
(3) Graphing the coordinate chart $\phi_\alpha : U_\alpha \longrightarrow X$ defines a section $\Phi_\alpha : U_\alpha \longrightarrow E_\alpha$ of the bundle E_α which is transverse to $\mathcal{F}_\alpha$ (because ϕ_α is nonsingular).

This local data pieces together globally as follows. Since the coordinate charts on overlapping coordinate patches differ by elements of G, the bundles E_α glue together to form a fiber bundle $\pi : E \longrightarrow M$ with fiber X and structure group G in the sense of Steenrod [64]. Since the action of G on X is *strongly effective,* (two elements of G which agree on a nonempty open set are identical), there is a foliation $\mathcal{F}$ of the total space E whose restriction to each E_α equals $\mathcal{F}_\alpha$. Since the foliation is defined by submersions to X in local foliation boxes E_α and the coordinate changes lie in G, the foliation has a *transverse (X,G)-structure* and the pair $(\pi : E \longrightarrow M, \mathcal{F})$ is a *flat (X,G)-bundle* over M. (In the terminology of Steenrod [64] a flat (X,G)-bundle is a bundle whose fiber is X and whose structure group is the group G^{discrete}, i.e. the group G endowed with the discrete topology. A flat structure on an (X,G)-bundle $\pi : E \longrightarrow M$, i.e. a reduction of the structure group of E to G^{discrete} is a foliation $\mathcal{F}$ as above transverse to the fibers of E whose holonomy lies in G such that for each leaf $L \subset E$, the restriction $\pi|_L : L \longrightarrow M$ is a covering space.)

Thus the "graph" of an (X,G)-structure on M consists of the following: a flat (X,G)-bundle $\pi : E \longrightarrow M$, called the *tangent flat (X,G)-bundle to M*, or the *(X,G)-holonomy bundle of M*, and a section $f : M \longrightarrow E$ which is transverse to the flat structure $\mathcal{F}$, called the *developing section.* We call the triple $(E, \mathcal{F}, f)$ the *graph* of the (X,G)-manifold and denote it by graph(M).

Suppose that $(E,\mathcal{F})$ is a flat (X,G)-bundle over M and let $f : M \longrightarrow E$ be a section which is transverse to $\mathcal{F}$. Choose a covering of the total space E by foliation boxes — open sets $U \subset E$ equipped with submersions $\psi_U : U \longrightarrow X$ defining the flat (X, G)structure $\mathcal{F}|_U$ on U. The inverse images $f^{-1}(U)$ define an open covering of M for which the compositions $\psi_U \circ f : f^{-1}(U) \longrightarrow X$ define coordinate charts for an (X, G)-structure on M. Thus we obtain:

LEMMA. *Let $(E, \mathcal{F})$ be a flat (X, G)-bundle over M. Then a section $M \longrightarrow E$ is a developing section for an (X, G)-structure on M if and only if it is transverse to $\mathcal{F}$.*

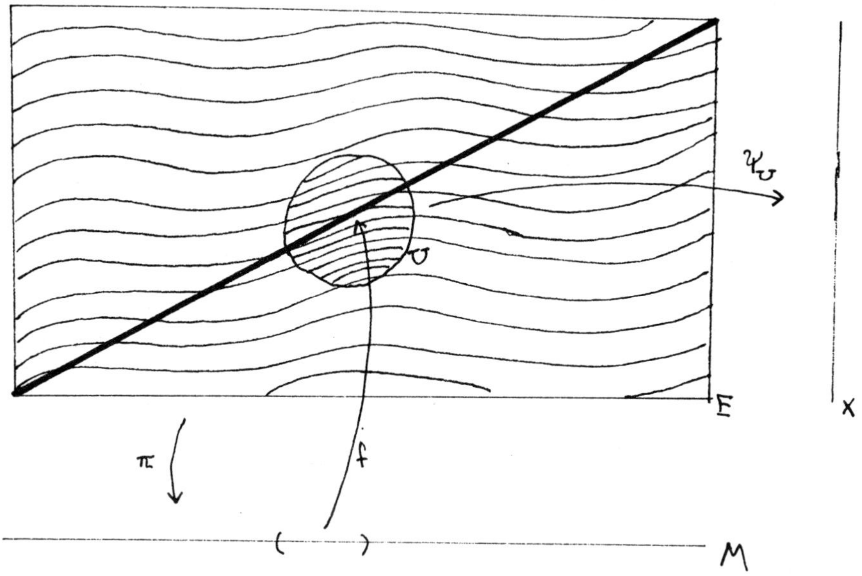

The graph of an (X, G)-structure

A flat (X, G)-bundle as above arises from a homomorphism of the fundamental group (the *holonomy homomorphism*) $\Gamma = \pi_1(M)$ into G as follows. Let $\rho : \Gamma \longrightarrow G$ be a homomorphism. Let $\mathbf{p} : \tilde{M} \longrightarrow M$ be a universal covering space of M with covering group Γ. Then Γ acts on the trivial (X, G)-bundle $\tilde{E} = \tilde{M} \times X$ over $\tilde{M}$ by

$$(1) \qquad \gamma : (\tilde{m}, x) \mapsto (\gamma\tilde{m}, \rho(\gamma) \cdot x)$$

where $\gamma \in \Gamma, \tilde{m} \in \tilde{M}, x \in X$. Since the action of Γ on $\tilde{M}$ is properly discontinous and free, so is the action on $\tilde{E}$ defined by (1) and it follows that the projection $\tilde{E} \longrightarrow \tilde{M}$ makes the quotient $E_\rho = \tilde{E}/\Gamma$ the total space of an (X, G)-bundle over $M = \tilde{M}/\Gamma$. The foliation of $\tilde{E}$ by leaves $\{\tilde{M} \times x\}_{x \in X}$ is a flat (X, G)-structure invariant under the Γ-action (1) and hence defines a flat structure $\mathcal{F}_\rho$ on E_ρ.

If $(\pi : E \longrightarrow M, \mathcal{F})$ is a flat (X, G)-bundle over M, then it is isomorphic to a flat (X, G)-bundle $(E_{\rho, \mathcal{F}_\rho})$ for a representation ρ as above. To see this, let $m_0 \in M$ be a basepoint and consider the fiber $E_0 = \pi^{-1}(m_0)$ over m_0. Let $\mathbf{p} : \tilde{M} \longrightarrow M$ be the universal covering space corresponding to m_0: a point of $\tilde{M}$ corresponds to a relative homotopy class of paths $\sigma : [0,1] \longrightarrow M$ with $\sigma(0) = m_0$ as above and the covering projection $\mathbf{p}$ is the evaluation map $\sigma \mapsto \sigma(1)$ at $t = 1$. We shall trivialize the pullback $\mathbf{p}^*E$ by the flat structure. Let $\sigma : [0,1] \longrightarrow M$ be a path with $\sigma(0) = m_0$. For each point $e_0 \in E_0$ there exists a unique path $\tilde{\sigma}^{e_0} : [0,1] \longrightarrow \tilde{M}$ with $\tilde{\sigma}^{e_0}(0) = e_0$ which is "$\mathcal{F}$-horizontal:" $\tilde{\sigma}^{e_0}(t)$ lies in a single leaf of $\mathcal{F}$. The correspondence

$$E_0 \times \tilde{M} \longrightarrow \mathbf{p}^*E$$
$$(e_0, \sigma) \mapsto \tilde{\sigma}^{e_0}(1)$$

defines a trivialization of $\mathbf{p}^*E$ as a flat (X,G)-bundle. Choose an identification $E_0 \longrightarrow X$; in other words, trivialize the bundle E over m_0 as an (X,G)-bundle. Then E is recovered from $\mathbf{p}^*E$ as a quotient of $\mathbf{p}^*E \cong \tilde{M} \times X$ by an action of $\pi_1(M; m_0)$ which extends the action of $\pi_1(M; m_0)$ on $\tilde{M}$ and comes from the action of G on the fibers. Such an action is given by (1) applied to a representation $\rho : \pi_1(M; m_0) \longrightarrow G$. Now G acts simply transitively on the trivializations of E_0; applying $g \in G$ to the trivialization $E_0 \longrightarrow X$ composes the holonomy representation ρ with the inner automorphism $G \longrightarrow G$ determined by g.

We may now interpret a section of the flat (X,G)-bundle $(E, \mathcal{F})$ associated to a representation ρ of the fundamental group as the graph of an equivariant map of the universal covering $\tilde{M}$ as follows. Since $\mathbf{p}^*E$ is the trivial (X,G)-bundle $\tilde{M} \times X$ over $\tilde{M}$, a section $\tilde{M} \longrightarrow \mathbf{p}^*E$ is just the graph of a map $\tilde{M} \longrightarrow X$. Suppose that $f : M \longrightarrow E$ is a section; then f pulls back to a section $\tilde{M} \longrightarrow \mathbf{p}^*E \cong \tilde{M} \times X$ which must be the graph of a map $\tilde{f} : \tilde{M} \longrightarrow X$ satisfying the equivariance condition

$$\tilde{f} \circ \gamma = \rho(\gamma) \circ \tilde{f}. \tag{2}$$

If f is the developing section for an (X,G)-structure, then the corresponding equivariant map $\tilde{f}$ is called a *developing map* for the (X,G)-structure. The condition that f is $\mathcal{F}$-transverse is just the condition that the developing map $\tilde{f}$ is a nonsingular smooth map (i.e. a local diffeomorphism).

A pair $(\tilde{f}, \rho)$ is called a *development pair* and is a useful globalization of an (X,G)-structure defined by local coordinates. The developing map pulls back the (X,G)-structure from X to $\tilde{M}$ and thus defines the (X,G)-structure on $\tilde{M}$. The holonomy homomorphism determines the action of $\pi_1(M)$ on $\tilde{M}$ by (X,G)-automorphisms. Thus a development pair completely determines the (X,G)-structure.

We have already seen that isomorphism classes of flat (X,G)-bundles over M correspond bijectively to G-orbits of representations $\pi_1(M) \longrightarrow G$; a well-defined holonomy representation is obtained once one chooses a basepoint $m_0 \in M$ and a trivialization τ_0 of the bundle $E_0 = E|_{m_0}$ over the base-point. Suppose that $f : M \longrightarrow E$ is a section; then the corresponding equivariant map $\tilde{f} : \tilde{M} \longrightarrow X$ is determined by the trivialization of $\mathbf{p}^*E$ corresponding to m_0 and τ_0. A given $g \in G$ changes τ_0 to $g \cdot \tau_0$ and the corresponding developing map is the composition $g \circ \tilde{f}$. Thus a development pair $(\tilde{f}, \rho)$ is determined up to the diagonal action of G obtained by composing $\tilde{f} : \tilde{M} \longrightarrow X$ with an element $g \in G$ and $\rho : \pi_1(M; m_0) \longrightarrow G$ with conjugation by g.

DEFORMATION SPACES

We wish to construct "a space of (X, G)-structures" on a compact smooth manifold S. To this end we consider pairs (ϕ, M) where $\phi : S \longrightarrow M$ is a diffeomorphism and M is an (X, G)-manifold. We call such a pair an *(X, G)-structure on S*; two (X, G)-structures (ϕ, M) and (ϕ', M') on S are *isotopic* if there is an (X, G)-map $h : M \longrightarrow M'$ (necessarily an (X, G)-isomorphism) such that ϕ' is isotopic to $h \circ \phi$. In other words, we consider the orbit of $\phi : S \longrightarrow M$ under the group $\mathrm{Diff}_0(S)$ acting by right-composition.

Before defining a topology on the space of isotopy classes of (X, G)-structures on S, we construct a larger space of "based" structures on S. Choose a base-point $s_0 \in S$ and let $\pi = \pi_1(S, s_0)$ and $\tilde{S} \to S$ be the corresponding fundamental group and universal covering spaces. We topologize the set $\mathrm{Hom}(\pi, G)$ with the compact-open topology, which (since π is finitely generated) equals the topology given by pointwise convergence of homomorphisms. If M is a (X, G)-manifold and $m_0 \in M$, we define an *(X, G)-germ at m_0* to be a germ of an (X, G)-structure at m_0, i.e. the germ of an (X, G)-map from a neighborhood N of m_0 into X. Clearly G acts simply transitively on the set of (X, G)-germs at m_0. Consider the set $\mathcal{D}'_{(X,G)}(S)$ of triples (ϕ, M, Ψ) where $\phi : S \longrightarrow M$ is a diffeomorphism to an (X, G)-manifold M and Ψ is an (X, G)-germ at $\phi(s_0)$. Let $\mathrm{graph}(M) = (E, \mathcal{F}, f)$ be the graph of the (X, G)-manifold M as above. Now the (X, G)-germ Ψ at $m_0 = f(s_0)$ defines an identification of the fiber E_0 over $m_0 = \phi(s_0)$ with X; thus there is a well-defined development pair $(\tilde{f}, \rho)$. The preceding discussion allows us to interpret elements $\mathcal{D}'_{(X,G)}(S)$ alternatively in terms of their graphs as well as their development pairs:

$$\begin{aligned}
\mathcal{D}'_{(X,G)}(S) = \{(\phi, M, \Psi) \mid \quad & \phi : S \to M \text{ is a diffeomorphism,} \\
& M \text{ is an } (X, G) - \text{manifold, and} \\
& \Psi \text{ is an } (X, G) - \text{germ at } \phi(s_0)\} \\
= \{(E, \mathcal{F}, f, \bar{\Psi}) \mid & (E, \mathcal{F}) \text{ is a flat } (X, G) - \text{bundle over } S, \\
& f : S \to E \text{ is a } \mathcal{F}\text{-transverse section,} \\
& \bar{\Psi} : E_0 \to X \text{ is a trivialization of the fiber } E_0 \text{ over } s_0\} \\
= \{(\tilde{f}, \rho) \mid \quad & \rho \in \mathrm{Hom}(\pi, G), \\
& \tilde{f} : \tilde{S} \to X \text{ is a } \rho\text{-equivariant nonsingular smooth map}\}
\end{aligned}$$

We can topologize $\mathcal{D}'_{(X,G)}(S)$ using the C^∞ topology either on the developing maps $\tilde{f}$, or in terms of the graphs $(E, \mathcal{F}, f)$ as follows. A neighborhood of $(E_1, \mathcal{F}_1, f_1)$ consists of all flat (X, G)-bundles $(E, \mathcal{F})$ over S such that the tangent spaces to the flat structure $\mathcal{F}$ are close in the C^∞ topology to those of $\mathcal{F}_1$ and $\mathcal{F}$-transverse sections f which are C^∞-close to f_1. It is clear these two topologies agree and that the map

$$\mathrm{hol}' : \mathcal{D}'_{(X,G)}(S) \longrightarrow \mathrm{Hom}(\pi, G)$$

associating to $(\tilde{f}, \rho)$ the holonomy representation ρ is continuous. Let $\mathrm{Diff}_0(\mathrm{S}, \mathrm{s}_0)$ denote the identity component in the group of all diffeomorphisms $S \to S$ which fix s_0. Moreover

$\mathrm{Diff}_0(\mathrm{S}, \mathrm{s}_0)$ acts properly and freely on $\mathcal{D}'_{(X,G)}(S)$ and hol$'$ is invariant under this action. Furthermore the group G acts on $\mathcal{D}'_{(X,G)}(S)$ by composition with Ψ and on $\mathrm{Hom}(\pi, G)$ by left-composition with innner automorphisms and hol$'$ is equivariant with respect to these G-actions.

DEFORMATION THEOREM. *The map* $\mathrm{hol}' : \mathcal{D}'_{(X,G)}(S) \longrightarrow \mathrm{Hom}(\pi, G)$ *is open and for each* $u \in \mathcal{D}'_{(X,G)}(S)$, *there exists a neighborhood* $U \subset \mathcal{D}'_{(X,G)}(S)$ *such that for each* $\rho \in \mathrm{hol}'(U)$ *and* $u_i = (E_i, \mathcal{F}_i, f_i, \Psi_i) \in U \cap \mathrm{hol}'^{-1}(\rho)$ *(where* $i = 1, 2$*), there exists* $h \in \mathrm{Diff}_0(\mathrm{S}, \mathrm{s}_0)$ *and* $g \in G$ *such that* $(h, g) \cdot u_1 = u_2$.

PROOF: Suppose that $u_0 = (E_0, \mathcal{F}_0, f_0, \bar{\Psi}_0) \in \mathcal{D}'_{(X,G)}(S)$ and let $\rho_0 = \mathrm{hol}'(u_0)$. We claim there exists a neighborhood V of $\rho_0 \in \mathrm{Hom}(\pi, G)$ such that the associated flat (X, G)-bundles $(E_v, \mathcal{F}_v)$ over M for $v \in V$ satisfy the following properties. First, by the covering homotopy property there are bundle isomorphisms $E_v \cong E$; indeed a smooth trivialization of the corresponding smooth family of (X, G)-bundles over V exists. Thus we assume that $E_v = E$; with respect to this identification the foliations $\mathcal{F}_v$ vary continuously in the C^∞ topology. By choosing V small enough, the original developing section f_0 remains transverse to $\mathcal{F}_v$ for all $v \in V$. By the previous lemma, this section is a developing section for a nearby (X, G)-structure on M. Thus there is a nearby structure on S with holonomy ρ_v; it follows that hol$'$ is an open map.

To prove that $\mathrm{Diff}_0(\mathrm{S}, \mathrm{s}_0)$ acts simply transitively on the nearby fibers of hol$'$, consider a tubular neighborhood W of the developing section $f_0(S) \subset E_0$ fibered by the intersections of leaves of $\mathcal{F}_0$ with W. Let U be a neighborhood of $u_0 \in \mathcal{D}'_{(X,G)}(S)$ such that $f(S) \subset W$ for $(E, \mathcal{F}, f, \bar{\Psi}) \in U$. For $u = (E, \mathcal{F}, f, \bar{\Psi}) \in U \cap \mathrm{hol}'^{-1}(\rho_0)$ we identify E with E_0 and since the holonomy $\mathrm{hol}'(u) = \rho_0$ of the flat (X, G)-bundle is constant, we identify $\mathcal{F}$ with $\mathcal{F}_0$. Then $u \in U \cap \mathrm{hol}'^{-1}(\rho_0)$ is completely determined by the section f_u and since $f(S) \subset W$, for each $s \in S$ let $h_u(s) \in S$ equal $f_0^{-1}(L_u(s))$ where $L_u(s)$ is the leaf of $\mathcal{F}$ containing $f_u(s)$. This defines the required isotopy joining f_u to f_0. ∎

This seems to be the most complete general statement one can make concerning the relationship of deformation spaces of geometric structures and spaces of homomorphisms of the fundamental group. This result seems to have been first observed in this generality by Thurston [66,§5.3.1] although numerous special cases seem to have been known previously (Weil [72], Hejhal [34], Earle [11], Hubbard [37]); a more detailed proof of this result, based on Weil [72], is given in Lok[50]. Proofs have been given by Dennis Johnson (unpublished) and Canary-Epstein-Green [7]. The proof given above was worked out with Moe Hirsch in 1978 and was also known to André Haefliger.

Let $\mathcal{D}_{(X,G)}(S)$ denote the quotient of $\mathcal{D}'_{(X,G)}(S)$ by $\mathrm{Diff}_0(\mathrm{S}, \mathrm{s}_0)$ and let $\mathrm{Hom}(\pi, G)/G$ denote the space of G-orbits of $\mathrm{Hom}(\pi, G)$ with the quotient topology. The map hol$'$ defines a local homeomorphism

$$\mathrm{hol} : \mathcal{D}_{(X,G)}(S) \longrightarrow \mathrm{Hom}(\pi, G).$$

We define the *deformation space* of (X, G)-structures on S to be the quotient

$$\mathfrak{T}_{(X,G)}(S) = (\mathcal{D}_{(X,G)}(S))/G$$

and there is a map

$$\mathrm{hol} : \mathfrak{T}_{(X,G)}(S) \longrightarrow \mathrm{Hom}(\pi, G)/G$$

which one would like to say is a local homeomorphism. In general this seems to be difficult, since one has little control over the orbits of the action of G.

Suppose that G is the group of $\mathbf{R}$-points of an algebraic group $\bar{G}$ defined over $\mathbf{R}$; then $\mathrm{Hom}(\pi, G)$ is the set of $\mathbf{R}$-points of an algebraic variety $\mathfrak{R}(\pi, \bar{G})$ defined over $\mathbf{R}$ (compare Johnson-Millson [38], Lubotzky-Magid [51]). Furthermore the action of G by conjugation on $\mathrm{Hom}(\pi, G)$ arises from an algebraic action of $\bar{G}$ on $\mathfrak{R}(\pi, \bar{G})$ and thus the orbits are locally closed. If $\bar{G}$ is reductive, then there is a distinguished open subset $\mathrm{Hom}(\pi, G)^{\mathrm{st}}$ of $\mathrm{Hom}(\pi, G)$ consisting of stable orbits, upon which the action of G is proper. (The algebraic-geometric quotient — denoted $X(\pi, G)$ and called the *character variety*, following Culler-Shalen [9] — is the closure of the space of stable orbits. When G is a linear $\mathbf{C}$-algebraic group, $X(\pi, G)$ may be identified with the space of all characters of representations $\pi \rightarrow G$.) Since G acts properly on any space from which there exists a G-equivariant map to a proper G-space, we obtain the following:

COROLLARY. *The restriction of* hol *to the subset* $\mathrm{hol}^{-1}(\mathrm{Hom}(\pi, G)^{\mathrm{st}}/G)$ *of* $\mathcal{D}_{(X,G)}(S)$ *is a local homeomorphism.*

Thus it follows that the local properties of $\mathrm{Hom}(\pi, G)/G$ translate into local properties of the deformation space $\mathfrak{T}_{(X,G)}(S)$ of (X, G)-structures on S. In the case considered above, $\mathrm{Hom}(\pi, G)^{\mathrm{st}}/G$ has the structure of an $\mathbf{R}$-algebraic space and therefore so does $\mathfrak{T}_{(X,G)}(S)$. If π is the fundamental group of a compact Kähler manifold, then by Goldman-Millson [29] in many cases the singularities of $\mathrm{Hom}(\pi, G)$ are described by systems of homogeneous quadratic equations in the Zariski tangent space. The corresponding quotient space has singularities which are quotients of such quadratic cones and one obtains similar information on the singularites of $\mathcal{D}_{(X,G)}(S)$ in this case. See Johnson-Millson [38] for more examples.

If π is the fundamental group of a compact Kähler manifold, $\mathrm{Hom}(\pi, G)/G$ often inherits a Kähler structure of its own (see Corlette [8]) and thus $\mathfrak{T}_{(X,G)}(S)$ also inherits such structure. Although the Kähler structure on $\mathfrak{T}_{(X,G)}(S)$ will depend on the choice of complex structure on the underlying manifold S, often the *symplectic geometry* is independent of the complex structure. In particular, if S is a closed surface, then the deformation space $\mathfrak{T}_{(X,G)}(S)$ admits a symplectic structure which will be invariant under the natural action of the mapping class group of S. This general construction includes as a special case the symplectic geometry arising from the Weil-Petersson Kähler form on Teichmüller space.

For constant negative curvature Lorentzian metrics on twisted S^1-bundles over closed hyperbolic surfaces, the deformation theorem was used in [22] to construct nonstandard complete examples. Indeed, the examples given in [22] demonstrate that the deformation space $\mathfrak{T}_{(X,G)}(S)$ of such structures is non-Hausdorff (even though G is semisimple, its isotropy group on X is semisimple and S is a compact quotient of a semisimple Lie group).

In [24] global properties of the holonomy map

$$\mathrm{hol} : \mathbf{CP}^1(S) \longrightarrow \mathrm{Hom}(\pi, \mathrm{PSL}(2, \mathbf{C}))/\,\mathrm{PSL}(2, \mathbf{C})$$

are investigated. In particular the fiberhol$^{-1}(\mathfrak{T}(S))$ over the set of equivalence classes of discrete embeddings $\pi \longrightarrow \mathrm{PGL}(2,\mathbf{R})$ is identified as the set $\Theta^{-1}(\mathfrak{T}(S) \times \mathfrak{ML}_{\mathbf{Z}}(S)$ corresponding to integral measured laminations (i.e. disjoint families of nontrivial simple closed curves) in Thurston's parametrization of complex projective structures. There is a similar statement for real projective structures on surfaces (although no analogue of the Thurston parametrization is currently known); $\mathbf{RP}^2$-structures on a compact surface S whose holonomy is Fuchsian can similarly classified in terms of disjoint families of simple closed curves weighted by elements of a certain elementary discrete semigroup (see [24] for details).

Using similar methods, one can show that the class of $\mathbf{CP}^1$-structures with holonomy representation $\pi_1(S) \longrightarrow \mathrm{PGL}(2,\mathbf{R})$ equals $\Theta^{-1}(\mathfrak{T}(S) \times \mathfrak{ML}_{\frac{1}{2}\mathbf{Z}}(S)$ where $\mathfrak{ML}_{\frac{1}{2}\mathbf{Z}}(S)$ is the space of *half-integral* points in $\mathfrak{ML}(S)$. (Laminations in $\mathfrak{ML}_{\frac{1}{2}\mathbf{Z}}(S)$ are disjoint families of nontrivial simple closed curves weighted by half-integers.) There is a map

$$H : \mathfrak{ML}_{\frac{1}{2}\mathbf{Z}}(S) \longrightarrow H_1(M;\mathbf{Z}/2)$$

obtained by summing the homology classes of these curves with twice the given weights. In [24] it is incorrectly stated the half-integral points correspond to representations into $\mathrm{PSL}(2,\mathbf{R})$. The correct statement is the following. Let M be a $\mathbf{CP}^1$-manifold whose holonomy lies in $\mathrm{Hom}(\pi, \mathrm{PGL}(2;\mathbf{R}))$ and suppose that its Thurston parameters are

$$\Theta(M) = (g, \lambda) \in \mathfrak{T}(S) \times \mathfrak{ML}(S)$$

where $\lambda \in \mathfrak{ML}_{\frac{1}{2}\mathbf{Z}}(S)$ is half-integral. Then the holonomy of M lies in $\mathrm{PSL}(2;\mathbf{R})$ if and only if $H(\lambda) = 0$. Furthermore the representations $\pi \longrightarrow \mathrm{PGL}(2;\mathbf{R})$ which occur as holonomy representations of $\mathbf{CP}^1$-structures correspond (under the isomorphism $\mathrm{PGL}(2;\mathbf{R}) \cong \mathrm{SO}(2,1)$) exactly to the representations $\pi_1(S) \longrightarrow \mathrm{SO}(2,1)$ which occur as holonomy representations of $\mathbf{RP}^2$-structures. In [19] it is shown that such representations are characterized as those whose image is not solvable.

Global properties of deformation spaces

In the above applications, the geometry of the variety of representations was used to deduce information on the deformation space of geometric structures. In the converse direction, geometric structures may be used to shed light on the geometry and topology of the spaces of homomorphisms. For example, if M is an (X,G)-manifold with graph $(E,\mathcal{F},f)$, then a tubular neighborhood of the developing section f inside E is isomorphic to the tangent bundle of M. This idea can be used to investigate the space of representations of a surface group inside $G = \mathrm{PSL}(2;\mathbf{R})$. Let X be the hyperbolic plane; then $\mathfrak{T}_{(X,G)}(S)$ equals the Teichmüller space of M and it is shown in [20] that a flat (X,G)-bundle $(E,\mathcal{F})$ over M admits a transverse section if and only if the Euler class of E equals $\pm\chi(M)$. Indeed as shown in [25] the Euler class of E characterizes the connected components of $\mathrm{Hom}(\pi, G)$. To understand the other components of $\mathrm{Hom}(\pi, G)$ one introduces hyperbolic structures with singularities as follows, (as first explained to me by W. Neumann). Instead of requiring all of the coordinate charts to be local diffeomorphisms, one allows charts which at isolated points look like the map

$$\begin{gathered} \mathbf{C} \longrightarrow \mathbf{C} \\ z \mapsto z^k \end{gathered}$$

(We say that such a chart defines a singularity of "cone angle" $\theta = 2k\pi$.) Such a singular hyperbolic structure may be alternatively described as a singular Riemannian metric whose curvature equals -1 plus Dirac distributions weighted by $2\pi - \theta$ at each singular point of cone angle θ. In particular a hyperbolic structure on M with isolated singularities at points $x_i, (i = 1, \ldots, k)$ of cone angle $\theta_i > 0$ each of which is a multiple of 2π has a graph $(E, \mathcal{F}, f)$ such that the Euler class is given by

$$e(E) = \chi(M) + \frac{1}{2\pi} \sum_{i=1}^{k} (\theta_i - 2\pi)$$

(It is convenient to assume that each $\theta_i = 4\pi$ and the points x_i are not necessarily distinct — a cone point of cone angle 4π with multiplicity m is then a cone point with cone angle $2(m+1)\pi$.) There is a uniformization theorem for such singular hyperbolic structures, due independently to McOwen [54], Troyanov [70], and Hitchin [35], that given a Riemann surface M, there exists a unique singular hyperbolic structure in the conformal class of M with cone angle θ_i at x_i for $i = 1, \ldots, k$ as long as

$$\chi(M) + \frac{1}{2\pi} \sum_{i=1}^{k} (\theta_i - 2\pi) \leq 0.$$

(Hitchin only considers the case when θ_i are multiples of 2π, while McOwen and Troyanov deal with arbitrary positive angles.) The resulting "uniformization map" then assigns to the collection of points $\{x_1, \ldots, x_k\}$ (where $0 \leq k < |\chi(S)|$) the singular hyperbolic structure with cone angles 4π (counted with multiplicity) at the x_i. The equivalence class of the holonomy representation in the component $e^{-1}(\chi(M) + k) \subset \mathrm{Hom}(\pi, G)/G$ defines a map from the symmetric product $\Sigma^k(M)$ to $e^{-1}(\chi(M) + k)$. The following remarkable result is due to Hitchin [35]:

THEOREM. *The above map*

$$\Sigma^k(M) \longrightarrow e^{-1}(\chi(M) + k)$$

is a homotopy equivalence.

Indeed, Hitchin obtains much more information; for a particular choice of complex structure on M, the component $e^{-1}(\chi(M) + k)$ has the structure of a holomorphic vector bundle over $\Sigma^k(M)$. See [35] for more details as well as a discussion of the rich geometry of these spaces.

Another striking use of geometric structures on surfaces with cone singularities has recently been given by Thurston [68]. For a certain class of singularities, Thurston has shown that the deformation space of Euclidean structures on the 2-sphere is a complex hyperbolic manifold, itself with cone singularities. The developing map for these complex hyperbolic manifolds is the holonomy map taking the space of developments of Euclidean structures on S^2 to a certain space of representations of the fundamental group of S^2 punctured at the various singular points. In certain cases these moduli spaces are complex hyperbolic orbifolds and some of the recent examples of complex hyperbolic manifolds discovered by Deligne-Mostow [10] are recovered in this way as moduli spaces of singular Euclidean structures on S^2. Geometric structures on 3-manifolds with cone singularities play a crucial role in Thurston's proof of the generalized Smith conjecture; see Hodgson [36] for details. Spherical structures with cone singularities on closed surfaces of higher genus have been considered recently by Calabi [6].

BUILDING DEFORMATION SPACES

One approach for understanding the global structure of spaces of representations and deformation spaces in general is to build them up via a decomposition of the manifold M. Since every closed surface can be decomposed into simpler subsurfaces, one might hope that deformation spaces of geometric structures on surfaces might be tractable from this point of view. This is the spirit of the Fenchel-Nielsen coordinates on Teichmüller space, which gives an explicit homeomorphism of the Teichmüller space of a compact orientable surface with an open cell. We shall outline proofs of the following theorems:

THEOREM. *Let S be a closed orientable surface of genus $g > 1$.*

(1) *The Teichmüller space $\mathfrak{T}(S)$ is diffeomorphic to a cell of dimension $6g-6$.*
(2) *The deformation space $\mathcal{P}(S)$ of convex $\mathbf{RP}^2$-structures on S is diffeomorphic to a cell of dimension $16-16$.*

Suppose that S is a compact manifold containing a hypersurface $F \subset S$ such that $S - F$ has two components S_1 and S_2. After appropriate base-points and connecting arcs are chosen, the fundamental group of S admits an amalgamated free product decomposition

$$\pi_1(S) = \pi_1(S_1) \amalg_{\pi_1(F)} \pi_1(S_2)$$

and we shall assume that the maps

$$\pi_1(F) \longrightarrow \pi_1(S_i) \longrightarrow \pi_1(S)$$

induced by the inclusions

$$F \hookrightarrow S_i \hookrightarrow S$$

are injective. There is a corresponding restriction map

$$\mathrm{Hom}(\pi_1(S), G) \longrightarrow \mathrm{Hom}(\pi_1(S_1), G) \times \mathrm{Hom}(\pi_1(S_2), G)$$

which is injective and whose image consists of pairs (ρ_1, ρ_2) of representations whose restriction to $\pi_1(F)$ coincide.

We shall be interested in the relationship between the character varieties $X(\pi_1(S), G)$ and $X(\pi_1(S_i), G)$, i.e. the spaces of conjugacy classes of stable representations. For that reason we shall restrict our attention to representations $\rho : \pi_1(S) \to G$ whose restrictions to $\pi_1(S_i)$ are stable. In the case when S is a compact surface and G is a reductive **R**-algebraic group, it is sufficient to assume that each image $\rho(\pi_1(S_i))$ has discrete centralizer. We henceforth assume this is the case; in all of our applications this condition will be satisfied. Passing to equivalence classes there is a restriction map

$$R : X(\pi_1(S), G) \longrightarrow X(\pi_1(S_1), G) \times X(\pi_1(S_2), G)$$

whose image consists of pairs of characters (χ_1, χ_2) of representations which coincide on $\pi_1(F)$. If (χ_1, χ_2) is such a pair, then its inverse image $R^{-1}(\chi_1, \chi_2)$ can be identified as follows. If ρ is a representation whose restriction ρ_i to $\pi_1(S_i)$ corresponds to χ_i, then the centralizer $\mathcal{Z}(\rho(\pi_1(F)))$ of $\rho(\pi_1(F))$ acts by conjugation on ρ_1 and can be amalgamated with ρ_2 to define a new representation $\pi_1(S) \to G$ as follows. Let $\zeta \in \mathcal{Z}(\pi_1(F))$; then

$$T_\zeta \rho : \gamma \mapsto \begin{cases} \zeta\rho_1(\gamma)\zeta^{-1} & \text{if } \gamma \in \pi_1(S_1) \\ \rho_2(\gamma) & \text{if } \gamma \in \pi_1(S_2) \end{cases}$$

is a representation whose character lies in $R^{-1}(\chi_1, \chi_2)$. One can show that this action is transitive on $R^{-1}(\chi_1, \chi_2)$. (See Goldman [23,25], Johnson-Millson [38] for more details of this construction.)

When $S - F$ is connected, there is a similar discussion. In this case $\pi_1(S)$ decomposes as an HNN-extension of $\pi_1(S-F)$ corresponding to the two embeddings $\iota_i : \pi_1(F) \to \pi_1(S)$, i.e. $\pi_1(S)$ admits a presentation with generators $\pi_1(S - F) \amalg \langle\eta\rangle$ with relations $\iota_1(\gamma) = \eta\iota_2(\gamma)\eta^{-1}$. Let $\theta : \pi_1(S) \longrightarrow \mathbf{Z}$ be the homomorphism containing $\pi_1(S - F) \subset \pi_1(S)$ in its kernel and $\theta(\eta) = 1$. The image of the restriction map

$$\mathrm{Hom}(\pi_1(S), G) \longrightarrow \mathrm{Hom}(\pi_1(S - F), G)$$

consists of representations $\rho \in \mathrm{Hom}(\pi_1(S - F), G)$ such that $\rho \circ \iota_1$ is conjugate to $\rho \circ \iota_2$. The centralizer $\mathcal{Z}(\rho(\pi_1(F)))$ acts on the set of representations whose character lies in the fiber of the restriction map

$$X(\pi_1(S), G) \longrightarrow X(\pi_1(S - F), G)$$

as follows. If $\zeta \in \mathcal{Z}(\rho(\pi_1(F)))$, then define

$$T_\zeta \rho : \gamma \mapsto \rho(\gamma)\zeta^{\theta(\gamma)}$$

If we restrict to representations ρ such that the restriction to $\pi_1(S_F)$ is stable, then $\mathcal{Z}(\rho(\pi_1(S_F)))$ acts transitively on the fiber of $R : X(\pi_1(S), G) \to X(\pi_1(S - F), G)$. In the cases we consider below, one can easily prove that this action is actually *simply transitive*.

Of particular interest is the case when S is a compact surface and F is a homotopically nontrivial simple closed curve. We consider two cases, both with $G = \mathrm{PGL}(n; \mathbf{R})$: hyperbolic structures on closed surfaces ($n = 2$) and convex real projective structures on closed surfaces ($n = 3$). In both cases the holonomy representation ρ is a stable point in

$\mathrm{Hom}(\pi_1(S-F),G)$ and the above discussion applies. Furthermore for each $\gamma \subset \pi_1(F)$, the image $\rho(\gamma)$ is a *positive hyperbolic* element of $\mathrm{PGL}(n;\mathbf{R})$ — represented by a real matrix with distinct positive real eigenvalues. In both cases conjugacy classes of hyperbolic elements are parametrized in terms of invariants arising from the coefficients of the characteristic polynomial and form a cell of dimension $n-1$. A hyperbolic element of $\mathrm{PGL}(2;\mathbf{R})$ has a lift to $\mathrm{SL}(2;\mathbf{R})$ whose trace lies in the interval $(-\infty,2)$; we use this interval to parametrize hyperbolic conjugacy classes. Similarly, the coefficients (x,y) of the characteristic polynomial t^3-xt^2+yt-1 of a positive hyperbolic element of $\mathrm{PGL}(3;R)\cong\mathrm{SL}(3;\mathbf{R})$ lies in the set

$$\mathcal{R}=\{(x,y)\in\mathbf{R}^2 \mid x^2y^2-4(x^3+y^3)+18xy-27>0, \quad x>0, y>0\}$$

and the conjugacy classes of positive hyperbolic elements in $\mathrm{PGL}(3;\mathbf{R})$ are parametrized by the open 2-cell $\mathcal{R}$.

Closely related are the centralizers of positive hyperbolic elements. If $A\in\mathrm{PGL}(n;R)$ is positive hyperbolic, then its centralizer corresponds to the corresponding conjugate of the subgroup of diagonal matrices ($\mathbf{R}$-split Cartan subgroup) containing A. Thus the centralizer of A is an $(n-1)$-dimensional vector group. In particular the space of positive hyperbolic conjugacy classes in $\mathrm{PGL}(n;\mathbf{R})$ and the centralizer of a positive hyperbolic element are each diffeomorphic to an open $(n-1)$-cell (a manifestation that the rank of $\mathrm{PGL}(n;\mathbf{R})$ equals $n-1$).

We consider the case of hyperbolic structures ($n=1$) first, deriving the Fenchel-Nielsen coordinates on Teichmüller space. Let $G=\mathrm{PGL}(2;\mathbf{R})$. Let S be a closed orientable surface of genus $g>1$ and choose a decomposition $\mathfrak{P}$ of S into pairs-of-pants (surfaces diffeomorphic to a sphere minus three discs); since $\chi(S)=2-2g$ and the Euler characteristic of a pair-of-pants equals -1, there are exactly $2g-2$ pants $P_1,\dots,P_{2g-2}$ in the decomposition. Since ∂P_i consists of 3 decomposition curves, and each decomposition curve abuts two pants, there are exactly $\frac{3}{2}(2g-2)=3g-3$ decomposition curves $C_1,\dots,C_{3g-3}$. Consider the restriction map

$$R_{\mathfrak{P}}: X(\pi_1(S),G)\longrightarrow \prod_{i=1}^{2g-2} X(\pi_1(P_i),G)$$

By the above discussion, there is an $\mathbf{R}$-action on each fiber corresponding to each C_j for $j=1,\dots,3g-3$; the disjointness of the curves implies that these $\mathbf{R}$-actions commute ([23]) and the resulting $\mathbf{R}^{3g-3}$-action on each fiber $R_{\mathfrak{P}}^{-1}(\chi_1,\dots,\chi_{2g-2})$ is simply transitive. Thus it suffices to determine the character variety for the fundamental group of a pair-of-pants, which is a free group of rank 2. This is accomplished by the following classical result, which is essentially due to Fricke-Klein [16] (compare Harvey [33], [25]).

PROPOSITION. *Let $\pi=\langle A,B,C \mid ABC=I\rangle$ be a presentation of the free group of rank two. Define $\chi:\mathrm{Hom}(\pi,\mathrm{SL}(2;\mathbf{C}))\longrightarrow\mathbf{C}^3$ by $\chi(\rho)=(\mathrm{tr}\,\rho(A),\mathrm{tr}\,\rho(B),\mathrm{tr}\,\rho(C))$. Then χ is invariant under the action of $\mathrm{SL}(2;\mathbf{C})$ by conjugation and if the image of $\rho_0\in\mathrm{Hom}(\pi,\mathrm{SL}(2;\mathbf{C}))$ is not solvable, then the inverse image $\chi(\chi^{-1}(\rho_0))$ equals the $\mathrm{SL}(2;\mathbf{C})$-orbit of ρ_0. For each $\chi_0\in(-\infty,-2)\times(-\infty,-2)\times(-\infty,-2)$ there exists a representation $\rho_0\in\mathrm{Hom}(\pi,\mathrm{SL}(2;\mathbf{R})$*

with $\chi(\rho_0) = \chi_0$ *and this representation is the holonomy representation of a hyperbolic structure of a pair of pants* P *with geodesic boundary. Conversely every hyperbolic structure on* P *with geodesic boundary has holonomy representation in* $\chi^{-1}((-\infty, 2)^3)$.

Geometrically, the last assertion means that a hyperbolic structure on a pair-of-pants P with geodesic boundary is determined up to isometry by the lengths of the boundary components (the length $l(A)$ of a boundary component is related to the holonomy $\rho(A) \in \mathrm{SL}(2; \mathbf{R})$ by $\operatorname{tr}\rho(A) = \pm 2\cosh(l(A)/2)$). Thus the restriction map is described by the map

$$\begin{aligned} \Theta_{\mathfrak{P}} : \mathfrak{T}(S) &\longrightarrow (-\infty, -2)^{3g-3} \\ [\rho] &\mapsto (\operatorname{tr}\rho(C_1), \dots, \operatorname{tr}\rho(C_{3g-3})) \end{aligned}$$

assigning to a representation the conjugacy-invariants of the image of the family of decomposition curves. Thus the Teichmüller space is expressed as a principal $\mathbf{R}^{3g-3}$ bundle over $\mathbf{R}^{3g-3}$. We note that the character map $\Theta_{\mathfrak{P}}$ above is a moment map for the action of $\mathbf{R}^{3g-3}$ on its fibers; see Goldman [23] for more details.

There is a similar description of the deformation space $\mathcal{P}(S)$ of convex $\mathbf{RP}^2$-structures over S. Once again we consider a pair-of-pants decomposition

$$S = \bigcup_{i=1}^{2g-2} P_i$$

where $C_1, \dots, C_{3g-3}$ are the decomposition curves. If M is a convex $\mathbf{RP}^2$-manifold diffeomorphic to S, then the family $\{C_1, \dots, C_{3g-3}\}$ is isotopic to a disjoint family of simple closed geodesic curves cutting M into a disjoint union of convex $\mathbf{RP}^2$-manifolds, each diffeomorphic to a pair-of-pants. Furthermore if γ is the holonomy around such a curve C, the development of $\tilde{C} \subset \tilde{M}$ is a segment joining the attracting fixed point of γ on $\mathbf{RP}^2$ to the repelling fixed point of γ on $\mathbf{RP}^2$; such a curve is said to be *principal* in [26]. Let $\mathcal{P}(P)$ denote the deformation space of convex $\mathbf{RP}^2$-structures on P with principal geodesic boundary. Recording the conjugacy class of the three boundary components defines a map $\Theta_{\partial P} : \mathcal{P}(P) \longrightarrow \mathcal{R}^3$. The following fact is proved in [26,§5].

PROPOSITION. *The map*

$$\Theta_{\partial P} : \mathcal{P}(P) \longrightarrow \mathcal{R}^3$$

is a fibration with fiber $\mathbf{R}^2$.

If M_1 and M_2 are two $\mathbf{RP}^2$-manifolds with principal geodesic boundary components C_1 and C_2 respectively and $f : N(C_1) \longrightarrow N(C_2)$ is a projective isomorphism between collar neighborhoods, then there is a unique $\mathbf{RP}^2$-structure on the identification space of $M = M_1 \cup M_2$ (where C_1 and C_2 are to be identified to give a curve $C \subset M$) such that there is a reflection in C defined in a tubular neighborhood of C realizing f. One can show ([26, 3.7]) that if M_1 and M_2 are convex, then so is M. Thus to build convex $\mathbf{RP}^2$-manifolds out of convex structures on pants, one needs only to know the structures on the pants and the projective maps needed to identify collar neighborhoods of their boundary components.

If $C \subset M$ is a principal simple closed geodesic, then the germ of the $\mathbf{RP}^2$ structure on a tubular neighborhood $N(C)$ of C (or equivalently on each component of $N(C) - C$) is completely determined by the conjugacy class of the holonomy $\rho(C)$. As noted above, the possible conjugacy classes are parametrized by an open 2-cell $\mathcal{R}$ and the possible identifications are described by the centralizer of a hyperbolic element of $\mathrm{PGL}(3;\mathbf{R})$, which is isomorphic to $\mathbf{R}^2$. Thus the restriction map

$$R_{\mathfrak{P}} : \mathcal{P}(S) \longrightarrow \prod_{i=1}^{2g-2} \mathcal{P}(P_i)$$

is a principal $(\mathbf{R}^2)^{3g-3}$-fibration onto its image, which can be determined as follows. The $\mathbf{RP}^2$-structures near the decomposition curves C_j are classified by the map

$$\begin{aligned} R_{\mathfrak{P}}(\mathcal{P}(S)) &\longrightarrow \mathcal{R}^{3g-3} \\ [\rho] &\mapsto ([\rho(C_1)], \dots, [\rho(C_{3g-3})]) \end{aligned}$$

whose fiber is a product of $2g-2$ cells of dimension 2, one for each P_i. Thus the image $R(\mathcal{P}(S))$ is a cell of dimension $2(3g-3) + 2(2g-2) = 10g - 10$ and $\mathcal{P}(S)$ is a cell of dimension $2(3g-3) + (10g-10) = 16g - 16$.

Once again the map $\mathcal{P}(S) \longrightarrow \mathcal{R}^{3g-3}$ is the moment map for an action of $(\mathbf{R}^2)^{3g-3}$. It seems likely that a refinement of the coordinates obtained in [26] for $\mathcal{P}(S)$ will be canonical coordinates, just as the Fenchel-Nielsen coordinates are canonical coordinates for the Weil-Petersson Kähler form on Teichmüller space as shown by Wolpert [73]. Furthermore there are various constructions of Riemannian metrics and almost complex structures on $\mathcal{P}(S)$ which restrict to the Weil-Petersson structure on $\mathfrak{T}(S) \subset \mathcal{P}(S)$, suggesting that there is at least one natural Kähler structure on $\mathcal{P}(S)$ which extends the Weil-Petersson Kähler geometry of Teichmüller space.

There are many interesting problems concerning the geometry of the space $\mathcal{P}(S)$. The holonomy map

$$\mathrm{hol} : \mathcal{P}(S) \longrightarrow \mathrm{Hom}(\pi, \mathrm{PGL}(3;\mathbf{R}))/\,\mathrm{PGL}(3;\mathbf{R})$$

is an embedding onto an open subset; if the image is also closed, then hol embeds $\mathcal{P}(S)$ as a connected component of $\mathrm{Hom}(\pi, \mathrm{PGL}(3;\mathbf{R}))/\,\mathrm{PGL}(3;\mathbf{R})$. One might try to compactify the image by projective classes of families of valuations à la Morgan-Shalen [58]. A closely related problem is to find alternate an alternate proof that $\mathcal{P}(S)$ is a cell, along the lines of the analytic proof that $\mathfrak{T}(S)$ is a cell: points of Teichmüller space are identified as quasiconformal deformations of a Fuchsian group and such quasiconformal deformations are parametrized by an open ball in a finite-dimensional complex vector space of Beltrami differentials. Alternatively one might try to parametrize points of $\mathcal{P}(S)$ in an analogous way to Thurston's description of $\mathbf{CP}^1(S)$ in terms of the "bending parameters" in $\mathfrak{T}(S) \times \mathfrak{ML}(S)$.

Triangle Groups

Finally we discuss the representations of Schwarz triangle groups in $G = \mathrm{PGL}(3;\mathbf{C})$ and their relation with geometric structures. The representations of these groups are easily computable and therefore one has an explicit description of the character varieties in this case.

Let $p, q, r \geq 3$ be integers satisfying

$$\frac{1}{p} + \frac{1}{q} + \frac{1}{r} < 1;$$

then there exists a triangle $\triangle$ (unique up to isometry) in the hyperbolic plane $\mathbf{H}^2$ with angles $\pi/p, \pi/q, \pi/r$. Let R_1, R_2, R_3 be the reflections in the sides of $\triangle$ and let Γ denote the group they generate. Then Γ is a discrete group of isometries of $\mathbf{H}^2$ with fundamental domain $\triangle$ and admits the presentation

$$\langle R_1, R_2, R_3 \,|\, (R_1)^2 = (R_2)^2 = (R_3)^2 = (R_1R_2)^p = (R_2R_3)^q = (R_2R_3)^r = I \rangle$$

(Although Γ is not the holonomy group of a hyperbolic structure on a manifold, any torsionfree subgroup of finite index is the holonomy group of a hyperbolic structure on a closed surface.)

The projective model for hyperbolic geometry identifies $\mathbf{H}^2$ with the region $\{[x, y, z] \in \mathbf{RP}^2 \,|\, x^2 + y^2 - z^2 < 0\}$ and the group of isometries with $\mathrm{PSO}(2,1) \subset \mathrm{PGL}(3; \mathbf{R})$. We obtain a representation $\rho_1 : \Gamma \longrightarrow \mathrm{PSO}(2,1) \subset \mathrm{PGL}(3; \mathbf{C})$ and we are interested in the component X_1 of $X(\Gamma, \mathrm{PGL}(3; \mathbf{C}))$ containing the equivalence class $[\rho_1]$. In particular, since SO(2,1) is the intersection of the two real forms SU(2,1) and $\mathrm{SL}(3; \mathbf{R})$ of $\mathrm{SL}(3; \mathbf{C})$, we are particularly interested in the deformations of ρ_1 in the real representation varieties $\mathrm{Hom}(\Gamma, \mathrm{SL}(3; \mathbf{R}))$ and $\mathrm{Hom}(\Gamma, \mathrm{SU}(2,1))$.

THEOREM. *The component X_1 of $X(\Gamma, \mathrm{PGL}(3; \mathbf{C}))$ containing the equivalence class $[\rho_1]$ is isomorphic to the curve $\mathbf{C}^*$. Indeed there is a map*

$$\mathbf{C}^* \longrightarrow \mathrm{Hom}(\Gamma, \mathrm{SL}(3; \mathbf{C}))$$
$$s \mapsto \rho_s$$

such that for each $\chi \in X_1$ there is a unique s such that $\chi = [\rho_s]$. Furthermore the component of $X(\Gamma, \mathrm{PGL}(3; \mathbf{R}))$ containing ρ_1 maps isomorphically onto $\mathbf{R}^* \subset \mathbf{C}^*$; the representations corresponding to the positive reals define convex $\mathbf{RP}^2$-structures on the Schwarz triangle orbifold. The component of $X(\Gamma, \mathrm{PU}(2,1))$ containing $[\rho_1]$ maps isomorphically onto the unit circle $\mathrm{S}^1 \subset \mathbf{C}^*$ and thus the component of $\mathrm{Hom}(\Gamma, \mathrm{PU}(2,1))$ containing ρ_1

Let $e_1, e_2, e_3 \in \mathbf{C}^3$ be the standard basis and let $s \in \mathbf{C}^*$. Consider the matrix

$$\mathbf{B}_s = \begin{bmatrix} 1 & -s^{-1}\cos\frac{2\pi}{p} & -\cos\frac{2\pi}{q} \\ -s\cos\frac{2\pi}{p} & 1 & -\cos\frac{2\pi}{r} \\ -\cos\frac{2\pi}{q} & -\cos\frac{2\pi}{r} & 1 \end{bmatrix}$$

When $s = 1$ this matrix is real, symmetric, and defines a nondegenerate quadratic form of signature (2,1). Define for $i = 1, 2, 3$

$$\rho_s(R_i) = -I + 2\mathbf{B}_s e_i e_i^\dagger$$

(where $A^\dagger$ denotes the transpose of A). A simple calculation shows that ρ_s defines a representation of Γ in $\mathrm{SL}(3;\mathbf{C})$ and that for $s = 1$ the representation ρ_s is conjugate to the standard representation $\rho_1 : \Gamma \to \mathrm{SO}(2,1) \subset \mathrm{SL}(3;\mathbf{C})$ defined above. When s is real, $\mathbf{B}_s$ is real and $\rho_s : \Gamma \longrightarrow \mathrm{SL}(3;\mathbf{R})$ is a real representation. When $|s| = 1$, the matrix $\mathbf{B}_s$ is Hermitian and the corresponding Hermitian form on $\mathbf{C}^3$ is nondegenerate and has signature (2,1); in that case representation $\rho_s : \Gamma \longrightarrow \mathrm{SL}(3;\mathbf{C})$ leaves invariant this Hermitian form and thus is conjugate to a representation $\Gamma \longrightarrow \mathrm{SU}(2,1)$.

For $s > 0$ the groups $\rho_s(\Gamma)$ preserve a convex domain $\Omega_s \subset \mathbf{RP}^2$ and for torsionfree finite-index subgroups $\Gamma' \subset \Gamma$, the quotient $\Omega_s/\rho_s(\Gamma')$ is a compact convex $\mathbf{RP}^2$-manifold. Pictured below are the tesselated domains Ω_s for $p = q = r = 4$ and various choices of s. These convex domains are bounded by C^1 curves which are never C^2 (unless they are conics); the first examples of such domains which cover closed surfaces were discovered by Kac-Vinberg [39]. In particular the family of representations $\rho_s : \Gamma \longrightarrow \mathrm{PGL}(3;\mathbf{R})$ for $s \in \mathbf{R}$ defines a noncompact family of properly discontinous actions on domains in $\mathbf{RP}^2$; one can check easily that the family $[\rho_s]$ does not converge in the quotient $\mathrm{Hom}(\Gamma, \mathrm{PGL}(3;\mathbf{R}))/\mathrm{PGL}(3;\mathbf{R})$.

For $|s| = 1$ we obtain a family of representations of Γ in the projective unitary group $\mathrm{PU}(2,1)$; for s sufficiently close to 1 the representations ρ_s will be discrete embeddings. These representations (restricted to torsionfree finite-index subgroups) correspond to non-compact complex hyperbolic surfaces on nontrivial 2-disc bundles over a closed Riemann surfaces F; alternatively they correspond to locally spherical CR-structures on a nontrivial oriented S^1-bundle over F. It would be extremely interesting to understand for what values of s these representations correspond to geometric structures and for which values of s are these representations discrete embeddings.

The predictions of the asymptotic theory of character varieties developed by Culler-Shalen [9], Morgan-Shalen [58,59] and Morgan [57] and Brumfiel [4] can be explictly verified for these examples. Since Γ is generated by elements of finite order, every action of Γ on an $\mathbf{R}$-tree must have a fixed point. When G has $\mathbf{R}$-rank 1, then the character variety $X(\Gamma, G)$ must be compact — as in the case $G = \mathrm{PU}(2,1)$ above. On the other hand, these examples of convex $\mathbf{RP}^2$-structures show that if G has rank 2, there is no such compactness and indeed there exist unbounded families of representations, which in this case are discrete embeddings.

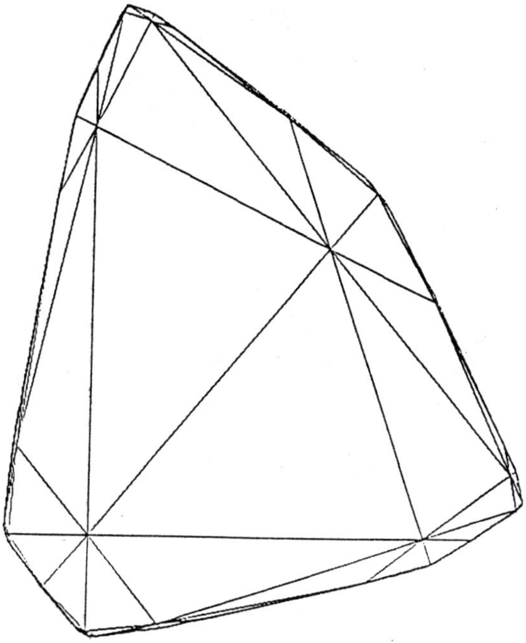

$s = \exp(1.6)$

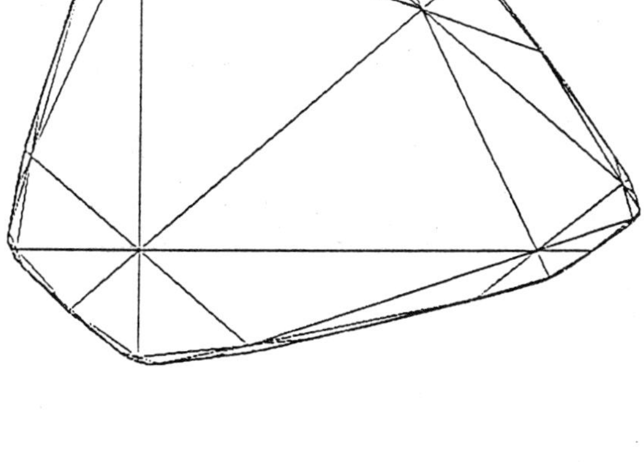

$s = \exp(1.3)$

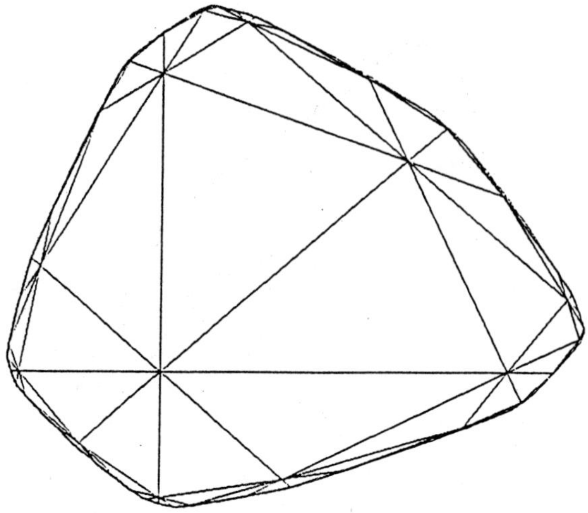

$s = \exp(1.0)$

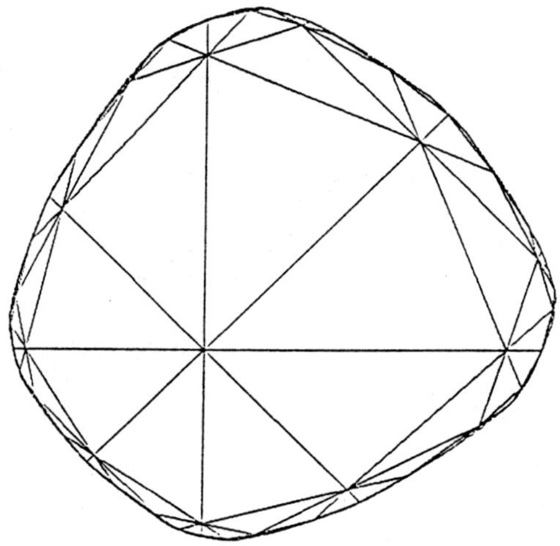

$s = \exp(0.6)$

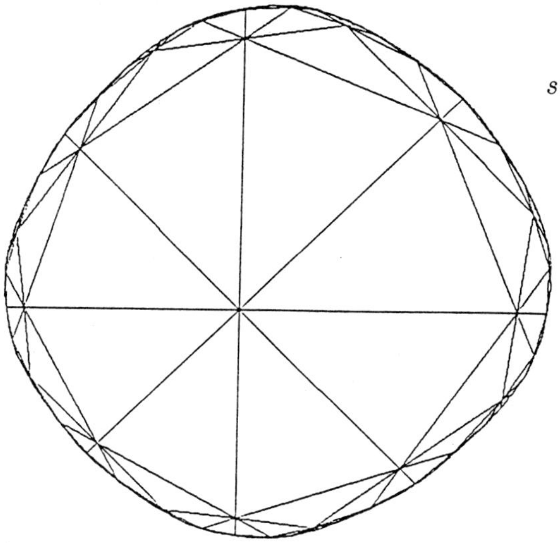

$s = \exp(0.3)$

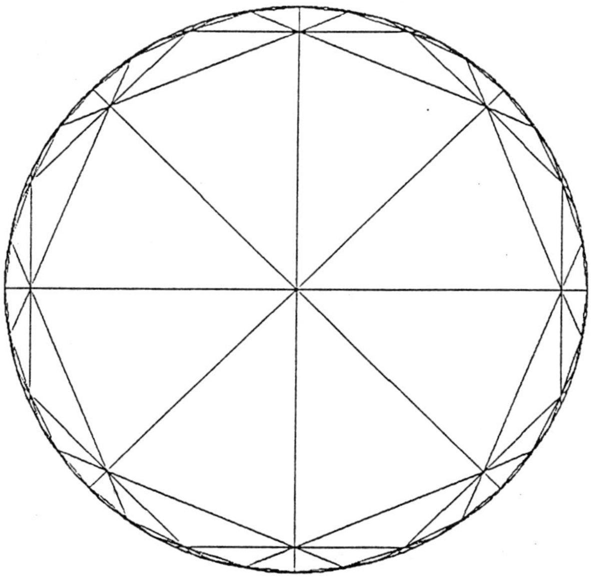

$s = 1$

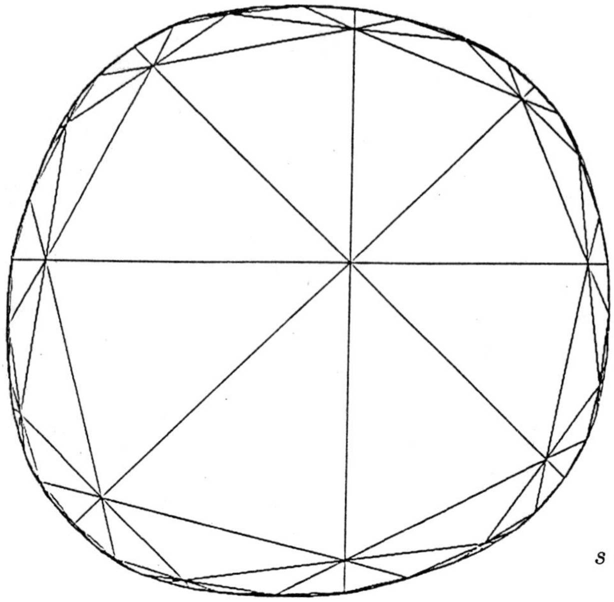

$s = \exp(-0.3)$

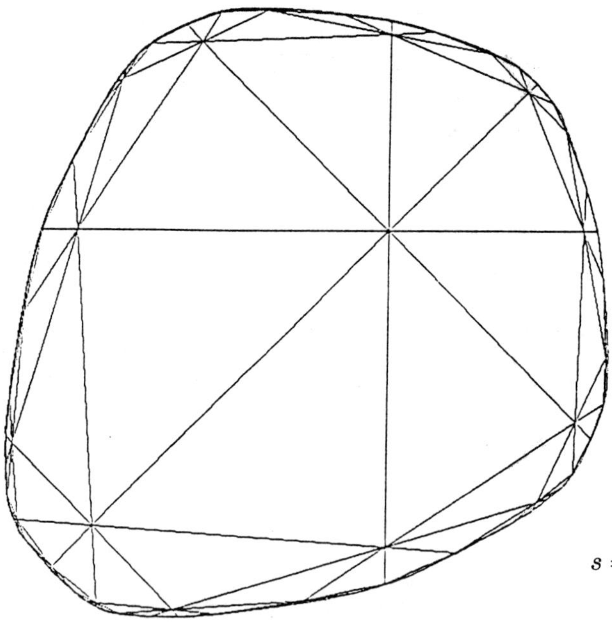

$s = \exp(-0.6)$

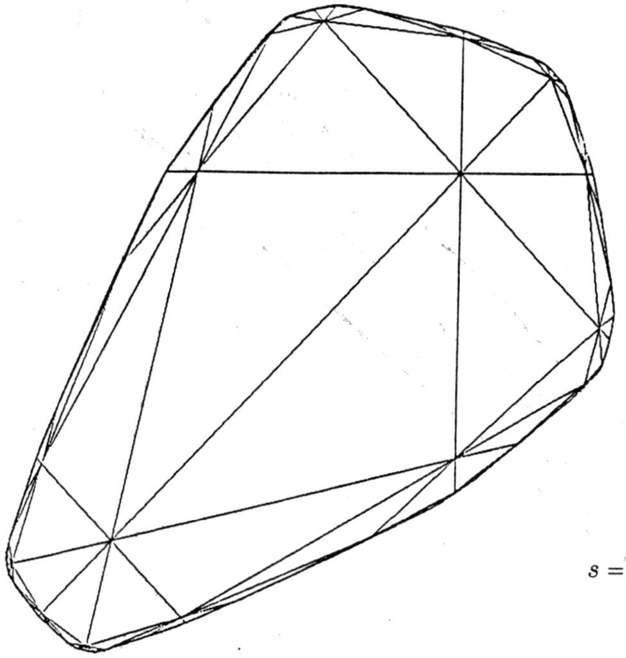

$s = \exp(-1.0)$

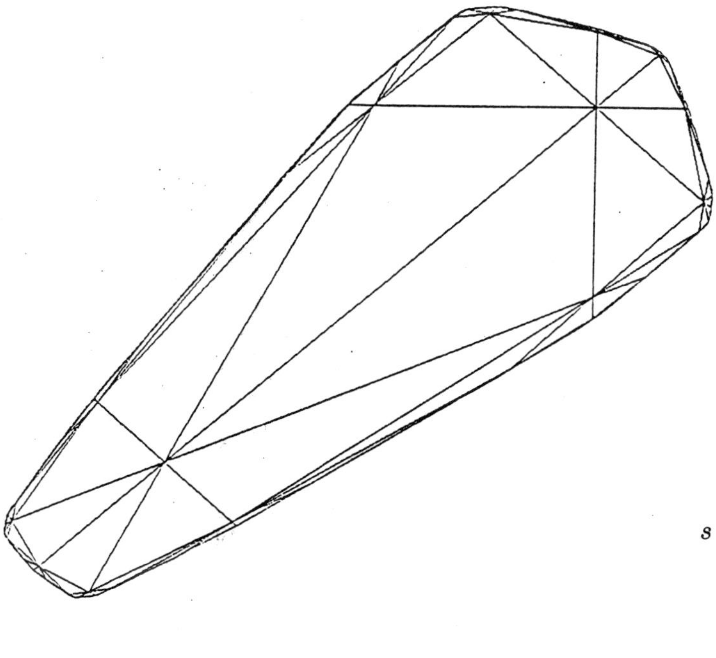

$s = \exp(-1.3)$

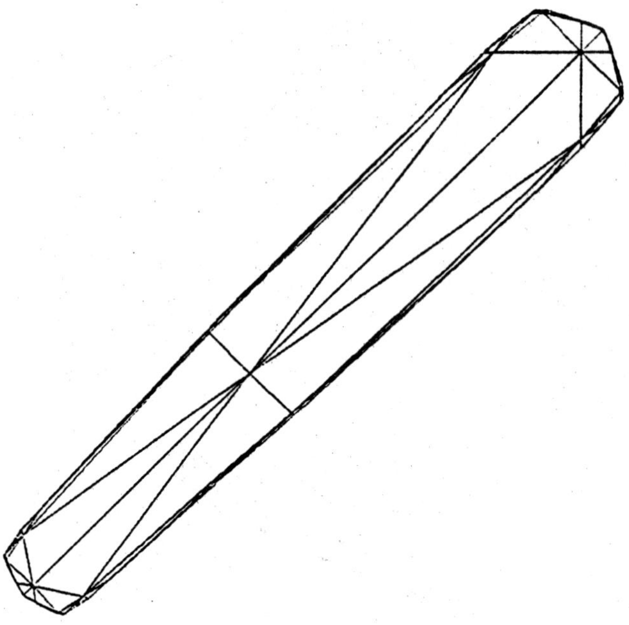

$s = \exp(-1.6)$

REFERENCES

[1] Apanosov, B., *Two kinds of deformation of conformal structure which are distinct from bending,* in "Proceedings of the Thirteenth Rolf Nevanlinna Colloquium and Teichmüller theory Conference, Jensuu, Helsinki August 1987

[2] Benzécri, J.P., *Variétés localement affines,* Sem. Topologie et Géom. Diff., Ch. Ehresmann (1958–60), No. 7 (mai 1959)

[3] Boyom, N.B., *The lifting problem for affine structures in nilpotent Lie groups,* (to appear)

[4] Brumfiel, G., *The real spectrum compactification of Teichmüller space,* these proceedings

[5] Burns, D., and Shnider,S., *Spherical hypersurfaces in complex manifolds,* Inv. Math. **33** (1976), 223-246

[6] Calabi, E., (in preparation)

[7] Canary, R.D., Epstein, D.B.A., and Green, P., *Notes on notes of Thurston,* in "Analytical and geometric aspects of hyperbolic space," London Mathematical Society Lecture Notes **111**, (D.B.A. Epstein, ed.), Cambridge University Press, Cambridge (1987), 3–92

[8] Corlette, K., *Gauge theory and representations of Kähler groups,* these proceedings

[9] Culler, M., and Shalen, P., *Varieties of group representations and splittings of 3-manifolds,* Ann. Math. **117** (1983),109–146

[10] Deligne, P. and Mostow, G.D., *Monodromy of hypergeometric functions and non-lattice integral monodromy,* Publ. Math. d'I.H.E.S. **63** (1986), 5–106

[11] Earle, C.J., *On variation of projective structures,* in "Riemann surfaces and related topics: Proceedings of the 1978 Stony Brook conference," (I. Kra and B. Maskit, eds.) Ann. of Math. Studies **97** (1980), 87–97

[12] Ehlers, F., Neumann, W. and Scherk, J., *Links of surface singularities and CR space forms,* (to appear)

[13] Ehresmann, C., *Sur les espaces localement homogenes,* L'ens. Math. **35** (1936), 317–333

[14] Epstein, D.B.A., and Marden, A., *Convex hulls in hyperbolic space, a theorem of Sullivan, and measured pleated surfaces,* in "Analytical and geometric aspects of hyperbolic space," London Mathematical Society Lecture Notes **111**, (D.B.A. Epstein, ed.), Cambridge University Press, Cambridge (1987), 114–253

[15] Fenchel, W., and Nielsen, J., "Discontinuous groups of non-Euclidean motions," (to appear)

[16] Fricke, R. and Klein, F., *Vorlesungen der Automorphen Funktionen,* Teubner, Leipzig, Vol. I (1897), Vol. II (1912)

[17] Fried, D., *Flat spacetimes,* J.Diff.Geo. **26** (1987), 385–296

[18] Fried, D. and Goldman, W., *Three-dimensional affine crystallographic groups,* Adv. Math. **47** (1983), 1–49

[19] D.Gallo, W. Goldman and R.M. Porter, *Projective structures with monodromy in* $\mathrm{PSL}(2,\mathbf{R})$, (submitted)

[20] Goldman, W., *Discontinuous groups and the Euler class,* doctoral dissertation, University of California, Berkeley (1980)

[21] Goldman, W., *Conformally flat manifolds with nilpotent holonomy and the uniformization problem for 3-manifolds,* Trans. Amer. Math. Soc. **278** (1983), 573–583

[22] Goldman, W., *Nonstandard Lorentz space forms,* J. Diff. Geom. **21** (1985), 301–308

[23] Goldman, W., *Invariant functions on Lie groups and Hamiltonian flows of surface group representations,* Inv. Math. **85** (1986), 263–302

[24] Goldman, W., *Projective structures with Fuchsian holonomy,* J. Diff. Geo. **25** (1987), 297–326

[25] Goldman, W., *Topological components of spaces of representations,* (to appear)

[26] Goldman, W., *Convex real projective structures on compact surfaces,* (to appear)

[27] Goldman, W. and Hirsch, M., *Affine manifolds and orbits of algebraic groups,* Trans. Amer. Math. Soc. **295** (1986), 175–198

[28] Goldman, W. and Kamishima, Y., *The fundamental group of a compact flat Lorentz space form is virtually polycyclic,* J. Diff. Geom. **19** (1984), 233–240

[29] Goldman, W. and Millson, J., *The deformation theory of representations of fundamental groups of compact Kähler manifolds,* Publ. I.H.E.S. (to appear)

[30] Gromov, M., Lawson, H.B. and Thurston, W., *Hyperbolic 4-manifolds and conformally flat 3-manifolds,* (preprint)

[31] Gunning,R.C., *On uniformization of complex manifolds: the role of connections,* Mathematical Notes **22**, Princeton University Press (1978), Princeton, New Jersey

[32] Gunning,R.C., *Affine and projective structures on Riemann surfaces,* in "Riemann surfaces and related topics: Proceedings of the 1978 Stony Brook conference," (I. Kra and B. Maskit, eds.) Ann. of Math. Studies **97** (1980), 225–244

[33] Harvey, W.J., *Spaces of discrete groups,* in "Discrete Groups and Automorphic Functions," Proceedings of a NATO Advanced Study Institute, Academic Press, London—New York—San Francisco, (1977), 295–348

[34] Hejhal, D., *Monodromy groups and linearly polymorphic functions,* Acta Math. **135** (1975) 1-55

[35] Hitchin, N.J. *The self-duality equations on Riemann surfaces,* Proc. Lond. Math. Soc. **55** (1987), 59–126

[36] Hodgson, C., (in preparation)

[37] Hubbard, J.H., *The monodromy of projective structures,* in "Riemann surfaces and related topics: Proceedings of the 1978 Stony Brook conference," (I. Kra and B. Maskit, eds.) Ann. of Math. Studies **97** (1980), 257–275

[38] Johnson, D. and Millson, J.J., *Deformation spaces associated to compact hyperbolic manifolds,* in "Discrete groups in geometry and analysis, Papers in honor of G.D. Mostow on his sixtieth birthday," Progress in Mathematics **67** (1987), 48–106 Birkhäuser, Boston—Basel

[39] Kac, V., and Vinberg, E.B., *Quasi-homogeneous cones,* Math. Notes **1** (1967), 231–235, (translated from Math. Zametki **1** (1967), 347-354)

[40] Kapovich, M. E., *Flat conformal structures on 3-manifolds,* (preprint)

[41] Kobayashi, S. and Ochiai, T., *Holomorphic projective structures on compact complex surfaces,* Math. Ann. **249** (1980), 75–94

[42] Kobayashi, S., *Projectively invariant distances for affine and projective structures,* in "Differential Geometry," Banach Center Publications, **12**, Polish Scientific Publishers, Warsaw, (1984), 127–152

[43] Korouniotis, C., *Deformations of hyperbolic manifolds,* Math. Proc. Cambridge Phil. Soc. **98** (1985), 247–261

[44] Kostant, B. and Sullivan, D., *The Euler characteristic of a compact affine space form is zero,* Bull. A.M.S. **81** (1975)

[45] Kra, I. and Maskit, B., *Remarks on projective structures,* in "Riemann surfaces and related topics: Proceedings of the 1978 Stony Brook conference," (I. Kra and B. Maskit, eds.) Ann. of Math. Studies **97** (1980), 343–358

[46] Kulkarni, R., *The principle of uniformization,* J. Diff. Geo. **13** (1978), 109–138

[47] Kulkarni, R., Lee, K.B., and Raymond, F., *Deformation spaces for Seifert manifolds,* in "Geometry and Topology, Proceedings, University of Maryland 1983–1984", J.

Alexander and J. Harer (eds.), Lecture Notes in Mathematics **1167**, Springer-Verlag Berlin—Heidelberg—New York (1985), 180–216

[48] Kulkarni, R., and Pinkall, U., *Uniformization of geometric structures with applications to conformal geometry* Diff. Geo. Peñiscola, Lecture Notes in Mathematics **1209** (1986) 190–209 Springer-Verlag

[49] Kulkarni, R. and Raymond, F., *3-dimensional Lorentz space-forms and Seifert fiber spaces,* J. Diff. Geo. **21** (1985),231–268

[50] Lok, W.L., *Deformations of locally homogeneous spaces and Kleinian groups,* Doctoral dissertation, Columbia University 1984

[51] Lubotzky, A. and Magid, A., *Varieties of representations of finitely generated groups,*, Memoirs A.M.S. **336** (vol. 5) (1985)

[52] Margulis, G.A., *Free properly discontinuous groups of affine transformations,* Dokl. Akad. Nauk SSSR **272** (1983), 937–940

[53] Margulis, G.A., *Complete affine locally flat manifolds with a free fundamental group,* J. Soviet Math. **134** (1987), 129–134

[54] McOwen, R., *Point singularities and conformal metrics on Riemann surfaces,* (preprint)

[55] Milnor, J.W., *On fundamental groups of complete affinely flat manifolds,* Adv. Math. **25** (1977), 178–187

[56] Millson, J.J., *Deformations of representations of finitely generated groups,* these proceedings

[57] Morgan, J.W., *Group actions on trees and the compactification of the space of classes of SO(n,1)-representations,* Topology **25** (1986), 1–33

[58] Morgan, J.W. and Shalen, P.B., *Valuations, trees and degenerations of hyperbolic structures I,* Ann. Math. **120** (1984), 401–476

[59] Morgan, J.W. and Shalen, P.B., *An introduction to compactifying spaces of hyperbolic structures by actions on trees,* in "Geometry and Topology, Proceedings, University of Maryland 1983–1984", J. Alexander and J. Harer (eds.), Lecture Notes in Mathematics **1167**, Springer-Verlag Berlin—Heidelberg—New York (1985), 228–240

[60] Schoen, R. and Yau, S.T., *Conformally flat manifolds, Kleinian groups and scalar curvature,* (to appear)

[61] Scott, G.P., *The geometries of 3-manifolds,* Bull. Lond. Math. Soc. **15** (1983), 401–487

[62] Shima, H., *Hessian manifolds and convexity,* in "Manifolds and Lie groups, Papers in honor of Yozô Matsushima," Progress in Mathematics **14** (1981), 385–392 Birkhäuser, Boston—Basel

[63] Smillie, J., *Affinely flat manifolds,* Doctoral dissertation, University of Chicago (1977)

[64] Steenrod, N.E., "The topology of fibre bundles," Princeton University Press (1951)

[65] Sullivan, D. and Thurston, W., *Manifolds with canonical coordinates: Some examples,* L'ens. Math. **29** (1983), 15–25

[66] Thurston, W., "The geometry and topology of three–manifolds," (preprint)

[67] Thurston, W., *Hyperbolic geometry, three-dimensional manifolds and Kleinian groups,* Bull. A.M.S. (New Series) **6** (1982), 357–381

[68] Thurston, W., *Shapes of polyhedra,* (preprint)

[69] Toledo, D., *Representations of surface groups in* PU$(n, 1)$ *with maximum characteristic number*, J. Diff. Geo. (to appear)

[70] Troyanov, M., *Les surfaces Riemannienes à singularites coniques*, (to appear)

[71] Wang, M., *The classification of flat compact complete space-forms with metric of signature (2,2)*, (to appear)

[72] Weil, A., *On discrete subgroups of Lie groups*, Ann. Math. **72** (1960), 369–384; **75** (1962), 578–602

[73] Wolpert, S., *The symplectic geometry of deformations of a hyperbolic surface*, Ann. of Math. **117** (1983), 207–234

[74] Yoshida, M. "Fuchsian differential equations," Aspects of Mathematics **11** (1987), Max Planck Institut für Mathematik, Bonn, Friedr. Vieweg & Sohn, Braunschweig—Wiesbaden

[75] Kamishima, Y., *Conformally flat manifolds whose development maps are not surjective*, Trans. A.M.S. **294** (1986), 607 -623

[76] Gusevskii, N. and Kapovich, M., *Conformal structures on three-dimensional manifolds*, Soviet Math. Dokl. **34** (1987)

[77] Kapovich, M., *Some properties of developments of conformal structures on three-dimensional manifolds*, Soviet Math. Dokl. **35** (1987)

Contemporary Mathematics
Volume **74**, 1988

Topological rigidity of developing maps with applications to conformally flat structures

WILLIAM M. GOLDMAN
AND
YOSHINOBU KAMISHIMA

18 April 1988 (revised)

Suppose that M^n is a closed n-manifold with a *geometric structure*, i.e. a structure with distinguished charts modeled on a 1-connected n-dimensional homogeneous space X of a Lie group G whose coordinate changes are restrictions of transformations from G. (We shall refer to a manifold with such a structure as a (G, X)-manifold.) Important examples of these types of structures include all locally homogeneous Riemannian structures as well as conformally flat, affinely flat, and projectively flat structures on manifolds. If G acts properly on X (as in the case when X is a G-homogeneous Riemannian manifold) it is an elementary fact (see Thurston [13]) that a closed manifold M with such a structure must be a quotient X/Γ, where $\Gamma \subset G$ is a discrete subgroup acting properly and freely on X. (A group is said to act *properly discontinuously* if its action is proper when given the discrete topology.) On the other hand, it is now well-known that without the properness assumption on the action of G on X, there are many pathological examples of closed (G, X)-manifolds which do not arise as quotients of domains (or even covering spaces of domains) in X by discrete groups. In general there is only a map

$$\mathbf{dev} : \tilde{M} \to X$$

defined on the universal covering $\tilde{M}$ of M which, although always locally a diffeomorphism, may be very far from *globally* being a diffeomorphism. To the best of our knowledge, Kuiper was the first to inquire whether, for a (G, X)-structure on a closed manifold M, the *developing map* **dev** must necessarily be a covering map onto its image. The first examples where **dev** is not a covering map onto its image were due to Maskit [9] and Hejhal [3] when $G = \mathrm{PSL}(2, \mathbf{C})$ and $X = \mathbf{CP}^1$, and occur even when the holonomy group is a discrete subgroup of G. These examples were rediscovered by Sullivan and Thurston [11] in the more general context of geometric structures, who found further examples in other geometries; independently Smillie [10] found examples of affine structures on the 3-torus whose developing maps are not covering maps onto their image. Further examples are given in Goldman [1], where developing maps for certain special holonomy homomorphisms are classified in terms of intrinsic topological data on M. It is the purpose of this note to show that if the holonomy group is "small enough," pathological developing maps will not arise.

The first author was supported in part by National Science Foundation Grant DMS-8613576 and an Alfred P. Sloan Foundation fellowship.

THEOREM 1. *Let M be a closed (G,X)-manifold. Suppose that $\Gamma \subset G$ is a (G,X)-holonomy group such that there exists a Γ-invariant open subset $\Omega \subset X$ upon which Γ acts properly discontinuously and freely with quotient a closed manifold. Suppose that the Hausdorff dimension of the complement $X - \Omega$ is less than $n-2$. Then the developing map*

$$\mathbf{dev} : \tilde{M} \to X$$

is a homeomorphism from $\tilde{M}$ onto Ω, i.e., M is isomorphic as a (G,X)-manifold to Ω/Γ.

A more general version of Theorem 1 has been obtained independently by Kulkarni-Pinkall [8], where it is shown that if the "domain of properness" of Γ is simply connected then a developing map must be injective. In our applications, the domain of properness must be the set Ω which is necessarily simply connected.

The basic examples which led to the above theorem are the following.

COROLLARY 2. *Let M^n be a closed n-dimensional manifold with a flat conformal structure whose holonomy group lies in a Schottky group $\Gamma \subset \mathrm{Conf}(S^n)$. Suppose that $n > 2$. Then M is conformally equivalent to a quotient Ω_1/Γ_1 where $\Gamma_1 \subset \Gamma$ is a subgroup and $\Omega_1 \subset S^n$ is the domain of discontinuity of Γ_1.*

Recall that a Kleinian group $\Gamma \subset \mathrm{Conf}(S^n)$ is said to be *convex cocompact* if the convex core of the corresponding hyperbolic $n+1$-dimensional manifold is compact. For example, Γ might be the fundamental group of a closed hyperbolic manifold of dimension $k < n$, acting on the sphere at infinity of hyperbolic $n+1$-space. In this case the limit set is a geometric $(k-1)$-sphere and the domain of discontinuity may be naturally identified with the product $S^{n-k} \times D^k$. The corresponding quotient manifold is the product $S^{n-k} \times D^k/\Gamma$.

COROLLARY 3. *Let M^n be a closed n-dimensional manifold with a flat conformal structure whose holonomy group Γ is convex cocompact and whose limit set has Hausdorff dimension $< n-2$. Then the developing map defines a homeomorphism of $\tilde{M}$ with the domain of discontinuity Ω and M is conformally equivalent to Ω/Γ.*

Thus, if Γ is the fundamental group of a closed hyperbolic k-manifold and $k < n-1$, then a closed conformally flat manifold with holonomy group Γ must be Ω/Γ. In Goldman [1] examples are given showing the conclusion of the above theorem is false when $k = n-1$.

One motivation for the above theorem is the recent work of Schoen-Yau [12] concerning conformally flat manifolds with positive scalar curvature. In [12] it is shown that if M is a closed conformally flat n-manifold of positive scalar curvature, then the holonomy group is discrete with limit set Λ having Hausdorff dimension $\leq \frac{n-2}{2}$ and the developing map is a homeomorphism of $\tilde{M}$ onto $S^n - \Lambda$. This paper gives a topological argument showing that the injectivity of the developing map actually follows from the results on Hausdorff dimension.

There are many results concerning tameness of developing maps for $\mathbf{CP}^1$-structures and flat conformal structures when the developing map is assumed not to be surjective, see Gunning [2], Kra [5], Kulkarni-Pinkall [8]. Here we obtain similar results with a different type of assumption: the holonomy group has an extremely "sparse" limit set.

Our use of Hausdorff dimension is the following "general position" property: If X is an n-dimensional manifold and $\Lambda \subset X$ has Hausdorff dimension δ, then the relative homotopy group $\pi_k(X, X - \Lambda) = 0$ for $k < n - \delta$. In particular if X is 1-connected and $\Lambda \subset X$ has Hausdorff dimension $< n - 2$, then the complement $X - \Lambda$ is also 1-connected.

We are grateful to Bob Edwards for an extremely helpful conversation concerning Hausdorff dimension and to Ser P. Tan and Michael Kapovich for pointing out an error in the original version.

For the proof of the theorem, we shall employ the following. Fix a universal covering $\mathbf{p} : \tilde{M} \to M$ with fundamental group π, and a developing map $\mathbf{dev} : \tilde{M} \to X$ with a compatible holonomy homomorphism $h : \pi \to G$. The holonomy group will be denoted $\Gamma = h(\pi)$. For definitions and further discussion of these notions the reader is referred to [1], [6], [7], or [11].

Our basic tool is the following:

LEMMA 4. *Let M be a closed (G, X)-manifold with holonomy group $\Gamma \subset G$ such that there exists a Γ-invariant open subset $\Omega \subset X$ with a Γ-invariant complete Riemannian metric g_Ω. Then $\mathbf{dev}^* g_\Omega$ is a complete metric on $\mathbf{dev}^{-1}(\Omega) \subset M$.*

PROOF: Let $\{\gamma(t) \mid 0 \leq t < t_0\}$ be a geodesic on $\mathbf{dev}^{-1}\Omega \subset \tilde{M}$ parametrized by arc length; it suffices to show that $\gamma(t)$ converges in $\mathbf{dev}^{-1}\Omega$ as $t \longrightarrow t_0$. Let $\{t_n\}$ be a sequence in the interval $[0, t_0)$ converging to t_0; since M is compact the sequence $\mathbf{p}\gamma(t_n)$ has an accumulation point as $n \longrightarrow \infty$. By passing to a subsequence we may assume that $\mathbf{p}\gamma(t_n)$ converges to $m_0 \in M$. Choose a neighborhood U of m_0 in M such that $\mathbf{p}^{-1}(U)$ is a disjoint union of components permuted freely by $\pi_1(M)$, each one of which is mapped diffeomorphically under $\mathbf{p} : \tilde{M} \longrightarrow M$. For sufficiently small $\epsilon > 0$, we may assume that if $x, y \in \mathbf{p}^{-1}(U) \cap \mathbf{dev}^{-1}(\Omega)$ satisfy $\rho(x, y) < \epsilon$, then x and y lie in the same component of $\mathbf{p}^{-1}(U)$ where ρ is the distance function. Since $\mathbf{dev} : \mathbf{dev}^{-1}(\Omega) \longrightarrow \Omega$ is a local isometry,

$$\rho(\mathbf{dev}\gamma(t_m), \mathbf{dev}\gamma(t_n)) \leq \rho(\gamma(t_m), \gamma(t_n)) \leq |t_m - t_n|$$

whence the sequence $\{\mathbf{dev}\gamma(t_n)\}$ is Cauchy in Ω and hence converges to a point $\omega_0 \in \Omega$. Furthermore $\{\gamma(t_n)\}$ is Cauchy in $\mathbf{dev}^{-1}(\Omega)$ and hence there exists N such that for $n > N$ all the $\gamma(t_n)$ lie in a single component U_0 of $\mathbf{p}^{-1}(U)$. It follows that $\gamma(t_n)$ converges to the unique point $\tilde{m_0}$ in $\mathbf{p}^{-1}(m_0) \cap U_0$. By the continuity of the developing map, $\mathbf{dev}(\tilde{m_0}) = \lim_{n \to \infty} \mathbf{dev}(\gamma(t_n)) = \omega_0$ and thus $\tilde{m_0} \in \mathbf{dev}^{-1}(\Omega)$. It follows easily that $\lim_{t \to t_0} \gamma(t) = \tilde{m_0}$ as claimed. ∎

PROOF OF THEOREM: Choose a Riemannian metric on the quotient manifold Ω/Γ and lift it to a Γ-invariant Riemannian metric g_Ω on Ω. Since Ω/Γ is compact, the original metric on Ω/Γ is complete, and hence so is g_Ω. We find Riemannian metrics $g_{\tilde{M}}$ on $\mathbf{dev}^{-1}(\Omega) \subset \tilde{M}$ and g_M on $\mathbf{p}\mathbf{dev}^{-1}(\Omega)$ such that

$$\mathbf{p}^* g_\Omega = g_{\tilde{M}} = \mathbf{dev}^* g_{\tilde{M}}$$

and by Lemma 4, both of these metrics are complete. Furthermore since the developing map $\mathbf{dev} : \mathbf{dev}^{-1}(\Omega) \to \Omega$ is a local isometry from the complete Riemannian manifold $\mathbf{dev}^{-1}(\Omega)$, we conclude that

$$\mathbf{dev} : \mathbf{dev}^{-1}(\Omega) \to \Omega$$

is a covering map.

LEMMA 5. $\mathbf{dev}^{-1}\Omega \subset \tilde{M}$ *is connected.*

PROOF OF LEMMA 5: Since $\Lambda \subset X$ has Hausdorff dimension $\delta < n-2$, the corresponding subset $M_\Lambda \subset M$ has Hausdorff dimension $\leq \delta < n-2$. Since M_Λ is compact, it can be covered by finitely many coordinate patches with compact closure which the coordinate charts map M_Λ homeomorphically into Λ. These coordinate charts are diffeomorphisms with bounded Lipschitz constant, and hence for every $\epsilon > \delta$ the ϵ-dimensional Hausdorff measure of M_Λ is zero.

Thus the Hausdorff dimension of M_Λ is $< n-2$ and the relative homotopy group $\pi_1(M, M_\Omega) = 0$. It follows from the exact homotopy sequence

$$\pi_1(M_\Omega) \longrightarrow \pi_1(M) \longrightarrow \pi_1(M, M_\Omega) \longrightarrow \pi_0(M_\Omega) \longrightarrow \pi_0(M)$$

that M_Ω is connected. Furthermore since $\pi_1(M_\Omega) \longrightarrow \pi_1(M)$ is surjective it follows that $\mathbf{p}^{-1}(M_\Omega) = \mathbf{dev}^{-1}(\Omega)$ is connected. ∎

Since Ω is simply connected, it follows that $\mathbf{dev} : \mathbf{dev}^{-1}(\Omega) \to \Omega$ is a homeomorphism, which is clearly equivariant with respect to the holonomy homomorphism $h : \pi \to \Gamma$. It follows that the induced map

$$M_\Omega = (\mathbf{dev}^{-1}(\Omega))/\pi \to \Omega/\Gamma$$

is a homeomorphism. Since Ω/Γ is compact, it follows that M_Ω is compact and hence closed in M. But $M_\Omega \subset M$ is open so M being connected implies $M_\Omega = M$. Thus $\mathbf{dev}^{-1}(\Omega) = \tilde{M}$ proving that

$$\mathbf{dev} : \tilde{M} \to \Omega$$

is a diffeomorphism. ∎

REFERENCES

[1] Goldman, W. *Projective structures with Fuchsian holonomy,* J. Diff. Geo. **25** (1987), 297–326

[2] Gunning, R., *Special coordinate coverings of Riemann surfaces,* Math. Ann. **170** (1967) 67-86

[3] Hejhal, D., *Monodromy groups and linearly polymorphic functions,* Acta Math. **135** (1975) 1-55

[4] Kamishima, Y. *Conformally flat manifolds whose development maps are not surjective,* Trans. A.M.S. **294** (1986) 607-623

[5] Kra, I., *Deformations of Fuchsian groups,* Duke Math. J. **36** (1969) 537-546

[6] Kulkarni, R., *The principle of uniformization,* J. Diff. Geo. **13** (1978) 109-138

[7] Kulkarni, R., *Geometric structures on manifolds and Kleinian groups,* (to appear)
[8] Kulkarni, R., and Pinkall, U., *Uniformization of geometric structures with applications to conformal geometry,* Diff. Geo. Peñiscola, Lecture Notes in Mathematics **1209** (1986) 190-209 Springer-Verlag
[9] Maskit, B., *On a class of Kleinian groups,* Ann. Acad. Sci. Fenn., A. I. **442** (1969) 1-8
[10] Smillie, J., *Affinely flat manifolds,* Doctoral dissertation, University of Chicago 1977
[11] Sullivan, D. and Thurston, W., *Manifolds with canonical coordinates: Some examples,* L'ens. Math. **29** (1983) 15-25
[12] Schoen, R. and Yau, S.T., *Conformally flat manifolds, Kleinian groups and scalar curvature,* (to appear)
[13] Thurston, W., "The geometry and topology of three-manifolds," (in preparation)
[14] Kapovich, M., *Flat conformal structures on 3-manifolds,* (to appear)

William M. Goldman, Department of Mathematics, University of Maryland, College Park, Maryland, USA 20742
Yoshinobu Kamishima, Department of Mathematics, Hokkaido University, Sapporo 060, JAPAN

Contemporary Mathematics
Volume **74**, 1988

MODULAR GROUPS AND REPRESENTATION SPACES

W.J. Harvey

Abstract

A brief and very selective survey is presented of recent work on modular groups, emphasizing geometric aspects. Some open questions are posed on realisation of geometric structures within the framework of Teichmüller Theory.

1. INTRODUCTION. This article considers the quest for geometric structure within the representation spaces associated with surface fundamental groups. It is well understood that the representation variety $R(\pi_1(X), PSL_2(\mathbb{R}))$ of a surface X (usually compact) has a connected component R(X) consisting of discrete faithful representations. Taking the quotient by conjugation with elements of $PSL_2(\mathbb{R})$ produces the Teichmüller space T(X) - also known as the character variety of $\pi_1(X)$ in some contexts. On this space the group $\Gamma(X)=\pi_0(\mathrm{Diff}\ X)$ of mapping classes operates via the Nielsen isomorphism, which identifies the outer automorphism group of $\pi_1(X)$ with $\Gamma(X)$. We refer to $\Gamma(X)$ as the modular group of X since when X is a torus $\Gamma(X)$ is $PSL_2(\mathbb{Z})$. Our purpose is to pursue aspects of the study of T(X) which have particular relevance to the modular group action, in the hope that this emphasis on $\Gamma(X)$ will encapsulate much of the interesting geometry residing in the space.

In section 2 we describe briefly the combinatorial framework and cohomological results of various authors. Here there is close analogy with other work in homology of discrete groups and with combinatorial group theory. In particular a closely parallel study of the outer automorphism group of a free group has been made along these lines by M. Culler and K. Vogtmann [4], as

---

Mathematics Subject Classification:- Primary 14H15, 20F32, 32G15, Secondary 20F36,57S30.
Key words:- Modular groups, Riemann surfaces, moduli spaces, geometric structures.

reported at these proceedings.

Later sections discuss a selection of geometric results on these moduli spaces. For lack of time, and also because excellent treatments occur in the literature, we omit discussion of the symplectic structure arising from the Weil-Petersson hermitian metric on T(X). The reader may consult S. Wolpert's article [17] for a detailed construction within hyperbolic plane geometry or the unified treatment by W.M. Goldman [6] which describes the symplectic structures intrinsic to all reasonable representation spaces of $\pi_1(X)$. Our aim has been to present a selection of facts and open questions which appear part of a developing pattern of geometry within a space which presents a bewildering range of different faces to the mathematical world, and which stands currently in a position of prominence amongst the growing tide of interactions with theoretical physics.

ACKNOWLEDGEMENTS

I am grateful to various institutes and universities for support during the academic year 1985-86, including the I.H.E.S. in Bures-sur-Yvette, the University of Maryland (College Park) and U.C.S.D. I am also grateful to the long-suffering British Tax-payer for a travel grant, administered through the Royal Society, to attend this conference, and to the N.S.F. for support during the conference.

2. SIMPLICIAL STRUCTURES ASSOCIATED WITH TEICHMÜLLER SPACES. Let X be a fixed surface of finite type, with g handles and n+k boundary components. There are several ways in which one can define simplicial decompositions for the Teichmüller spaces $T^k_{g,n}$ = T(X), representing marked Riemann surfaces of type g, n, k (n boundary points, k boundary curves). When $n+k > 0$, there are triangulations for which the operation of the modular group is simplicial, and this has the effect of rendering the group accessible to standard algebraic methods of study; thus the homological properties and presentation by generators and relations can be worked out with reference to this combinatorial pattern. We discuss this briefly now, partly in the hope that an outline may suggest how to proceed in a general analysis of automorphism groups for geometric structures on other manifolds. Detailed

accounts may be found in excellent comprehensive survey articles [8,12].

The simplicial complexes $\mathfrak{A} = \mathfrak{A}(X)$ constructed by Harer involve curve systems on the reference surface. These curves begin and end at a designated point (or at any of a given subset of the designated n+k points in the general case, which we suppress for simplicity). A maximal set of curves determines a simplex of the subdivision; barycentric co-ordinates specify the weighted lengths of a dual set of curves for reconstructing a hyperbolic structure (marked), or conformal structure if one employs complex analytic techniques. Allowing one curve to have zero weight corresponds to a frontier point abutting another simplex; this identification of common faces in the tesselation of T(X) can be understood in terms of a Whitehead move on the dual graph.

In order to gain a more complete picture of this we need not just T(X) but also a bordification of it, which is a partial compactification

$$T(X) \cup \partial T(X) = \hat{T}(X)$$

having the property that the action of the modular group $\Gamma = \Gamma(X)$ extends continuously from T to $\hat{T}$ and produces a (Hausdorff) compact quotient space . The ideal boundary ∂T comprises a disjoint union of connected pieces, each parametrising the singular surfaces corresponding to a degeneration of the hyperbolic structure by shrinking one or more homotopy classes of loop in X to a point. By use of Fenchel-Nielsen co-ordinates (see for instance Abikoff [1] and Wolpert [17],[18]) this process can be put into precise analytic form; the group Γ operates on ∂T by permuting the components of it, with the stabiliser of a given component acting as an appropriately extended modular group for that boundary subspace. If one loop ℓ in X is shrunk to a point, the corresponding degenerate surface has a single node; when ℓ is homologically non-trivial, the stabiliser in Γ is an extension, by the infinite cyclic group $\langle\tau_\ell\rangle$ generated by the Dehn twist about ℓ, of the modular group $\Gamma(X_\ell)$ for the surface X_ℓ obtained by splitting X open along ℓ. In general, there will be a finite set $\Lambda = (\ell_1,\ldots,\ell_n)$ of homotopically distinct loops to become node points; the stabilising subgroup of Γ will be the subgroup normalising the free abelian rank-n group generated by the twists τ_j about the ℓ_j, expressible simply in terms of the modular groups of the components of X_Λ, the complementary surface.

It was pointed out in [11] that this boundary action of Γ is homotopically equivalent to a purely simplicial action of Γ on the geometric realisation $|\mathfrak{S}(X)|$ of the poset $\mathfrak{S}(X)$ of simple loop systems Λ in X. The proof of this fact involves the procedures developed by Borel and Serre [3] in their fundamental work on symmetric spaces with arithmetic group actions; it is possible, using Fenchel-Nielsen hyperbolic length and twist parameters, to set up a collection of real analytic actions by the group $\mathbb{R}_+$ of positive real numbers on T(X) which perform the task of pushing the space in the directions which lead to ∂T. Provided that one operates sufficiently far out; that is where some geodesic loop has small length, these flows may be considered independent of each other in the sense that either they commute because they are operating on co-ordinates for disjoint loops or they do not stabilise the same end (boundary component) of ∂T. Each flow preserves the twist parameters and so there is a collection of vestigial co-ordinates at infinity, on which Dehn twists act by translation. For a maximal collection of loops $\Lambda = \{\ell_1,\ldots,\ell_{3g-3+n}\}$, the surface X_Λ with nodes has each component a 3-holed sphere, so that there are no moduli apart from the vestigial twist space $\mathbb{R}^{3g-3+n}$ endowed with the stabilising action of a Euclidean discrete group G_Λ which is of the form

$$1 \longrightarrow \bigoplus_j \mathbb{Z}\tau_j \longrightarrow G_\Lambda \longrightarrow A \longrightarrow 1,$$

where A is the finite group of automorphisms of the surface X_Λ.

Now Harer has shown how to incorporate this picture of the boundary into the simplicial complex $\mathfrak{A}$; there is a subcomplex $\mathfrak{A}_\infty$ such that T(X) is homeomorphic to $|\mathfrak{A} - \mathfrak{A}_\infty|$. Furthermore $\mathfrak{A}_\infty$ is Γ-equivariantly homotopy equivalent to $\mathfrak{S}$ and, since it is possible to show that this has the homotopy type of a wedge of spheres with dimension d = 2g-3+n, a standard application of the results of Borel and Serre shows that the modular group $\Gamma(X)$ is a <u>virtual duality group</u> in the sense of Bieri-Eckmann with virtual cohomology dimension 4g-4+n; here we are assuming that $n \geq 1$ and $g \geq 1$. It follows easily that $\Gamma_{g,0}$ is of the same ilk, with v.c.d. 4g-5.

There is much more to be gleaned here about the algebraic topology of the modular group. Notably, it is possible to introduce a stabilisation process by adding on handles to the reference surface X_g and comparing the resulting complexes and cohomologies. A stable modular group $\Gamma_\infty = \lim_{g\to\infty} \Gamma(X^1_{g,0})$ results, which has much in common with the stabilised groups $SL(\mathbb{Z})$, $Sp(\mathbb{Z})$. An important result of Harer's is that the homology groups stabilise under the addition of handles:

$$H_k(\Gamma^1_{g,n}) \text{ is independent of } g,n \text{ if } g \geq 3k+1.$$

Thus we have a powerful algebraic tool for understanding the structure of $\Gamma(X)$.

Results on the cohomology of the modular groups are important in algebraic geometry because they are reflected, since $\Gamma(X_g)$ operates properly discontinuously on $T(X_g)$ with all fixed subsets of the action having large codimension, in the algebraic topology of the quotient moduli space of curves M_g. In fact the bordification process above corresponds under a suitably defined exponential map, to the inclusion of M_g in the stable compactification $\hat{M}_g$ comprising Riemann surface structures with nodes, or stable curves, of genus g. Although we shall not discuss the complex analytic geometry of $T(X)$ or M_g here, it is possible to present at least the almost complex structure within the context of hyperbolic structures on X and its real representation space; see for this the article [19] of Scott Wolpert. The complement $\hat{M}_g - M_g$ turns out to be a divisor in the projective variety $\hat{M}_g$, whose irreducible components $D_0, D_1, \ldots, D_k$, $k = [g/2]$, correspond bijectively to Γ-orbits of boundary components in $\partial T(X_g)$. Mumford's article [15] makes this the basis for initiating a detailed study of the algebraic cohomology of $\hat{M}_g$ via its Chow ring. See also [7],[20] for discussion of the homology of M_g. At the same time, however, we should note that there are subtle facts lying concealed in the algebraic geometry of M_g that are inaccessible to our present procedures. For instance, there are no compact complex subvarieties of M_g with $\mathbb{C}$-dimension greater than g-2, by a theorem of S. Diaz. This needs to be reconciled with the fact that there are readily constructible compact submanifolds with dimension 4g-5; it would be interesting to study this discrepancy between the topological and analytic/algebraic K-theories of M_g more closely. Other sources to consult on this work are cited in the bibliography [14],[15].

We close this section with a comment on the combinatorial group theory. As the modular group acts on the complexes $\mathfrak{A}, \mathfrak{T}$

simplicially with quotient a finite complex, there results a simplicial presentation of the various $\Gamma^k_{g,n}$. I intend in a future publication to describe a particularly simple version of this which takes account of the stabilisation process; the infinite genus modular group has in many ways a more straightforward theory than the finite genus groups.

3. GEOMETRY OF THE TEICHMÜLLER SPACES. It is known that Teichmüller space T(X) with the Teichmüller metric is not a symmetric space in the sense of differential geometry; it has no isometries apart from the modular group when X has negative Euler characteristic. This might of course mean that the metric is wrong; however a theorem of Ivanov ([8],[12]) shows that there is no faithful representation of $\Gamma(X)$ to an arithmetic subgroup of any linear algebraic group, and one can extend the reasoning to any cofinite volume lattice subgroup by results of Margulis. The main point in the argument is an analysis of the dimension, which quickly outlaws any rank 1 symmetric space structure for T(X), since then any cusp (which must occur since T/Γ is not compact) would have an associated abelian stabiliser in Γ of rank (dim T)-1 contradicting the fact that the maximal rank of an abelian group in Γ is $3g-3+n$. Higher rank symmetry is ruled out for T(X) by certain elementary properties of the group Γ of an algebraic nature which violate the properties of lattice subgroups in the corresponding Lie automorphism group.

This leaves open several directions in which we can seek geometric structures within T(X) and Γ. First of all, we can restrict attention to subspaces and ask for a geometric structure on a submanifold $Y \subseteq T(X)$ such that the stability group of Y in Γ is commensurable with a discrete subgroup of a transformation group which acts transitively on Y. For instance, here is an intriguing problem which I have been unable to resolve. Let $\gamma \in \Gamma$ be a pseudo-Anosov element, which means that the diffeomorphism γ leaves fixed no closed loop in X. See for instance [1] for a discussion of this notion. Then γ has an axis $L_\gamma \subseteq T(X)$, which is a geodesic line preserved by γ and on which γ acts as a translation. If one forms the fibre space over T(X) consisting of hyperbolic planes obtained by reduction of the structure group $G = PSL_2(\mathbb{R})$ for the trivial bundle R(X) to the subgroup $SO(2,\mathbb{R})$, then the resulting total space F is isomorphic to $T(\hat{X})$ where $\hat{X}$

denotes the surface X punctured at a point. Passing to the part F_γ of F lying over the line L_γ produces a 3-dimensional ball on which the group $\langle\gamma\rangle$ acts by isometry, forming with the standard fibre action of $\pi_1(X)$ an extension $G\gamma$, with

$$1 \longrightarrow \pi_1(X) \longrightarrow G\gamma \longrightarrow \langle\gamma\rangle \longrightarrow 1;$$

here G_γ is the fundamental group of the 3-manifold which is the mapping torus of the homeomorphism γ. Now a famous result of W. Thurston states that the space F_γ carries a structure of hyperbolic manifold H/G, with $G \approx G_\gamma$.

PROBLEM 1. Is there a way to construct the hyperbolic structure on $F\gamma$ which derives from a standard structure on $\hat{T}(\dot{X})$?

Note that the restriction of the Weil-Petersson metric is complete on $F\gamma$, although it is incomplete on $T(\dot{X})$ itself. Recently a fresh definition of metric on T-spaces has been introduced by Thurston, and this turns out from Scott Wolpert's work [19] to coincide with the W-P distance, thereby giving it a description which fits intrinsically into the structure of $R(\pi_1(X), PSL_2(\mathbb{R}))$. In any case, the group $G\gamma$ defined above is a subgroup of the modular group $\Gamma(\dot{X})$ and so preserves all of the standard metrics on $T(\dot{X})$. This same viewpoint on the Weil-Petersson metric emerges from work of F. Bonahon [The geometry of Teichmüller space via geodesic currents, preprint from U.S.C. (1986)] where a picture of T_g analogous to the hyperboloid model of hyperbolic space is developed.

A similar question arises in connection with Kleinian groups which represent non-degenerate limits (in the sense of pointwise convergence) of quasi-fuchsian groups isomorphic to $\pi_1(X)$. Such groups have been studied by Maskit [13] by considering pairs of loop systems Λ_1, Λ_2 which may be shrunk to nodes simultaneously at opposite ends of a quasi-fuchsian representation family.

PROBLEM 2. Characterise the Kleinian groups in the closure of $R_{q.f}(\pi_1(X), PSL_2(\mathbb{C})$ by using the geometry of $T(\dot{X})$.

Much of the interest in T(X) and $\Gamma(X)$ stems from the link with moduli of algebraic curves, and it is always fascinating to discover inter-relations between the algebraic and analytic theories, particularly results which are relatively elementary in one aspect but inaccessible in the other. For instance one can recognise moduli spaces of curves with an automorphism (that is a holomorphic self-mapping) or more generally a given group H of symmetries by considering inclusions of Teichmüller representation spaces arising from an inclusion of Fuchsian groups

$$\pi_1(X) \hookrightarrow G$$

with quotient group H. The group G corresponds to the quotient surface X/H and any deformation of it determines a deformation of X. Now in certain cases there is an explicit description available for the corresponding moduli space in terms of a parametrisation by special values of theta functions. This is explained for the hyperelliptic involution in Mumford's book (Tata Lectures on Theta II). Recently a parallel description has been given for the class of surfaces with an automorphism of prime order [7] where the quotient surface is a sphere. In these cases, there is a representation into the transformation group T(X), Γ(X) of a braid group operating on the configuration space of k-tuples of points on the sphere, with k the number of fixed points of the automorphism. It seems to be an interesting question to determine the contribution to the cohomology of Γ(X) arising from that of these braid groups. Also it appears likely that the quotient moduli space of surfaces with such a symmetry is often a rational variety, as is the case for the hyperelliptic moduli space.

Along these lines we may pose a further problem of a general nature about the geometrisation of surface moduli spaces.

PROBLEM 3. Determine all homogeneous spaces Y with subgroups $G \subset \mathrm{Aut}\, Y$ such that a mapping $f\colon Y \longrightarrow T(X)$ exists with

$$f \circ g(y) = \rho(g) f(y)$$

for some representation $\rho\colon G \longrightarrow \Gamma(X)$.

Of course the constituent pieces should be non-trivial and one may need to impose further restrictions to make the answers interesting. Also it is probably appropriate to extend the framework to encompass the possibility of an infinitesimal linear model of the geometric structure at a cusp boundary point of T(X). A recent article of M. Yoshida provides such a pattern in genus 2 in terms of a hyperbolic Kac-Moody algebra [21].

4. INFINITE DIMENSIONAL GEOMETRY. The Krichever construction (see for instance [2]) defines a map from a certain bundle over Teichmüller space into the Grassmann manifold of a Hilbert space; this depends on a reformulation by Segal and Wilson [16] of results by the Japanese school (notably M. Sato) on soliton solutions to the K.-P. and related equations. There is a very beautiful infinite dimensional geometry intrinsic to the Grassmannian which interacts closely with the geometry of the moduli space and with that of the moduli spaces of vector bundles too. For instance, in this way one can give a unified description for <u>all</u> Riemann surfaces of the Hodge bundle, which is the fundamental line bundle λ_1 obtained by taking the (rank g) vector bundle of holomorphic 1-forms on a surface and forming the gth exterior power; the important factor here is that it comes as a holomorphic bundle over the entire family of all Riemann surfaces. When restricted to each moduli space $T(X_g)$, λ_1 determines by its first Chern class a generator of the cohomology group $H^2(\Gamma_g)$ for $g \geqslant 3$.

As a final comment, it now appears clear that this opens up a new direction for Teichmüller theory, bringing in the action of the Virasoro algebra of holomorphic vector fields on the circle S^1 as a vital new ingredient to deformation theory of Riemann surfaces. Briefly, if we consider S^1 as the boundary of a normalised co-ordinate disc U_∞ at the reference point P_∞ on a marked surface X, then introducing a vector field on S^1 determines a way to deform the Riemann surface because the vector field defines a 1-cocycle on X with respect to the covering of X by U_0, U_1 for some pair of discs U_0, $X-U_1$ with $X-U_1 \subset U_\infty \subset U_0$ (so that $U_0 \cap U_1$ is an annular neighbourhood of S^1) - the space $Z^1(X, \mathcal{O}(K^{-1}))$ of such cocycles is, by Kodaira-Spencer theory, the tangent space to moduli space, recognisable to Teichmüller theorists via Serre duality as the quadratic differential forms on X. Furthermore, this procedure lends itself to constructions on suitable universal Teichmüller spaces, including infinite genus surfaces; the model presented at this conference by Bob Penner is particularly intriguing.

1. W. Abikoff, Real analytic Theory of Teichmüller space, L. Notes in Math. 820, Springer Verlag (1980).

2. L. Alvarez-Gaumé, C. Gomez and C. Reina, New methods in string theory, C.E.R.N. preprint TH4775/1987.

3. A. Borel and J.P. Serre, Corners and arithmetic groups, Comm. Math. Helv. 48 (1973) 436-491.

4. M. Culler and K. Vogtmann, Moduli of graphs and automorphisms of free groups, Inv. Math. 84 (1986), 90-120.

5. S. Giddings and S. Wolpert, A triangulation of moduli space from light cone string theory, Comm. Math. Phys. 1987 to appear.

6. W.M. Goldman, The symplectic nature of fundamental groups of surfaces, Adv. in Math. 54 (1984) 200-225.

7. G. Gonzalez-Diez, Studies in the complex analytic geometry of moduli space for curves. Ph.D. Thesis at King's College London, 1987.

8. J. Harer, The Cohomology of the moduli space of curves, notes of C.I.M.E. Lectures (1985).

9. J. Harer, Stability of the homology of mapping class groups, Ann. of Math. 121 (1985) 215-249.

10. J. Harer, Virtual cohomological dimenion of the mapping class group, Inv. Math. 84 (1986) 157-176.

11. W.J. Harvey, Boundary structure of the modular group, in Ann. of Math. Studies no.97, Princeton U. Press (1981) 245-251.

12. N.V. Ivanov, Algebric properties of the mapping class groups, LOMI preprint (1985). Also a survey article to appear in Russ. Math. Surveys 42 (3), 1987.

13. B. Maskit, Parabolic elements in Kleinian groups, Ann. of Math. 117 (1983) 659-668.

14. E.Y. Miller, Homology of the mapping class group, J. Diff. Geom. 24 (1986), 1-14.

15. D. Mumford, Enumerative geometry of moduli space of curves, in Arithmetic and Geometry, Progress in Math. Vol.36, (1983) Birkhauser 271-328.

16. G. Segal and G. Wilson, Loop groups and equations of KdV type, Publ. Math. I.H.E.S. 61 (1985), p.1.

17. S.A. Wolpert, Symplectic geometry of deformations of a hyperbolic surface, Ann. of Math. 117 (1983); 207-234.

18. S.A. Wolpert, Weil-Petersson geometry of moduli space of curves, Amer. J. Math. 107 (1985), 959-997.

19. S.A. Wolpert, Thurston's Riemannian metric on Teichmüller space, J. Diff. Geom. 23 (1986) 143-174.

20. S. Wolpert, Homology of the moduli space of stable curves, Ann. of Math. 118 (1983), 491-523.

21. M. Yoshida, Discrete reflection groups in a parabolic subgroup of Sp(2,$\mathbb{R}$), J. Math. Soc. Japan 36 (1984), 243-258.

Contemporary Mathematics
Volume 74, 1988

LOCAL STRUCTURE OF REPRESENTATION VARIETIES:[1] EXAMPLES

Alexander Lubotzky*
and
Andy R. Magid*

ABSTRACT. The isomorphism classes of irreducible complex representations of a finitely generated discrete group D form a complex algebraic variety $S_n(D)$ whose Zariski tangent spaces are subspaces of cohomology groups. The tangents to formal curves (the curve tangents) are given by certain Lie algebra homomorphisms. If the group is abelian-by-finite, these sets of curve tangents are computable as equivariant Lie homomorphisms inside the cohomology space. Other general computations for the varieties $S_n(D)$, based on the inductive stucture of simple representations of abelian-by-finite groups, are also made.

I. Tangents, curve tangents, cohomology

1. INTRODUCTION. Throughout, we let D denote a finitely generated group with presentation

$$D = \langle g_1, \ldots, g_d \mid s_a \ a \text{ in } A \rangle.$$

An n-dimensional representation of D is a homomorphism

$$f: D \dashrightarrow Gl_n(\mathbf{C})$$

and we denote the set of all such by $R_n(D)$ and call it the <u>variety of representations</u> of degree n. It is indeed a variety: we can identify $R_n(D)$ with the n-tuples (A_i) of invertible matrices satisfying $s_a((A_i)) = I$ for all a. It is also useful to view $R_n(D)$ as the **C**-points of the scheme $Hom(D, Gl_n(*))$. A representation is simple it its image spans $M_n(\mathbf{C})$. The set of simple representations

1. 1980 Mathematics Subject Classification 20C15
* Both authors supported by NSF and BSF

forms an open subvariety of the representation variety which we denote by $R_n(D)^s$. Two representations f and f' are equivalent if there is an invertible matrix A with $A^{-1}fA = f'$. This conjugation action defines an action of $PGl_n(\mathbf{C})$ on $R_n(D)$ which has trivial stabilizers on $R_n(D)^s$. The resulting geometric quotient variety

$$p: R_n(D)^s \dashrightarrow R_n(D)^s/PGl_n = S_n(D)$$

is then locally trivial for the etale topology. The points of $S_n(D)$ correspond to the equivalence classes of simple representations, the class of the representation f being denoted [f] (or p(f)). The fibre over [f] is the PGl_n orbit of f. As a general reference to representation varieties we cite [6]

To analyze the varieties $S_n(D)$ and $R_n(D)$ we need to consider both their underlying point sets and their local geometric structure. In [6] we study these two aspects for nilpotent groups, proving, among other things, that $S_n(D)$ is always non-singular. It is important, however, to recognize that even when the qualitative geometric structrure of $S_n(D)$ is known, as for example in the case of nilpotent groups [6, Thm. 6.6, p.97], the quantitative structure (for example, number of irreducible components) may require separate analysis for each D. The exposition [9] of $S_n(D)$ for D the integral Heisenberg group is a good illustration.

Once one drops the assumption of nilpotency, things become

much more complicated, even for solvable groups. An example of Rudnick [10] shows that $S_n(D)$ can then have singular points. The theory of solvable groups can be easily reduced to the case of abelian-by-finite groups (see section 4 below). Abelian-by-finite groups can be studied by the methods of [8] (see section 3 below). We will illustrate this in section 4 with examples. However, we have not succeeded in settling the general question of singularities even for abelian-by-finite groups, despite the fact that the set-theoretic structure of their representation varieties is completely known by Mackey theory (see [11] for the split case and [5] in general). The only examples we know of singularities in this case are those which, like those of [10], come from the intersection of two irreducible components. We have thus been led to the following hypothesis:

<u>Conjecture</u> If D is an abelian-by-finite group, then every irreducible component of $S_n(D)$ is non-singular.

Sections 2 and 3 of part I of this paper study the local structure of representation varieties in general from the point of view of tangents and curve tangents. Section 4 studies the construction of curve tangents from a cohomological viewpoint.

In part II we illustrate how the methods described in part I can be "working tools" when one wants to study $S_n(D)$ for a specific group D. These methods are especially suitable for abelian-by-finite groups, so most of our examples are of this kind. As a by product, we obtain more examples of singularities, as well as some "non-singularity" results for quite special families of groups.

Support for the research reported here was provided by the National Science Foundation and the United States-Israel Binational Science Foundation.

2. TANGENT SPACES. The Zariski tangent space $T_f(R_n(D))$ to $R_n(D)$ at the representation f can be described as follows: informally speaking, we pass a curve through f and take its derivative at zero. Speaking precisely, this means to consider representations $f_e: D \to Gl_n(\mathbf{C}[e])$ into the dual numbers $\mathbf{C}[e]$ (generated over $\mathbf{C}$ by the relation $e^2 = 0$), such that when composed with the map $e \to 0$ they give f. One can identify $Gl_n(\mathbf{C}[e])$ with the semi-direct product $M_n(\mathbf{C}) \rtimes Gl_n(\mathbf{C})$ where the action is by conjugation. Then f_e is a map into the semi-direct product which becomes f with projection on the $Gl_n(\mathbf{C})$ factor. It readily follows that f_e is given in the other factor by a cocycle $D \to M_n(\mathbf{C})$ (D acts on $M_n(\mathbf{C})$ through f composed with conjugation, an action we denote by Adf). This defines a linear embedding of $T_f(R_n(D))$ into $Z^1(D,\mathrm{Adf})$. We can regard the orbit of f under the $PGl_n(\mathbf{C})$ action as the image in $R_n(D)$ of the conjugation map $Gl_n(\mathbf{C}) \to R_n(D)$ sending A to

$A^{-1}fA$. The tangent space at I of $Gl_n(\mathbf{C})$ is $M_n(\mathbf{C})$ and its image under the differential of the conjugation map is $B^1(D,Adf)$.

Because p: $R_n(D)^s \dashrightarrow S_n(D)$ is etale locally trivial, with its fibres being the orbits, if f is simple we can identify $T_{[f]}(S_n(D))$ with a subspace of $Z^1(D,Adf)/B^1(D,Adf) = H^1(D,Adf)$. Even if f is not simple, we can still say that $H^1(D,Adf)$ indicates how far the orbit tangent space is from the full tangent space. For instance, if every representation of D is semisimple, so that $H^1(D,Adf) = 0$, the the orbit is necessarily open in $R_n(D)$. This result is due to A. Weil, who introduced the methods of representation varieties and used the connection between their tangent spaces and group cohomology.

3. CURVE TANGENTS. The varieties $R_n(D)$ can have singularities. At a singular point, not every element of the Zariski tangent space is tangent to a curve in the representation variety. Indeed, singular means that the tangent cone is a proper subvariety of the tangent space, and curve tangents are necessarily in the tangent cone. In fact, they can also form a proper subset of the tangent cone, also. Since every tangent vector <u>is</u> tangent to a curve at a non-singular point, we can use the space of curve tangents to detect some singularities. For example, the space of curve tangents being non-linear is a sufficient condition for singularity.

A curve tangent can be thought of as either tangent to an analytic or a formal curve. By the Artin Approximation Theorem, both notions give the same set of curve tangents. For our purposes, the definition based on formal curves is most convenient. A formal curve in $R_n(D)$ can be viewed as a representation

$$f_t : D \dashrightarrow Gl_n(\mathbf{C}[[t]])$$

of D by matrices of formal power series. We say it lies over f if $f_t = f \pmod t$ and that its tangent at f is $f_t \pmod{t^2}$. We denote the set of curve tangents at f (or [f]) by $CT_f(R_n(D))$ (or $CT_{[f]}(S_n(D))$).

To properly describe the set of curve tangents over f, it is necessary to make some observations about the group $Gl_n(\mathbf{C}[[t]])$: this is a pro-algebraic group whose pro-unipotent radical is $I+tM_n(\mathbf{C}[[t]])$ and which has $Gl_n(\mathbf{C})$ as a maximal reductive subgroup. Any homomorphism from D to a pro-algebraic group factors through the pro-algebraic completion A(D). (A(D) is the inverse limit of the Zariski closures of all homomorphisms from D to algebraic groups.) This group is itself a product of its pro-unipotent radical U(D) and a maximal pro-reductive subgroup; we fix one of these and denote it P(D).

Let f be a simple representation in $R_n(D)^s$, and let F

denote the induced map $A(D) \to Gl_n(\mathbf{C})$. If $g : U(D) \to I+tM_n(\mathbf{C}[[t]])$ is a homomorphism, we can define a function gF from $A(D) = U(D)P(D)$ to $Gl_n(\mathbf{C}[[t]])$ by $gF(up) = g(u)F(p)$. This map will be a homomorphism precisely when g commutes with the D actions on its domain and range, and up to replacing $F(P(D))$ by a conjugate, all formal curves arise this way.

A similar analysis applies to the representations of D in $Gl_n(\mathbf{C}[e])$ used in the description of tangents at f. Such representations arise by the above process from D-homomorphisms from $U(D)$ to $I+eM_n(\mathbf{C})$ (this is the unipotent radical of $Gl_n(\mathbf{C}[e])$). Together, these descriptions allow us to describe the spaces of curve tangents:

<u>Theorem</u> [8] Let $f : D \to Gl_n(\mathbf{C})$ be a simple representation. Let V denote the image in $Hom_D(U(D),I+eM_n(\mathbf{C}))$ of $Hom_D(U(D),I+tM_n(\mathbf{C}[[t]]))$. Then

$$CT_f(R_n(D)) = V \times \dot{B}^1(D,Adf)$$

$$CT_{[f]}(S_n(D)) = V$$

The first part of the theorem also applies to semi-simple representations

As the theorem makes clear, to understand curve tangents, one must understand the homomorphism spaces from $U(D)$ to

$I+tM_n(\mathbf{C}[[t]])$. Homomorphisms of pro-unipotent groups are determined by the corresponding homomorphisms on Lie algebras. What must be studied, therefore, is the space of Lie algebra homomorphisms $Lie_D(LieU(D),tM_n(\mathbf{C}[[t]]))$ and its image in the space $Hom_D(LieU(D),eM_n(\mathbf{C}))$ of D-module homomorphisms. Note that $eM_n(\mathbf{C})$ is an abelian Lie algebra, and it is not the same Lie algebra as $M_n(\mathbf{C})$ with the commutator bracket.

To begin to analyze the space of D equivariant Lie homomorphisms, it is convenient to note the following proposition on Lie homomorphisms in the nilpotent case:

Proposition [8] Let L be a finite-dimensional nilpotent Lie algebra with minimal generating set $x_1, \ldots, x_d$. Let $g:L \to tM_n(\mathbf{C}[[t]])$ be a Lie homomorphism and suppose $g(x_i) = tA_i$ + higher degree. Then the Lie subalgebra of $M_n(\mathbf{C})$ generated by the A_i is a homomorphic image of the associated graded Lie algebra gr(L). If L is graded, then every homomorphism from L to $M_n(\mathbf{C})$ arises in this way.

The proposition yields the following applications to spaces of curve tangents:

Corollary [8] Assume that (1) U(D) is finite dimensional. Then the image of $Lie_D(LieU(D),tM_n(\mathbf{C}[[t]]))$ in $Hom_D(LieU(D),eM_n(\mathbf{C}))$ is contained in $\mathrm{Lie}_D(\mathrm{Lie}U(D), eM_n(\mathbf{C}))$. If further (2) U(D) is abelian, then the two sets are equal.

4. COHOMOLOGY. The problem of determining which tangents are curve tangents can be approached inductively: we start with a representation $r=r_0$: D --> $Gl_n(\mathbf{C})$ and a tangent vector in $T_r(R_n(D)$ viewed as a representation r_e: D --> $Gl_n(\mathbf{C}[e])$. Suppose we can inductively determine representations $r^{(m)}$: D --> $Gl_n(\mathbf{C}[t]/t^{m+1})$ such that $r^{(1)} = r_e$ and $r^{(m+1)} = r^{(m)}$ (modulo t^m). Then we obtain in the projective limit a representation r_t: D --> $Gl_n(\mathbf{C}[[t]])$ with $r_t = r_e$ (modulo t^2). The inductive step in this procedure may not always be possible (in fact it has a two cohomological obstruction, as we see below), and of course it's conceivable that some sequences starting from r_e are eventually obstructed and some aren't. By hypothesizing that all are unobstructed we obtain a sufficient condition for non-singularity, which applies in particular to surface groups.

In this section it is convenient to deal with representations into the **C**-points of an affine algebraic group scheme G. It is clear how the corresponding representation scheme $\underline{R}(D,G)$ (with A-points Hom(D,G(A)) in the **C**-algebra A) and representation variety R(D,G) are defined. For r in R(D,G), $T_r(\underline{R}(D,G) = Z^1(D, Ad_{Lie(G)}r)$ and so $T_r(R(D,G))$ embeds in this 1-cocycle space.

From the map $\mathbf{C}[t]/t^{m+2} \to \mathbf{C}[t]/t^{m+1}$ we have the exact sequence

$$0 \to \mathrm{Lie}(G) \to G(\mathbf{C}[t]/t^{m+2}) \to G(\mathbf{C}[t]/t^{m+1}) \to 1$$

Given $r^{(m)}: D \to G(\mathbf{C}[t]/t^{m+1})$, the obstruction to lifting $r^{(m)}$ to $r^{(m+1)}: D \to G(\mathbf{C}[t]/t^{m+2})$ lies, by a simple group cohomology calculation, in $H^2(D,\mathrm{Lie}(G)) = H^2(D,\mathrm{Ad}\ r)$. Hence we conclude:

<u>Proposition</u> Let r be in R(D,G) and assume $H^2(D,\mathrm{Ad}\ r) = 0$. Then every scheme tangent at r is a curve tangent. In particular, <u>R</u>(D,G) is reduced at r and R(D,G) is non-singular at r.

The vanishing of 2-cohomology is a strong condition. We can use the technique of cohomological obstruction in more general cases also, as we now show:

<u>Proposition</u> (Goldman [4]) Let D be a surface group. Then $S_n(D)$ is non-singular, and for [r] in $S_n(D)$, $T_{[r]}(S_n(D)) = H^1(D,\mathrm{Ad}\ r)$.
Proof: We need the Poincaré duality property that surface groups have, namely that $H^2(D,V) = H^0(D,V)^*$ for any linear D-module V [2, p.36]. For a simple $Sl_n(\mathbf{C})$-valued representation r of such a group, we have $H^2(D,\mathrm{Lie}(Sl_n(\mathbf{C}))) = H^0(D,\mathrm{Lie}(Sl_n(\mathbf{C})))^* = (\mathrm{End}_D(r)/\mathbf{C})^* = 0$.

For any simple r in $R_n(D)$, we can multiply r by the linear character 1/det(r) to make r(D) contained in $Sl_n(\mathbf{C})$; character multiplication is an operation [6, Prop. 5.2, p.81] and preserves non-singularity. The scheme tangents at r are $Z^1(D,\mathrm{Lie}(Gl_n(\mathbf{C})))$ = $Z^1(D,\mathrm{Lie}(Sl_n(\mathbf{C})))$ x $Z^1(D,\mathbf{C})$. And $Z^1(D,\mathbf{C})$ = Hom(D,**C**). We claim that all these homommorphisms are curve tangents: if f is in Hom(D,**C**), define X_f: D --> 1+t**C**[[t]] by $X_f(d)$ = exp(td). It's clear that X_f:D --> $Gl_n(\mathbf{C}[[t]])$ is a curve tangent at r through f. Now take f in the other factor $Z^1(D,\mathrm{Lie}(Sl_n(\mathbf{C})))$. By the preceding proposition, there is a curve F: D --> $Sl_n(\mathbf{C}[[t]])$ through f. We can regard F as a mapping to $Gl_n(\mathbf{C}[[t]])$ and hence conclude that f is also a curve tangent as an element of $T_r(R_n(D))$. Thus r is non-singular on $R_n(D)$ and, since r is simple, [r] is non-singular on $S_n(D)$.

II. Abelian-by-finite groups

We focus in this section on abelian-by-finite and related groups. This is, admittedly, a very special part of discrete group theory. We adopt it primarily because the construction of the <u>sets</u> (as opposed to the <u>varieties</u>) of isomorphism classes of simple representations of such groups are known: the Mackey theory [11, Thm. 17, p. II-18] in the case of semi-direct products, and the work of Hanrath and Plesken [5, Thm. IV.7, p.136] in the general case. As we shall see, however, the geometry of the varieties is far from being completely understood. By considering in detail some examples here, we hope to point the way toward a better understanding.

In defense of abelian-by-finite groups, we should point out that the crystallographic groups are of this type [1] and hence studied for their applications. Interestingly enough, there may be a connection between scheme singularities of representation varieties and certain physical phenomena. (Daniel Farkas called our attention to the physical concept of Brillouin zone and its connection with our work.) The phenomena are described in [3] and the work on possible connections is in progress.

Moreover, if D is a solvable-by-finite group with solvable normal subgroup K of index m, then any irreducible representation r: D --> $Gl_n(\mathbf{C})$ must have the Zariski closure H of r(K) a reductive solvable group, whose identity component H^0 is a torus. H^0 is normalized by r(D), and it follows that r(K) contains an abelian normal subgroup of r(D) of finite index, the index uniformly bounded depending only on n. (To see this, we note that H can be written as TS, where T is a torus and S is finite (S can meet T non-trivially). By the Jordan theorem on the existence of normal abelian subgroups, S has an abelian normal subgroup A of index bounded by a function of n which we denote $f_1(n)$. A acts on T and since T is at most n-dimensional, and the automorphism group of a k-dimensional torus is $Gl_k(Z)$, A has a subgroup B of index bounded by a function of n which we denote $f_2(n)$ which acts trivally on T. Then TB is an abelian normal subgroup of H of index bounded by $f_1(n)f_2(n)$.) If b is this bound, and L the intersection of all subgroups of K of index b, then $S_n(D) = S_n(D/(L,L))$, and of course D/(L,L) is abelian-by-finite. Thus the simple representation theory of virtually solvable groups and of abelian-by-finite groups is equivalent.

We fix in this section some special notation:
D is the semi-direct product of the torsion-free abelian normal subgroup N, and the finite subgroup W.

We now recall the Mackey theory of representations of abelian-by-finite semi-direct products. Our reference is [11, 9.2] and we continue to establish notation:
$X(N) = Hom(N, Gl_1(\mathbf{C}))$ is the character group of N; X(N) is a W-module. For V a subgroup of W we let $X(N,V) = \{f \text{ in } X(N) \mid Stab_W(f) = V\}$. If $r: D \to Gl_m(\mathbf{C})$ and V is contained in $Stab_W(f)$ then $fr: NV \to Gl_m(\mathbf{C})$ is a simple representation of the subgroup NV of D. We let $S(V,[r]) = \{ [fr] \mid f \text{ is in } X(N,V) \}$. If $n = m[W:V]$ then $Ind^D(fr)$ is in $R_n(D)$ and again simple [11, Prop. 16, p. II-11]. We let $I(V,[r]) = \{ [Ind^D(fr)] \mid f \text{ in } X(N,V) \}$. We use F(V,[r]) to denote the closure of I(V,[r]) in $S_n(D)$. We now recall the main result of the Mackey theory:

<u>Theorem</u> [11, Thm. 17, p.II-18] $S_n(D)$ is the union of the subsets I(V,[r]), where $\dim(r)[W:V] = n$. If representatives $X = \{ f_a \mid a$

in A } are selected for the W orbits on X(N) and if V_a denotes the stablizer of f_a then $[\mathrm{Ind}^D(f_a r)] = [\mathrm{Ind}^D(f_b s)]$ if and only if $f_a = f_b$ and $[r] = [s]$.

It is important to realize that the sets I(V,[r]) in $S_n(D)$ are not closed. The theorem, which gives a complete description of $S_n(D)$ as a point set, makes no assertions about its topology.

To begin to analyze the geometrical consequences of the theorem, we note the following:
{ f in X(N) ¦ $\mathrm{Stab}_W(f)$ contains V } is a closed subset of X(N) which we denote $X_V(N)$. Since $X_V(N)$ is closed, in X(N) and X(N,V) = $X_V(N) - \bigcup\{ X_U(N)$ ¦ U properly contains V }, we see that X(N,V) is contained in the closure of $X_V(N)$ (in general, properly). We use this observation to identify points in the closures of the I(V,[r]):

<u>Proposition</u> Let r be a simple representation of the subgroup V of W and let f be in the closure of X(N,U). Let $U \supset V$ be the stabilizer of f and assume that $s = \mathrm{Ind}^U(r)$ simple. Then $[\mathrm{Ind}^D(fs)]$ is in the closure of I(V,[r]).
Proof: There is a sequence of characters f_t in X(N,V) converging to f. The representations $\mathrm{Ind}^D(f_t r)$ converge to $\mathrm{Ind}^D(fr)$ (induction from NV), which equals $\mathrm{Ind}^D(fs)$.

We would like to assert that the proposition describes completely the closure F(V,[r]) of I(V,[r]) in $S_n(D)$. What we have is the following partial converse:

<u>Proposition</u> Assume V $\subset$ U are subgroups of W and that for some f in X(N,U) and simple representations s of U and r of V that [Ind^D(fs)] is in F(V,[r]). Then:

a) f is in $\bigcup$ { $X_Y(N)$ | Y = gVg^{-1} some g in W }

b) [Ind^W(s)] = [Ind^W(r)]

Proof: Restriction to N defines a map from $R_n(D)^s$ to X(N) with the isomorphism class of the representation of the form Ind^D(gt) determining the W-orbit of the charcter g. Part a) of the proposition follows from this fact and the theorem above. If we select a sequence of representations in I(V,[r]) converging to Ind^D(fs), and restrict to W, then each representation in the sequence produces [Ind^W(r)] while the limit produces [Ind^W(s)]. This establishes b).

It seems plausible that the converse proposition is also true, but we do not have a proof.

The above propositions allow us to use the methods of part I in some special cases. We first consider the case that W is abelian; our conclusions will follow from the following general proposition:

Proposition Let r be a one-dimensional representation of V. Then $I(V,[r])$ is an open subset of $S_n(D)$ and all its point are non-singular on the associated scheme.

Proof: Let $I = \mathrm{Ind}^D(fr)$. The scheme tangents at [I] are identified with $\mathrm{Hom}_D(N,\mathrm{End}(I)) = \mathrm{Hom}_D(N,\mathrm{End}_N(I))$ and the curve tangents with $\mathrm{Lie}_W(N,\mathrm{End}_N(I))$. It follows from [11, Prop. 15, p. II-10] that $\mathrm{Res}_N(I) = \bigoplus \{ \dim(r)wf \mid w \text{ in } W/V \}$ in general, so that, since $\dim(r) = 1$, $\mathrm{Res}_N(I)$ is the direct sum of the conjugates of f, which are all distinct since $V = \mathrm{Stab}_W(f)$. It follows that $\mathrm{End}_N(I)$ is an abelian Lie algebra, so all homomorphisms to it from N are Lie homomorphisms. Thus every scheme tangent is a curve tangent.

Since every simple representation of an abelian group is one-dimensional, we have:

Corollary If W is abelian then $S_n(D)$ is non-singular for every n.

The non-singularity of $I(V,[r])$ for r one-dimensional is also relevant to certain non-abelian W. First, we need lots of representations induced from one-dimensional ones; for example, that every simple representation of every subgroup of W is induced from a one-dimensional representation of it (that is, W is an M-group). Also, we need lots of closure relations on character sets. It is easy to see that the algebraic group $X_V(N)$ has the same dimension as the **C**-vector space $(N \otimes \mathbf{C})^V$. If $(N \otimes \mathbf{C})^V$ strictly contains $(N \otimes \mathbf{C})^U$ whenever U strictly contains V then we call N a <u>strict</u> module. For example, **Z**[W] is strict. In the strict case, we know that X(N,V) is dense in $X_V(N)$ because its complement is a union of smaller dimensional subsets. With these definitions, we obtain the following:

<u>Proposition</u> Assume that W is an M-group and that N is a strict module. Then

$$S_n(D) = \bigcup \{ F(V,[r]) \mid [W:V] = n, \dim(r) = 1\}$$

Moreover,

$$\dim(F(V,[r]) = \dim(\mathrm{Hom}_D(N,\mathrm{End}_N(I))) \text{ for any } [I] \text{ in } I(V,[r]).$$

Proof: Let I = $\mathrm{Ind}^D(fr)$ (induction from NU) be a simple representation of D. By assumption, r = $\mathrm{Ind}^U(s)$ (induction from some subgroup V of U) for a suitable one-dimensional s. The closure of X(N,V) is $X_V(N)$ which contains $X_U(N)$ and hence f. Then a proposition above shows that I is in the closure of I(V,[r]).

N is strict, for example, in case D is the semi-direct product of **Z**[W] by W (with the obvious multiplication action). We should note, however, that in the situation of the proposition $S_n(D)$ can have singularities, as the following example shows:

<u>Example</u>: [8] Let D be the semi-direct product $\mathbf{Z}[D_4] \rtimes D_4$, where

D_4 denotes the dihedral group of order 8. Let f in $R_2(D)$ be the simple representation of D_4 preceded by projection on D_4. We let N denote $\mathbf{Z}[D_4]$. It is possible to show that $M_2(\mathbf{C})$ is the sum of the four linear characters of D_4. Thus $\mathrm{Lie}_D(N, M_2(\mathbf{C}))$ is the union of three planes. There are three index 2 subgroups of D which contain N. Induction of linear characters from them to D yield locally closed subsets of $S_2(D)$ whose closures are two dimensional and which meet at [f], thus yielding the three planes in $CT_{[f]}(S_2(D))$. Of course the non-linearity of the space of curve tangents shows that [f] is a singular point.

This example shows that our assertion of non-singularity for abelian W in the corollary above, as weak as it sounds, is the best possible. An attempt to extend it in a different direction, to the case where the extention of the (abelian) W by N is not semi-direct also fails, as the following example of Rudnick's [10] shows:

Example: $D = \langle P, Q, T \mid T^2 = 1, P^2 = Q^2, TPT^{-1} = Q^{-1} \rangle$ [10, Example 3.3, p. 271]. Let $X = P^{-1}Q$, $Y = P^2$, and $N = \langle X, Y \rangle$. Also let $H = \langle P, Q \rangle$, $K = \langle PT, X, Y \rangle$ and $F = \langle T, X, Y \rangle$. Then Rudnick shows that $S_2(D)$ is the union of the two one-dimensional spaces $S_2(D/(H,H))$ and $S_2(D/(K,K))$ while $S_2(D/(H,H)) \cap S_2(D/(K,K)) = S_2(D/(F,F))$ is a finite set with two elements. Let $W = D/N$. (Note: D is not the semidirect product of W and N). $W = (\mathbf{Z}/2\mathbf{Z})^{(2)}$ with generators t =

TN and p =PN. X(W) has four elements: the identity f_0, f_1 which sends t to -1 and p to -1, f_2 which sends t to -1 and p to 1, and $f_3 = f_1 f_2$. One can check that **CN** = $f_1 + f_2$ (the basis weight vectors are X and Y). Rudnick shows that N becomes central in D/(F,F) so that it acts trivially on Adr for simple D/(F,F)-modules. If r is one of the simple two-dimensional D/(F,F) modules, Adr = $f_0 \times f_1 \times f_2 \times f_3$. It follows that $\mathrm{Lie}_W(N,\mathrm{Adr})$ is the union of the two one-dimensional subspaces $\mathrm{Hom}_W(f_i,f_i)$, i =1,2. Thus $CT_{[r]}(S_2(D))$ is the union of these two lines. This reflects the combinatorial structure previously noted.

We conclude with two other classes of groups whose representation varieties can be understood in terms of our characterization of curve tangents. First, nilpotent groups: it was shown in [6] and again in [10] that $S_n(D)$ is non-singular for D nilpotent. As noted above, any solvable D can be replaced by a suitable abelian-by-finite quotient to calculate $S_n(D)$. For nilpotent D, we can replace D by a central-by-finite quotient, so, in our current notation, assume that D acts trivially on N (We do not assume semi-direct product). Thus for a D-module M we have $\mathrm{Lie}_D(N,\mathrm{End}(M)) = \mathrm{Lie}_D(N,\mathrm{End}_D(M))$. If M is simple, or even just multiplicity free, $\mathrm{End}_D(M)$ is abelian, so all D-homomorphisms from N to End(M) are Lie homomorphisms and every tangent is a curve tangent. Thus we obtain:

Proposition [6, 6.1, p. 93] [10, 2.3, p.267] Let D be nilpotent and let r in $R_n(D)$ be semi-simple and multiplicity free. Then r is non-singular on $R_n(D)$. In particular, $S_n(D)$ is non-singular

For our final case, we depart from our establised notation for this section and consider groups D which have a free nornal subgroup F of finite index (so D is free-by-finite). Then U(D) = U(F) is a free prounipotent group [7,Thm 4.3, p.93]. It follows that a homomorphism from U(D) to $I+tM_n(\mathbf{C}[[t]])$ corresponds to a set map from a set of generators of U(F) to $I+tM_n(\mathbf{C}[[t]])$ [7, Prop 2.2, p.83]. As a consequence, $Hom_D(U(D),I+tM_n(\mathbf{C}[[t]]))$ maps onto $Hom_D(U(D),eM_n(\mathbf{C}))$, so every tangent is again a curve tangent. Thus we obtain:

Proposition If D is free-by-finite, then every semi-simple representation in $R_n(D)$ is non-singular. In particular, $S_n(D)$ is non-singular.

This proposition applies, for example, to the group $SL_2(Z)$, where the corresponding fact was obtained in [6, 3.12, p.64]

Bibliography

1. Altmann,S. Induced Representations in Crystals and Molecules,

Academic Press, London, 1977

2. Eckmann, B. "Poincare duality groups of dimension two are surface groups" in Gersten, S. and Stallings, J. Combinatorial Group Theory and Topology, Annals of Math Studies 111, Princeton University Press, Princeton, 1987

3. Frech, R. "Oblique phonons in uniaxial crystals", The Journal of Chemical Physics 67(1977), 952-956

4. Goldman, W. "The symplectic nature of fundamental groups of surfaces", Advances in Math 54(1984), 200-225

5. Hanrath, W. and Plesken, W. "Group rings and irreducible representations of space groups", J. fur die reine und angewandte Math., 362(1985), 130-140

6. Lubotzky, A. and Magid, A. Varieties of Representations of Finitely Generated Groups, Mem. Amer. Math. Soc. 336 (1985)

7. Lubotzky, A. and Magid, A. "Cohomology of unipotent and prounipotent groups", J. of Alg. 74(1982), 76-95

8. Magid, A. "Curve tangents to representation varieties", preprint

9. Nunley, C. "Simple representations of the integral Heisenberg group", in preparation

10. Rudnick, Z. "Representation varieties of solvable groups", J. Pure Applied Algebra 45(1987), 261-272

11. Serre, J.-P. Representations Lineaires des Groupes Finis, Hermann, Paris 1967

Department of Mathematics
Hebrew University
Givat Ram, Jerusalem, ISRAEL

Department of Mathematics
University of Oklahoma
Norman, OK 73019

Contemporary Mathematics
Volume **74**, 1988

Deformations of Representations of Finitely Generated Groups

John J. Millson*

This paper is a slightly expanded version of our conference lecture by the same title. Although it is the lattice subgroups of Lie groups that are of primary importance, for many reasons it is important to widen the class of subgroups considered to include all finitely generated ones (not necessarily discrete and not necessarily of finite covolume). There are several reasons for doing this. First there are many examples of this broader class of groups occurring in geometry. For example, in algebraic geometry it is now known that the monodromy representation of a variation of polarized Hodge structure is not necessarily a lattice, [DM], [N]. In topology, the holonomy groups of locally homogeneous structures are usually not lattices and are frequently not discrete. Above all by removing the restrictions of discreteness and finite covolume one often obtains continuous families. This allows the introduction of the techniques of deformation theory developed in the last three decades. In order to obtain a good deformation space it is important to fix the isomorphism class of the finitely generated group. Thus we are led to consider the set $\mathrm{Hom}(\Gamma,G)$ where Γ is a fixed finitely-generated group and G is a fixed semi-simple Lie group which we assume henceforth is real algebraic. It is then a standard fact that there is an algebraic variety $R(\Gamma,G)$ defined over $\mathbb{R}$ such that $\mathrm{Hom}(\Gamma,G)$ is the associated set of real points. We will try to describe the neighborhood of an element $\rho \in R(\Gamma,G)$ for Γ a finitely-generated group and G a Lie group in the following three cases:

---

*Partially supported by NSF grant DMS-85-01742.

(a) Γ is a lattice in a semi-simple Lie group H of real rank 1 and any G.

(b) Γ is the fundamental group of a compact Kahler manifold M and G is compact.

(c) Γ is the fundamental group of a compact Kahler manifold M and ρ is the monodromy representation of a variation of polarized Hodge structure parametrized by M.

The point of studying the examples from (1) is two-fold. First, one would like to ascertain the limits of the Margulis Super-Rigidity Theorem (see section 2), i.e., to determine to what degree it carries over to the rank 1 case. Of course, there are counterexamples to the naïve hope that the theorem would hold without modification, we give a family of such here. However, these counterexamples depend on the existence of totally-geodesic hypersurfaces or at least the non-vanishing of the first Betti number of the associated locally symmetric space. The examples in (a) are also of interest from the point of view of locally homogeneous structures on manifolds. We give new information concerning the space of flat conformal and projective structures on hyperbolic manifolds containing totally geodesic hypersurfaces.

The point of studying the examples of (b) is that their conjugacy classes parametrize semi-stable bundles over M for those C^{∞} bundles admitting flat connections by Uhlenbeck-Yau [UY]. In particular, one could use algebraic geometry techniques to test the rigidity properties of such representations. Similar considerations apply to (c).

In the last section of the paper we discuss a recent theorem [GM3] obtained in collaboration with William Goldman which proves that in cases (b) and (c) ρ has a neighborhood real analytically equivalent to a neighborhood of 0 on the quadratic cone in $Z^1(\Gamma,\mathfrak{g})$ (the Eilenberg-MacLane 1-cocycles) determined by the vanishing of the bracket square (see section 2). The main lines of the proof were suggested to us by Pierre Deligne. It seems possible that representations of fundamental groups of Kahler manifolds satisfy surprising local rigidity properties. K. Corlette [C] has obtained promising results along these lines.

1. *Stable Points of* $R(\Gamma,\underline{G})$

In what follows Γ will be a finitely generated group and $\underline{G}$ will be a reductive algebraic group defined over $\mathbb{R}$ with the associated Lie group of real points denoted G. We will identify $\underline{G}$ with its group of complex points. We obtain in the usual way *non-reduced* varieties, the representation variety $R(\Gamma,\underline{G})$, and its algebraic geometry quotient $X(\Gamma,\underline{G})$, the character variety, see [LM] or [JM]. We let $R(\Gamma,G)$ and $X(\Gamma,G)$ be the associated sets of real points. We obtain quotient maps

$$\underline{\pi} : R(\Gamma,\underline{G}) \longrightarrow X(\Gamma,\underline{G})$$

$$\pi : R(\Gamma,G) \longrightarrow X(\Gamma,G).$$

It is easily proved by using a Reynolds operator that $\underline{\pi}$ is onto but π is not even locally onto. We will usually identify a variety with its set of complex points but it is good to keep in mind that there is more structure here – there is a non-reduced "coordinate ring," see [LM]. We now make two definitions which are important in the discussion of the character variety $R(\Gamma,\underline{G})$.

Definitions.

(i) A representation ρ in $\mathrm{Hom}(\Gamma,\underline{G})$ is *stable* if the $\underline{G}$ orbit of ρ is closed and the isotropy $Z(\rho)$ of ρ in $\mathrm{Ad}\,\underline{G}$ is finite.

(ii) A representation ρ in $\mathrm{Hom}(\Gamma,\underline{G})$ is *good* if $Z(\rho)$ is trivial.

(iii) A representation ρ in $\mathrm{Hom}(\Gamma,G)$ is stable (good) if it is stable (good) when considered as a point of $\mathrm{Hom}(\Gamma,\underline{G})$.

We remark that the G-orbit of ρ in $\mathrm{Hom}(\Gamma,G)$ is closed if and only if the $\underline{G}$-orbit of ρ in $\mathrm{Hom}(\Gamma,\underline{G})$ is closed. This follows from [B]. Clearly the isotropy of ρ in $\underline{G}$ is finite if and only if the isotropy of ρ in G is. Thus stability of a real representation could be defined directly as in (i) with $\underline{G}$ replaced by G. We have the following criterion for stability, [JM], Theorem 1.1. We take this opportunity to correct a typographical error in [JM]. The G at the end of the statement in Theorem 1.1 should be underlined. The uncorrected statement is false.

Theorem. ρ is stable if and only if $\rho(\Gamma)$ is not contained in a proper parabolic subgroup of $\underline{G}$.

Corollary. If G is a complex Lie group the ρ is stable if and only if $\rho(\Gamma)$ is not contained in a parabolic subgroup of G.

Examples.

(i) If $\underline{G} = SL(n,\mathbb{C})$ then ρ is stable if and only if ρ is irreducible.

(ii) If $\underline{G} = SO(n,\mathbb{C})$ resp. $\underline{G} = Sp(n,\mathbb{C})$ then ρ is stable if and only if $\rho(\Gamma)$ leaves no non-zero totally isotropic subspace of $\mathbb{C}^n$ resp. $\mathbb{C}^{2n}$ invariant.

We see that a Fuchsian representation, [JM], section 3, of a lattice in $SO(1,n)$ into $SO(1,n+1)$ is reducible but stable.

Let $R(\Gamma,\underline{G})'$ denote the set of good representations and $X(\Gamma,\underline{G})'$ be the image of $\mathrm{Hom}(\Gamma,\underline{G})'$ under $\underline{\pi}$. We define

$$R(\Gamma,G)' = R(\Gamma,G) \cap R(\Gamma,\underline{G})'$$

$$X(\Gamma,G)' = X(\Gamma,G) \cap X(\Gamma,\underline{G})'.$$

The next theorem combines Theorem 1.2, Proposition 1.3 and the corollary to Lemma 3.7 of [JM].

Theorem.

(i) $R(\Gamma,\underline{G})'$ and $X(\Gamma,\underline{G})'$ are Zariski open subsets of $R(\Gamma,G)$ and $X(\Gamma,\underline{G})$ respectively.

(ii) The restriction of $\underline{\pi}$ (to be denoted $\underline{\pi}'$)

$$\underline{\pi}' : R(\Gamma,\underline{G})' \longrightarrow X(\Gamma,\underline{G})'$$

is a principal fiber bundle with group $\mathrm{Ad}\,\underline{G}$.

(iii) The restriction of π' (to be denoted π)

$$\pi' : R(\Gamma,G)' \longrightarrow X(\Gamma,G)'$$

is onto a connected component (in the classical topology) of $X(\Gamma,G)'$.

Remark. Let $\rho \in R(\Gamma,G)'$ and S be a local analytic slice to G through ρ. Then $\pi|S$ is an analytic equivalence between S and a neighborhood if $\pi(\rho)$ in $X(\Gamma,G)$. The same picture holds if ρ is stable except that you have to divide out S by the isotropy of ρ to get a neighborhood of $\pi(\rho)$ in $X(\Gamma,G)$.

Problem. Let ρ be a stable point in $R(\Gamma,G)$.

(i) Describe a neighborhood U of ρ in $R(\Gamma,G)$.

(ii) Describe $U \cap S$.

2. *Representations of Lattice Subgroups of* SO(1,n) *and* SU(1,n)

We consider the diagram

$$\begin{array}{ccc} H & & G \\ \uparrow & \nearrow & \\ \Gamma & & \end{array}$$

where H and G are non-compact semi-simple Lie groups with the real rank of H greater than 1, the vertical arrow is the inclusion of Γ as an irreducible lattice in H and the diagonal arrow is a homomorphism ρ with Zariski dense image. We recall the statement of the Margulis Super-Rigidity Theorem [Z], 5.1.2.

Theorem. Assume G has trivial center and ρ and H are as above, then ρ extends to a representation of H in G.

As a consequence of the Super-Rigidity Theorem we see that there are only a finite number of ρ as above, in particular any such ρ is locally rigid.

In fact, we can deduce the local rigidity of ρ even when G is compact by using Raghunathan's Vanishing Theorem [R], see [JM], Theorem 1.2. Moreover, Raghunathan's theorem implies that if H is simple and ρ is not locally rigid then either $H = SO(1,n)$ or $H = SU(1,n)$. We will now discuss the results obtained for these groups in [JM] and [GM1].

In [JM] we showed that super rigidity fails for lattices in $SO(1,n)$. We described a large number of choices for Γ, G and ρ so that the slice S is not reduced to a point. The basic idea there was to deform a rational representation of Γ, that is a representation of Γ which extends to H. In [JM] we concentrated on the two basic families of ρ which follow. These examples are the most interesting from the point of view of geometry. However, our basic existence result [JM], Proposition 5.1, applies to a large number of other G.

Examples.

(i)

$$\begin{array}{ccc} SO(1,n) & \xrightarrow{\ \tau\ } & SO(1,n+1) \\ \uparrow & \nearrow \rho & \\ \Gamma & & \end{array}$$

(ii)

$$\begin{array}{ccc} SO(1,n) & \xrightarrow{\ \tau\ } & PGL_{n+1}(\mathbb{R}) \\ \uparrow & \nearrow \rho & \\ \Gamma & & \end{array}$$

In both cases τ is the natural representation and ρ is the restriction of τ to Γ.

Remarks. In case (i) ρ is stable but not good and in case (ii) ρ is good.

In both cases ρ is the holonomy representation of a locally homogeneous structure. In case (i) ρ is the holonomy representation of the flat conformal structure associated to the hyperbolic structure on M and in case (ii) ρ is the holonomy representation of the flat projective structure associated to the hyperbolic structure.

We recall that there exist many examples of lattices $\Gamma \subset SO(1,n)$ which are congruence subgroups of the unit groups of suitable quadratic forms [JM]. We will call these the "standard arithmetic examples." These exist in all dimensions and are all the arithmetic lattices if n is even. We use these examples because they have lots of totally geodesic hypersurfaces. The following theorem is Theorem 7.1 of [JM].

Theorem A. Let Γ be a standard arithmetic lattice in $SO(1,n)$. Then by passing to congruence subgroups of Γ we can make the dimension of the slice S through ρ arbitrarily large in both examples (i) and (ii).

Theorem A follows from a general existence theorem [JM], Proposition 5.1, asserting the existence of deformations of Lie group representations of fundamental groups of graphs of groups. This result is an algebraic version of the Thurston bending deformation. This construction assigns a curve in S through ρ to an embedded totally geodesic hypersurface N. We say that we have bent ρ along N. One can bend independently along collections of disjoint embedded totally geodesic hypersurfaces. See [K] for a geometric description justifying the term bending.

Now that we have established that the dimension of S is greater than zero we may ask what it looks like. We obtained partial information on this in [JM], Theorem 7.6.

Theorem B. Let Γ be a standard arithmetic lattice in $SO(1,n)$. Assume $n \geq 4$. Then there exists a congruence subgroup $\Gamma' \subset \Gamma$ such that the component of the singular set passing through ρ has dimension larger than zero in both (i) and (ii).

Corollary. In both (i) and (ii) $R(\Gamma,G)$ has singularities at irreducible representations.

Proof. This is automatic in (ii) because ρ is irreducible. It is true for (i) because all representations in S sufficiently close to ρ but not equal to ρ are irreducible [JM], Theorem 3.1.

We will go into some detail concerning the proof of Theorem B since we will need the ideas involved in the rest of this paper. These ideas are specializations of general notions from deformation theory namely "infinitesimal deformation" and the "first obstruction to integrability."

Let ρ_t be a curve in $R(\Gamma,G)$ with $\rho_0 = \rho$. We consider the function $\dot{\rho} : \Gamma \longrightarrow T_{\rho(\gamma)}(G)$ given by

$$\dot{\rho}(\gamma) = \frac{d}{dt}\rho_t(\gamma)\big|_{t=0}.$$

We let $\mathfrak{g} = T_e(G)$ denote the Lie algebra of G and define $c : \Gamma \longrightarrow \mathfrak{g}$ by

$$c(\gamma) = \dot{\rho}(\gamma)\rho(\gamma)^{-1}.$$

Let $\mathrm{Ad}\,\rho$ denote the composition $\mathrm{Ad}\circ\rho$ where $\mathrm{Ad} : G \longrightarrow \mathrm{Aut}(\mathfrak{g})$ is the adjoint representation.

Lemma. c is a cocycle with values in $\mathrm{Ad}\,\rho$; that is,

$$c(\gamma\eta) = c(\gamma) + \rho(\gamma)c(\eta)\rho(\gamma)^{-1}.$$

Proof. We differentiate $\rho_t(\gamma\eta) = \rho_t(\gamma)\rho_t(\eta)$ to obtain

$$\dot{\rho}(\gamma\eta) = \dot{\rho}(\gamma)\rho(\eta) + \rho(\gamma)\dot{\rho}(\eta)$$

and hence

$$\dot{\rho}(\gamma\eta)\rho(\gamma\eta)^{-1} = \dot{\rho}(\gamma)\rho(\gamma)^{-1} + \rho(\gamma)\dot{\rho}(\eta)\rho(\eta)^{-1}\rho(\gamma)^{-1}.$$

The lemma follows.

We let $Z^1(\Gamma,\mathfrak{g})$ denote the space of 1-cocycles. Since a 1-cocycle is determined by its values on a set of generators for Γ the space $Z^1(\Gamma,\mathfrak{g})$ is finite dimensional. We call a 1-cocycle an ***infinitesimal deformation.***

A cocycle c coming from a curve ρ_t via the above procedure is said to be tangent to ρ_t or is said to be integrable. Not every cocycle is integrable. There is an infinite sequence of obstructions (polynomial functions of c with values in $H^2(\Gamma,\mathfrak{g})$) that must vanish in order for c to be integrable. We derive the first of these, it is a quadratic function of c. We define $[c,c]$ a 2-cochain on Γ with values in $\mathfrak{g}$ by

$$[c,c](\gamma,\eta) = [c(\gamma),\rho(\gamma)c(\eta)\rho(\gamma)^{-1}].$$

We leave to the reader the verification that $[c,c]$ is a 2-cocycle. The next lemma is of great importance to us.

Lemma. If c is integrable then $[c,c]$ is a coboundary.

Proof. Suppose c is tangent to ρ_t. We define a curve of cocycles c_s by the formula

$$c_s(\gamma) = \dot{\rho}_s(\gamma)\rho_s(\gamma)^{-1}$$

where

$$\dot{\rho}_s(\gamma) = \frac{d}{dt}\rho_t(\gamma)\Big|_{t=s}$$

We let $\dot{c}$ be the 1-cochain on Γ with values in $\mathfrak{g}$ defined by

$$\dot{c}(\gamma) = \frac{d}{ds}c_s(\gamma)\Big|_{s=0}.$$

Let δ be the Eilenberg-MacLane coboundary operator from $C^1(\Gamma,\mathfrak{g})$ to $C^2(\Gamma,\mathfrak{g})$. We claim

$$\delta\dot{c} = [c,c].$$

Indeed we have for all s

$$c_s(\gamma\eta) = c_s(\gamma) + \rho_s(\gamma)c_s(\eta)\rho_s(\gamma)^{-1}.$$

Differentiating and evaluating at $s = 0$ we obtain

$$\begin{aligned}\dot{c}(\gamma\eta) = \dot{c}(\gamma) &+ \rho(\gamma)\dot{c}(\eta)\rho(\gamma)^{-1} \\ &+ \dot{\rho}(\gamma)c(\eta)\rho(\gamma)^{-1} \\ &- \rho(\gamma)c(\eta)\rho(\gamma)^{-1}\dot{\rho}(\gamma)\rho(\gamma)^{-1}.\end{aligned}$$

Hence

$$\delta\dot{c}(\gamma,\eta) = c(\gamma)\rho(\gamma)c(\eta)\rho(\gamma)^{-1} - \rho(\gamma)c(\eta)\rho(\gamma)^{-1}c(\gamma)$$

$$= [c(\gamma),\rho(\gamma)c(\eta)\rho(\gamma)^{-1}].$$

Definition. The (reduced) tangent cone $TC_\rho \subset Z^1(\Gamma,\mathfrak{g})$ is the set of all infinitesimal deformations which are actually tangent to (i.e., leading coefficients of) curves. Let Q denote the quadratic cone

$$Q = \{c \in Z^1(\Gamma,\mathfrak{g}) : [c,c] = 0 \text{ in } H^2(\Gamma,\mathfrak{g})\}.$$

We have just seen

$$TC_\rho \subset Q.$$

Remark. This inclusion is usually strict, there are higher degree polynomial functions of c with values in quotients of $H^2(\Gamma,\mathfrak{g})$ which must vanish in order that c be in TC_ρ. These functions are analogous to Massey products.

We can now explain the proof of Theorem B. Suppose M contains two embedded totally geodesic hypersurfaces N and N′ which intersect obliquely (i.e., not at 90°). Let ρ_t be the curve obtained by bending ρ along N and c the corresponding infinitesimal deformation. Let ρ'_t be the curve obtained by bending ρ along N′ and c' the corresponding infinitesimal deformation. A great deal of [JM] is devoted toward proving the following result.

Proposition. Let Γ be a standard arithmetic lattice in $SO(1,n)$. Assume $n \geq 4$. Then in both (i) and (ii) there exists a congruence subgroup $\Gamma' \subset \Gamma$ such that the corresponding hyperbolic manifold contains hypersurfaces N and N′ as above with

$$[\lambda c+\mu c',\lambda c+\mu c'] \neq 0 \text{ unless } \lambda = 0 \text{ or } \mu = 0.$$

Thus only the axes in the plane in $Z^1(\Gamma,\mathfrak{g})$ spanned by c and c' can be integrated. We remark that if N and N′ intersect orthogonally then the entire λ, μ plane can be integrated in the projective case.

Finally, in order to obtain non-isolated singularities we find another embedded totally geodesic hypersurface N'' disjoint from N and N'. The infinitesimal bendings of ρ along N, N' and N'' span a three-dimensional subspace of $Z^1(\Gamma, \mathfrak{g})$. Only those cocycles in the coordinate planes corresponding to N, N'' and N', N'' are integrable. We obtain a curve of singularities passing through ρ in $R(\Gamma, G)$ corresponding to bending ρ along N''. Indeed bending ρ along N'' does not change the restriction of ρ to N and N'. It follows that c and c' propagate to cocycles c_t and c'_t with values in the deformed representation ρ_t. However, it requires considerable care to prove that $[c_t, c'_t] \neq 0$ for all t. Note that the non-vanishing of the cup-product at $t = 0$ does not imply the non-vanishing near $t = 0$.

It is a natural question to ask whether there are results analogous to those just detailed for the case of complex hyperbolic space; that is, for lattices in $SU(1,n)$. At first glance, it would seem that an analogue of Theorem A would hold for the diagram

$$\begin{array}{ccc} SU(1,n) & \xrightarrow{\ \tau\ } & SU(1,n+1) \\ \uparrow & \nearrow_{\rho} & \\ \Gamma & & \end{array}$$

Here τ is the natural representation and Γ is a standard arithmetic lattice (a congruence subgroup of the group of units of a hermitian form defined over a totally imaginary extension of a totally real field). For such Γ, Borel and Wallach [BW] showed that the space of infinitesimal deformations $H^1(\Gamma, V)$ was non-zero, here V is the standard module for $SU(1,n)$ restricted to Γ. In [GM1] we proved that none of these deformations were integrable. We showed this by establishing that the bracket $[c,c]$ was never zero for $c \in H^1(\Gamma, V)$. In fact, as we will see in the next section the non-vanishing of $[c,c]$ was not only sufficient to prove local rigidity, it was also necessary, any zero of the bracket is automatically integrable. The above rigidity result is particularly striking because in the real hyperbolic case outlined previously the infinitesimal deformations are again $H^1(\Gamma, V)$ where V is now the standard representation of $SO(1,n)$. As we have seen many of these are in fact integrable. We conclude this discussion by observing that if we complexify and consider the diagram

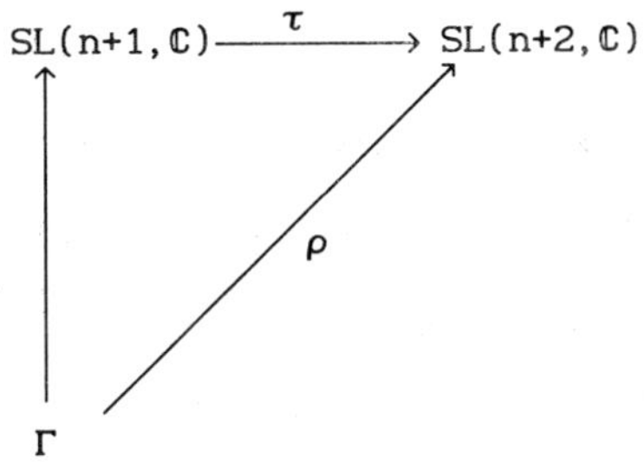

then ρ is no longer rigid. The point is that the image of ρ lies in $SL(n+1,\mathbb{C})$ and each element of $H^1(\Gamma, V)$ gives a class of deformations of ρ in the affine group $SL(n+1,\mathbb{C}) \ltimes V \subset SL(n+2,\mathbb{C})$. There deformations correspond to (unstable) deformations of the holomorphic structure on the bundle over M associated to the action of Γ on the Lie algebra of $SL(n+2,\mathbb{C})$. They do not correspond to deformations of the underlying flat (indefinite) unitary structure.

3. *Representations of Fundamental Groups of Compact Kahler Manifolds*

In this section we describe how the quadratic cone Q of the previous section is the universal deformation of a representation $\rho \in R(\Gamma,G)$ for Γ the fundamental group of a compact Kahler manifold and ρ as in (ii) or (iii) of the introduction. We have the following theorem, [GM3].

Theorem. Let Γ be the fundamental group of a compact Kahler manifold. Then $\rho \in R(\Gamma,G)$ has a neighborhood real analytically equivalent to a neighborhood of 0 in Q if one of the following conditions is satisfied:

(i) $\rho(\Gamma)$ lies in a compact subgroup of G.

(ii) ρ is the monodromy of a variation of Hodge structure over M.

(iii) G is the group of automorphisms of a hermitian symmetric space X and the associated flat X-bundle over M possesses a holomorphic section.

This theorem can be broken into two parts. The first part is an assertion that $TC_\rho = Q$, that is, all the higher obstructions to integrability mentioned in section 2 vanish. This is a direct consequence of the Hodge theory for Kahler manifolds though in cases (ii) and (iii) it is necessary to use Deligne's Hodge theory for local systems which underlie real variations of polarized Hodge structure. The point here is that the assumption (ii) (or (iii)) implies that the flat $\mathfrak{g}$-bundle associated to $\operatorname{Ad}\rho$ underlies a real variation of polarized Hodge structure of weight zero. This first part follows from [GM2] though only case (i) was stated and proved there.

The second part of the theorem is that ρ has a neighborhood in $R(\Gamma,G)$ which is analytically equivalent to a neighborhood of 0 in TC_ρ. That this is not generally the case for algebraic varieties is seen by considering the case of the curve $y(y-x^2) = 0$, a parabola together with its tangent line at the vertex. The tangent cone coincides with the tangent line. To effect the second step for the case in hand we establish isomorphisms of nth-order neighborhoods of ρ and 0 for each n. This is accomplished by giving a description of each nth-order neighborhood as a set of gauge equivalence classes of flat connections with nilpotent parameters then applying the previous Hodge theory techniques. Passing to the limit we obtain a formal isomorphism and consequently by Artin's Theorem [A] an analytic isomorphism.

Our description of the universal deformation is as concrete as could be expected in the general case. It depends only upon knowing the bracket product on the first cohomology group with local coefficients equal to the $\mathfrak{g}$-bundle associated to $\mathrm{Ad}\,\rho$. However, it is important to obtain more concrete information in special cases and especially to construct examples of representations which are not rigid.

We conclude with a discussion of an interesting example from [LM]. Let Γ be a $(3,3,3)$ triangle group. Then Γ is a Bieberbach group. We have an exact sequence

$$\mathbb{Z}^2 \longrightarrow \Gamma \longrightarrow S_3.$$

Let ρ be the irreducible two-dimensional representation of S_3 in $GL(2,\mathbb{C})$ pulled back to Γ. In [LM] it is established that

(i) $\dim H^1(\Gamma,\mathfrak{g}) = 1$,

(ii) ρ is locally rigid.

It is shown in [GM3] that the quadratic singularity theorem (case (i)) stated previously for fundamental groups of Kahler manifolds holds also for Bieberbach groups. Combining this with (i) and (ii) of [LM] we obtain the following theorem.

Theorem. $R(\Gamma,G)$ is isomorphic to the non-reduced scheme $PGL(2,\mathbb{C}) \times \mathrm{Spec}\ \mathbb{C}[t]/(t^2)$.

References

[A] M. Artin, On solutions to analytic equations, Invent. Math. 5 (1968), 277-291.

[B] D. Birkes, Orbits of linear algebraic groups, Annals of Math. 93 (1971), 459-475.

[C] K. Corlette, Gauge theory and representations of Kahler groups, these proceedings.

[BW] A. Borel and N. Wallach, *Continuous Cohomology, Discrete Subgroups, and Representations of Reductive Groups*, Annals of Math. Studies No. 94, Princeton University Press (1980).

[DM] P. Deligne and G.D. Mostow, Monodromy of hypergeometric functions and non-lattice integral monodromy, Publ. Math. I.H.E.S., no. 63 (1986), 5-89.

[G] W.M. Goldman, Geometric structures on manifolds and varieties of representations, these proceedings.

[GM1] W.M. Goldman and J.J. Millson, Local rigidity of discrete groups acting on complex hyperbolic space, Invent. Math. 88 (1987), 495-520.

[GM2] W.M. Goldman and J.J. Millson, Deformations of flat bundles over Kahler manifolds, in "*Geometry and Topology, Manifolds, Varieties and Knots,*" C. McCrory and T. Shifrin (eds.), Lecture Notes in Pure and Applied Mathematics 105, Marcel Dekker, New York-Basel (1987), 129-145.

[GM3] W.M. Goldman and J.J. Millson, The deformation theory of representations of fundamental groups of compact Kahler manifolds, submitted to Publ. Math. of the I.H.E.S.

[JM] D. Johnson and J.J. Millson, Deformation spaces associated to compact hyperbolic manifolds, in "*Discrete Groups in Geometry and Analysis, Papers in Honor of G.D. Mostow on His Sixtieth Birthday,*" R. Howe (ed.), Progress in Math. 67, Birkhauser, Boston-Basel-Stutgart (1987), 48-106.

[K] C. Kourouniotis, Deformations of hyperbolic structures on manifolds of several dimensions, Math. Proc. Comb. Phil. Soc. 98 (1985), 247-261.

[LM] A. Lubotzky and A. Magid, Varieties of representations of finitely generated groups, Memoirs A.M.S. 336 (vol. 5) (1985).

[N] M. Nori, Non arithmetic monodromy, C.R. Acad. Sci. Paris, to appear.

[R] M.S. Raghunathan, On the first cohomology of discrete subgroups of semi-simple Lie groups, Amer. J. Math. 87 (1965), 103-139.

[UY] K. Uhlenbeck and S.T. Yau, On the existence of Hermitian-Yang-Mills connections in stable vector bundles, Comm. Pure and Appl. Math. 39 (1986), 257-293.

[Z] R. Zimmer, *Ergodic Theory and Semi-simple Groups*, Monographs in Mathematics, Birkhauser (1984).

Department of Mathematics
UCLA
Los Angeles, CA 90024

Contemporary Mathematics
Volume **74**, 1988

CONNECTED COMPONENTS OF REPRESENTATION VARIETIES

Kent Morrison[1]

ABSTRACT. This is primarily a survey of results known about the connected components of varieties of representations of groups and algebras. The examples discussed are finite groups, finite dimensional algebras, commutative algebras, finite dimensional Lie algebras, finitely generated nilpotent groups, fundamental groups of compact orientable surfaces.

Let Γ be a finitely generated discrete group or let A be a finitely generated associative k-algebra over a field k. The set of n-dimensional representations of Γ or A, that is, either $\mathrm{Hom}_{grp}(\Gamma,GL_n(k))$ or $\mathrm{Hom}_{k\text{-}alg}(A,M_n(k))$, is a variety defined over k. The basic problem of interest in this article is the characterization of the connected components of these varieties in terms of algebraic data coming from Γ or A, or, as we shall see with surface groups, in terms of topological data -- characteristic classes -- of flat bundles on the surface. Since these spaces of representations are algebraic varieties, for each n there are only a finite number of connected components.

Let $R_n(\Gamma)$ and $R_n(A)$ denote these algebraic varieties. (This is the notation of [LM] which will be used in this article.) They are affine varieties over k, which we take to be algebraically closed. We will generally use $R_n(A)$ in what follows. The reader should realize that setting $A = k\Gamma$, the group algebra of Γ, will give us $R_n(\Gamma)$. Two representations are isomorphic when they are in the same orbit of the conjugation action of $GL_n(k)$ on $R_n(A)$. Each orbit is connected since it is the image of the connected group $GL_n(k)$ under the orbit map, so we need only consider the connected components of the orbit space $R_n(A)/GL_n(k)$. This, however, is not usually a nice space but there is a better quotient; it is the variety associated to the ring of invariants $\mathcal{O}(R_n(A))^{GL_n(k)}$. Its points are in one-to-one correspondence with

[1]1980 Mathematics Subject Clasification (1985 Revision). 16A64, 20C99, 14D20, 19L99, 53C05, 55R25, 55R50.

the isomorphism classes of semi-simple representations, these being precisely the closed orbits of the action, and is denoted $SS_n(A)$. See [LM, Pr]. When Γ is a fundamental group topologists as in [CS] are calling $SS_n(\Gamma)$ the "character variety" of n-dimensional representations, a suggestive name because the traces of the elements $\rho(\gamma)$, for $\gamma \in \Gamma$ fixed and ρ varying over representations, generate the coordinate ring of $SS_n(\Gamma)$. For this quotient the connected components are also in one-to-one correspondence with those of $R_n(\Gamma)$, the reason being that a component of $R_n(\Gamma)$ is determined by a minimal idempotent in its coordinate ring and idempotents are $GL_n(k)$-invariant since $GL_n(k)$ is connected. So we have reduced the main question to this: when are two semi-simple representations in the same component of $SS_n(A)$?

When $k = \mathbb{C}$ we can consider the usual (strong) topology on $R_n(A)$ and $SS_n(A)$ and we can ask about the strongly connected components and the path components. Happily all three notions -- the Zariski components, the strong components, the path components -- coincide for complex algebraic varieties. It is easy to see that path connected $\Rightarrow$ strongly connected $\Rightarrow$ Zariski connected. To show Zariski connected $\Rightarrow$ path connected, first note that a Zariski connected component is the union of irreducible components with a connected incidence graph. Vertices represent irreducible components and two vertices are connected by an edge if the corresponding irreducible components meet. Then use the fact that two points in an irreducible variety lie in an irreducible curve, which over $\mathbb{C}$ is a Riemann surface, possibly singular, but certainly path connected.

Now the problem of classifying the components should be easier than that of classifying all finite dimensional representations and even easier, as we have seen, than classifying the semi-simple representations. In the classical moduli problems of algebraic geometry there are two aspects: first to describe the discrete invariants (genus of a curve, degree and rank of a vector bundle, Hilbert polynomial) and then to describe the continuous invariants (construction of a good moduli space). The discrete invariants are usually easy to see and understand while the continuous invariants require all the work. But for general representation varieties we have a good moduli space $SS_n(A)$, although we do not know too much about it, but we do not even know the discrete invariants, i.e. the connected components.

In order to put some algebraic structure into the picture one may collect together the connected components of $R_n(A)$ for all n in $\mathbb{N}$ and make them into an abelian monoid using direct sum for addition. Then form the associated group by adjoining formal inverses and denote this group $D(A)$. In the case of representations of a group Γ the tensor product provides a multiplication to make $D(\Gamma)$ a commutative ring, and $D(\Gamma)$ is a quotient of the usual representation ring $R[\Gamma]$. In fact, by topologizing $R[\Gamma]$ with the natural topology obtained by constructing $R[\Gamma]$ as the group associated to the monoid $\bigcup_{n\geq 0} R_n(\Gamma)/GL_n(k)$ (disjoint union), we see that $D(\Gamma)$ is the ring of connected components, $R[\Gamma]$ modulo the connected component containing 0.

Intermediate between $R[\Gamma]$ and $D(\Gamma)$ is $K_0(\Gamma\text{-}\underline{\text{mod}})$ where Γ-mod is the category of finite dimensional representations of Γ. There are surjective homomorphisms $R[\Gamma] \twoheadrightarrow K_0(\Gamma\text{-}\underline{\text{mod}}) \twoheadrightarrow D(\Gamma)$. Now $K_0(\Gamma\text{-}\underline{\text{mod}})$ as a group is free abelian on the classes of simple modules and the class of ρ in $D(\Gamma)$ only depends on the image of ρ in $SS_n(\Gamma)$, which is the isomorphism class of the semi-simplification of ρ, that is, the sum of the simple modules in a composition series for ρ. Therefore $K_0(\Gamma\text{-}\underline{\text{mod}})$ maps surjectively onto $D(\Gamma)$. Likewise the corresponding statements hold for the groups $R[A]$, $K_0(A\text{-}\underline{\text{mod}})$ and $D(A)$.

A word of warning: it is not known in general that the cancellation property holds for the monoid of components, so it is not known whether we lose any information in the construction of the group $D(A)$. We do not know whether it is possible for $\rho \oplus \sigma$ to be in the same component as $\rho \oplus \tau$ while σ and τ are in different components. In all known examples, however, cancellation does in fact hold.

The basic data are collected in the following examples.

Example 1 Let A be a finite dimensional k-algebra. Each component of $R_n(A)$ contains a unique closed orbit of semi-simple representations, so that the connected components are bijective with the isomorphism classes of semi-simple representations. Two representations are in the same component if and only if they have isomorphic simple factors in their composition series. If A is semi-simple then the components are exactly the orbits and also the irreducible components; this is the situation for representations of a finite group Γ when the characteristic of k does not divide the order of Γ. We have $D(A) = R[A]$ and $D(\Gamma) = R[\Gamma]$; the quotient varieties $SS_n(A)$ and $SS_n(\Gamma)$ are finite sets of isolated points [Ar, Ga, Pr].

__Example 2__ A is a finitely generated commutative k-algebra, k algebraically closed. A splits into a product of connected algebras (i.e. having no non-trivial idempotents or equivalently a connected spectrum) $A = A_1 \times ... \times A_s$. An A-module decomposes into a sum of A_i-modules and $SS_n(A)$ is the disjoint union of $SS_{n_1}(A_1) \times ... \times SS_{n_s}(A_s)$ as $(n_1, ..., n_s) \in \mathbb{N}^s$ ranges over s-tuples with $n_1 + ... + n_s = n$. Then $SS_m(A_i) = (\text{Spec } A)^m/\Sigma_m$ because the simple A_i-modules are one-dimensional and $R_1(A_i) = SS_1(A_i) = \text{Spec } A_i$; Σ_m is the symmetric group acting on the m-fold product of Spec A by permuting the factors. Then $SS_m(A_i)$ is connected so that the components of $SS_n(A)$ are identified by $(n_1, ..., n_s)$ which is a dimension vector. Cancellation holds in the monoid and $D(A) = \mathbb{Z}^s$ [Mo 1].

__Example 3__ $A = U(\mathfrak{g})$, the enveloping algebra of a finite dimensional Lie algebra $\mathfrak{g}$ over a field k, algebraically closed of characteristic zero. In this case we write $R_n(\mathfrak{g})$, and so on, in place of $R_n(U(\mathfrak{g}))$. The components of $R_n(\mathfrak{g})$ are bijective with the components of $R_n(\mathfrak{g}/\text{rad } \mathfrak{g})$, via the natural map $R_n(\mathfrak{g}/\text{rad } \mathfrak{g}) \to R_n(\mathfrak{g})$, and these are precisely the isomorphism classes of $\mathfrak{g}/\text{rad } \mathfrak{g}$ - modules since $\mathfrak{g}/\text{rad } \mathfrak{g}$ is semi-simple. Again cancellation holds in the monoid of connected components since it is inherited from the category of $\mathfrak{g}/\text{rad } \mathfrak{g}$-modules. $D(\mathfrak{g})$ as a group is free abelian on the classes of simple $\mathfrak{g}/\text{rad } \mathfrak{g}$-modules. [Mo 1].

Examples 1-3 have an underlying similarity. In each case there is a subalgebra $B \subset A$ such that $R_n(A) \to R_n(B)$ induces a bijection on connected components __and__ B is a semi-simple algebra so that the components of $R_n(B)$ are just the orbits. Thus $SS_n(B)$ is a finite set of points and the inverse images of the map $SS_n(A) \to SS_n(B)$ are the connected components. In example 1, B is a semi-simple algebra isomorphic to A/N where N is the nilpotent radical. In example 2, $B = k \times ... \times k$, s factors, embedded in

$A = A_1 \times \dots \times A_s$. In example 3, $B = U(\mathfrak{s})$ where $\mathfrak{s}$ is the semi-simple factor in the Levi decomposition $\mathfrak{g} = \mathfrak{s} \oplus \text{rad}\,\mathfrak{g}$ [Mo 1].

Example 4 Γ is a nilpotent group and Γ^{ab} is torsion free (Γ^{ab} is the abelianization $\Gamma/[\Gamma,\Gamma]$.) The components of $SS_n(\Gamma)$ are the twist-isomorphism classes and they are the irreducible components, too. Let $\rho = \rho_1 \oplus \dots \oplus \rho_m$ be a semi-simple representation decomposed into simple ones. Let $\chi = (\chi_1, \dots, \chi_m)$ be a multi-character; i.e. each χ_i is a one-dimensional representation. Then $\chi \bullet \rho = \chi_1\rho_1 \oplus \dots \oplus \chi_m\rho_m$ is the twist of ρ by χ and is said to be "twist-isomorphic" to ρ. The twist of a simple representation remains simple so $D(\Gamma)$ is generated by the twist-isomorphism classes of simple representations, and in fact freely generated by them, the argument being that the monoid of connected components is the free abelian monoid on the twist-isomorphism classes of simple representations. If $\rho = \rho_1 \oplus \dots \oplus \rho_m$ and $\sigma = \sigma_1 \oplus \dots \oplus \sigma_l$ are in the same component then each simple factor ρ_i is twist isomorphic to some simple σ_j and $l = m$. The monoid is $\mathbb{N}^{(S)}$ with S the set of twist-isomorphism classes of simple representations and $D(\Gamma) = \mathbb{Z}^{(S)}$ [LM, Ru].

Whereas the components in these four examples are given by algebraic data in the rest of the examples topological data -- characteristic classes or classes of vector bundles -- prove effective. We also consider representations into groups other than $GL_n(k)$ such as $PSL_2(\mathbb{R})$ and $U(n)$.

Let M be a manifold. For each homomorphism ρ from the fundamental group of M to a Lie group G there is a principal G-bundle P over M with a flat connection and also flat connections on any associated bundles. Thus $GL_n(\mathbb{C})$ representations give flat complex vector bundles. The construction of the principal bundle is this: $\pi_1(M)$ acts on $\tilde{M} \times G$ by $\gamma \bullet (z,g) = (z\gamma^{-1}, \rho(\gamma)g)$, where $\tilde{M}$ is viewed as a principal $\pi_1(M)$-bundle, and $P = (\tilde{M} \times G)/\pi_1(M)$. The flat connection on P, viewing it as a horizontal distribution, descends from the trivial flat connection on $\tilde{M} \times G$ whose leaves are $\tilde{M} \times \{g\}$, and ρ is the holonomy or the monodromy of the connection. The term holonomy is preferred by differential geometers and monodromy is preferred by algebraic geometers. If ρ is changed to an isomorphic representation the resulting flat connection is isomorphic, using the appropriate notion of morphism of connections, to the flat connection arising from ρ, and every flat connection on a principal G-bundle can be constructed from its holonomy (monodromy). Therefore there is a bijection between $\mathrm{Hom}(\pi_1(M),G)/G$, the isomorphism classes of representations, and the isomorphism classes of flat connections on all principal G-bundles over M.

Key point If ρ_0 and ρ_1 are in the same component of $\mathrm{Hom}(\pi,G)$ then there is a homotopy of the associated principal bundles so that the bundles must be isomorphic (disregarding the flat connections). Thus any topological invariant of G-bundles is a discrete invariant of representations into G.

First and foremost among these invariants is simply the topological type of the bundle and is the main theme of the remaining examples. The topological type may be expressed as an Euler class or a Stiefel-Whitney class or a Chern class, particularly when the base space is a surface, in which case it can also be seen as the obstruction to lifting the representation $\pi_1 \to G$ to the universal cover $\tilde{G}$.

When G is an algebraic group (real or complex), as in all our examples, $\mathrm{Hom}(\pi,(M),G)$ has a finite number of components, since it is an algebraic variety, so there are only a finite number of distinct principal G-bundles that have flat connections. Understanding which G-bundles have flat connections is part of the problem of classifying the connected components of $\mathrm{Hom}(\pi,G)$ using topological information. The components (finite in number) are grouped into classes with isomorphic underlying G-bundles so that the classification problem has two parts:

(1) Identify the G-bundles admitting flat connections;

(2) For each such G-bundle indentify the components associated to it.

Example 5 We will see this program explicitly here for M a compact orientable surface of genus $g \geq 2$ and $G = PSL_2(\mathbb{R})$ or $SL_2(\mathbb{R})$. Goldman [Go 1] and Jankins [Ja] determined that the connected components of $Hom(\pi_1(M),PSL_2(\mathbb{R}))$ are exactly the fibers of the continuous map $e : Hom(\pi_1(M),PSL_2(\mathbb{R})) \rightarrow \mathbb{Z}$, the "Euler number" map defined so that $e(\rho)$ is the Euler number of the $\mathbb{R}\mathbb{P}^1$-bundle associated to the principal $PSL_2(\mathbb{R})$-bundle arising from ρ where $PSL_2(\mathbb{R})$ acts on $\mathbb{R}\mathbb{P}^1 = S^1$ in the natural way as linear fractional transformations on $\mathbb{R} \cup \{\infty\}$, or equivalently as automorphisms of $\mathbb{R}\mathbb{P}^1$ as the set of lines in $\mathbb{R}^2$. It was already known by work of Wood [Wo] that an S^1-bundle with structure group $Diff^+ S^1$ has a flat connection if and only if its Euler number e satisfied $|e| \leq |\chi(M)| = 2g - 2$. (Beware that these are not principal S^1-bundles under consideration.) Earlier Milnor had shown that a principal bundle with structure group $GL_2^+(\mathbb{R})$, the plus denoting positive determinant, has a flat connection if and only if its Euler number (that of the associated plane bundle) e satisfies $|e| \leq g - 1$ [Mi]. Now for $GL_2^+(\mathbb{R})$ we may just as well use $SL_2(\mathbb{R})$. The factor of 2 present in the first inequality can be explained by the fact that $q : SL_2(\mathbb{R}) \rightarrow PSL_2(\mathbb{R})$ is a double cover.

We have the commutative diagram

$$\begin{array}{ccc} Hom(\pi_1(M),SL_2(\mathbb{R})) & \xrightarrow{q_*} & Hom(\pi_1(M),PSL_2(\mathbb{R})) \\ e' \downarrow & & \downarrow e \\ \mathbb{Z} & \xhookrightarrow{2} & \mathbb{Z} \end{array}$$

where e and e' are the Euler class maps and 2 denotes multiplication by 2. Goldman [Go 2] has also determined the components of $Hom(\pi_1(M),SL_2(\mathbb{R}))$. For $1 - g < k < g - 1$, $(e')^{-1}(k)$ is a component and is a covering space of $e^{-1}(2k)$ of index 2^{2g}. For $k = 1 - g$ and $k = g - 1$, $(e')^{-1}(k)$ is 2^{2g} components, each one of which is diffeomorphic to $e^{-1}(2k)$. The difference for the extreme value of k is due to the fact that the components of $Hom(\pi_1(M),PSL_2(\mathbb{R}))$ with Euler number $\pm(2 - 2g)$ correspond to Fuchsian representations of $\pi_1(M)$ and their orbit spaces correspond to Teichmüller space which is contractible. Thus we see that there are $2^{2g+1} + 2g - 3$ components of $\mathrm{Hom}(\pi_1(M), SL_2(\mathbb{R}))$ partitioned into $2g - 1$ topological types.

More can be said about the topology of the other components of $\mathrm{Hom}(\pi_1(M), PSL_2(\mathbb{R}))$. Hitchin [Hi] has recently shown that the moduli space $e^{-1}(k)/PSL_2(\mathbb{R})$ is diffeomorphic to a complex vector bundle of $\mathbb{C}$-rank $g - 1 + k$ over the symmetric product $S^{2g-2-k}M$.

The Euler number can be seen more directly for a representation $\rho : \pi_1(M) \to G$ as an obstruction to lifting ρ to $\tilde{\rho} : \pi_1(M) \to \tilde{G}$, where $\tilde{G}$ is the universal covering group of G. Let $\pi_1(M)$ have the standard presentation $\langle \alpha_1, \ldots, \alpha_g, \beta_1, \ldots, \beta_g : \Pi [\alpha_i, \beta_i] = 1 \rangle$. Let $\rho(\alpha_i) = A_i$, $\rho(\beta_i) = B_i$ in G. Pick any lifts of A_i and B_i in $\tilde{G}$ and call them $\tilde{A}_i$ and $\tilde{B}_i$. Then $\Pi [\tilde{A}_i, \tilde{B}_i] \in \mathrm{Ker}\, p = \pi_1(G)$, $p : \tilde{G} \to G$. As $\pi_1(PSL_2(\mathbb{R})) = \mathbb{Z}$ we get an integer which is the Euler number. For an arbitrary manifold M we consider the exact sequence of Čech cohomology sets

$$H^1(M, \tilde{G}) \to H^1(M, G) \to H^2(M, \pi_1(G))$$

arising from the exact sequence of constant sheaves over M

$$0 \to \pi_1(G) \to \tilde{G} \to G \to 1,$$

and we interpret $H^1(M,G)$ as the isomorphism classes of principal G-bundles with flat connections, i.e. $H^1(M,G) = \mathrm{Hom}(\pi_1(M), G)/G$. The obstruction class map o_2 is the composition

$$\mathrm{Hom}(\pi_1(M), G) \to \mathrm{Hom}(\pi_1(M), G)/G \xrightarrow{\approx} H^1(M,G) \to H^2(M, \pi_1(G)).$$

For $G = U(n)$ or $GL_n(\mathbb{C})$ we get the first Chern class, an integral cohomology class since $\pi_1(U(n)) = \mathbb{Z}$. For $G = SO(n)$, $n \geq 3$, we get the second Stiefel-Whitney class in $H^2(M, \mathbb{Z}_2)$, whose vanishing means the structure group lifts to $\mathrm{Spin}(n)$.

In 1976 Heitsch posed the problem of finding the supremum of the absolute value of the Euler number for real vector bundles of rank $2n$ with structure group a discrete subgroup of $SL_{2n}(\mathbb{R})$ i.e. flat orientable bundles, over a manifold M of dimension $2n$ and to do so in terms of topological invariants of M such as the Euler and Pontryagin classes [DT]. Earlier Sullivan showed that an upper bound is the number of top dimensional simplices of a triangulation of M [Su]. See also [Du, He].

Example 6 M is a compact surface of genus g just as in the previous example. The space of equivalence classes of unitary representations $\mathrm{Hom}(\pi_1(M),U(n))/U(n)$ is the moduli space of semi-stable holomorphic vector bundles of degree 0 (i.e. topologically trivial) and rank n on a Riemann surface with M as the underlying differentiable manifold [NS, Se]. A priori $\mathrm{Hom}(\pi_1(M),U(n))/U(n)$ has no complex structure but in fact it is a projective variety and the complex structure depends on the complex structure of M, an aspect described in greater generality in [AB, H]. When $n = 1$, the U(1) conjugation action is trivial, and $\mathrm{Hom}(\pi_1(M),U(1))$ is the Jacobian variety of the Riemann surface. Topologically, $\mathrm{Hom}(\pi_1(M),U(1)) = \mathrm{Hom}(H_1(M,\mathbb{Z}),U(1)) = \mathrm{Hom}(\mathbb{Z}^{2g},U(1)) = U(1)^{2g}$, but of course this ignores the complex structure [Gu 1, Gu 2].

$\mathrm{Hom}(\pi_1(M),U(n))$ is connected as shown by the work of Narasimhan and Seshadri. The quotient space fibers over the Jacobian variety of M; the projection is the map that sends each unitary representation to its determinant. The fibers are irreducible varieties. It should be noted that since U(n) is compact each representation is semi-simple and so each orbit is closed and $\mathrm{Hom}(\pi_1(M),U(n))/U(n)$ is automatically Hausdorff.

Example 7 For bundles of non-zero degree Atiyah and Bott [AB] have shown that there is a central extension of $\pi_1(M)$ by $\mathbb{R}$ that they denote $\Gamma_{\mathbb{R}}$ which generalizes the role of the fundamental group in degree zero. The orbit space $\mathrm{Hom}(\Gamma_{\mathbb{R}},U(n))/U(n)$ is the moduli space of all semi-stable holomorphic vector bundles of rank n over M. The degree (first Chern class) is a discrete invariant and there is exactly one connected component for each degree. Thus there are an infinite number of components, but $\mathrm{Hom}(\Gamma_{\mathbb{R}},U(n))$ is not an algebraic variety since $\Gamma_{\mathbb{R}}$ is not finitely generated. It is defined by the exact sequence $0 \to \mathbb{R} \to \Gamma_{\mathbb{R}} \to \pi_1(M) \to 1$ with $\prod [\alpha_i,\beta_i] = 1 \in \mathbb{R}$. For $n = 1$ every homomorphism $\Gamma_{\mathbb{R}} \to U(1)$ factors through $\Gamma_{\mathbb{R}}^{ab} \cong U(1) \times \mathbb{Z}^{2g}$ so $\mathrm{Hom}(\Gamma_{\mathbb{R}},U(1)) = \mathrm{Hom}(U(1) \times \mathbb{Z}^{2g},U(1)) \cong \mathbb{Z} \times U(1)^{2g}$ with the $\mathbb{Z}$-factor specifying the degree. This space, which is an abelian group, is the Picard group of M; the connected component containing 0 is the Jacobian variety.

Example 8 We know that $\mathrm{Hom}(\pi_1(M),GL_2(\mathbb{C})) = R_2(\pi_1(M))$ is connected for M a compact orientable surface of genus $g \geq 0$, and for $g = 0$ and 1, $R_n(\pi_1(M))$ is connected, but it is still unknown for $g \geq 2$ and $n \geq 3$ whether or not $R_n(\pi_1(M))$ is connected. The question can be reduced to the special linear representations:

If $\mathrm{Hom}(\pi_1(M),SL_n(\mathbb{C}))$ is connected for M a compact orientable surface then $\mathrm{Hom}(\pi_1(M),GL_n(\mathbb{C}))$ is connected. That is because the tensor product (twist) map $t : \mathrm{Hom}(\pi_1(M),\mathbb{C}^x) \times \mathrm{Hom}(\pi_1(M),SL_n(\mathbb{C})) \to \mathrm{Hom}(\pi_1(M),GL_n(\mathbb{C}))$ is surjective, since the equation $\chi \otimes \sigma = \rho$ can be solved for χ and σ, when ρ is given, by $\chi = (\det \rho)^{1/n}$ and $\sigma = \chi^{-1} \otimes \rho$. The nth roots of $\det \rho$ exist because $\det \rho$ factors through the free abelian group $H_1(M,\mathbb{Z})$ and maps to the divisible group $\mathbb{C}^x$. The proposition follows since $\mathrm{Hom}(\pi_1(M),\mathbb{C}^x) \cong (\mathbb{C}^x)^{2g}$ is connected, so the domain of t is connected.

In [Go 1, Go 2] it is shown that $\mathrm{Hom}(\pi_1(M),SL_2(\mathbb{C}))$ is connected by rather explicit computations and arguments that will not work for $SL_n(\mathbb{C})$ in general. However the proof of connectedness would follow easily from this stronger conjecture: the fibers of the commutator map

$$SL_n(\mathbb{C}) \times SL_n(\mathbb{C}) \to SL_n(\mathbb{C})$$
$$(A,B) \to ABA^{-1}B^{-1}$$

are connected. Then it would be possible to deform a representation σ considered as a 2g-tuple $(A_1,A_2, \ldots, A_g,B_1, \ldots, B_g)$ to the trivial representation by arbitrarily moving $A_1, \ldots, A_{g-1},B_1, \ldots, B_{g-1}$ each to I and then deforming A_g and B_g so as to satisfy the defining relation. For $SL_2(\mathbb{C})$ it is true that these commutator fibers are connected, but the proof in [AM] does not generalize. There is only one isomorphism class of complex vector bundle arising from representations of $\pi_1(M)$ in $GL_n(\mathbb{C})$, namely the topologically trivial bundle, as contrasted with real vector bundles arising from $GL_2(\mathbb{R})$ representations. One explanation is that the Chern classes of a complex vector bundle with a flat connection are all torsion classes because as elements of $H^*(M,\mathbb{R})$ they can be constructed from the curvature form of a connection. Over an orientable surface there is no torsion and the Chern classes completely describe the bundles. The connectedness conjecture is equivalent to the claim that there is only one connected component of gauge-equivalence classes of flat connections on the trivial bundle $M \times \mathbb{C}^n$ when M is a surface.

The stable version of the conjecture is that $D(\pi_1(M)) = \mathbb{Z}$, and although weaker it is also unknown.

__Example 9__ This is not an example of known connected components but rather a general setting for the stable version of the connectedness conjecture for surfaces. For a finite group Γ there is a beautiful theorem of Atiyah [At 1] and for Γ a compact Lie group a generalization of Atiyah and Segal [AS] that relates the representation ring $R[\Gamma]$ with the ring of vector bundles $K(B\Gamma)$ over the classifying space of Γ. In particular the completion $\widehat{R[\Gamma]}$ with respect to the augmentaton ideal defined by the dimension map is isomorphic to $K(B\Gamma)$, which is complete by its construction as a limit of $K(B\Gamma_n)$ where $B\Gamma_n$ is the n-skeleton. When Γ is finite then $D(\Gamma)$ is the same as $R[\Gamma]$, but when Γ is discrete but not finite then the natural map $\alpha : R[\Gamma] \to K(B\Gamma)$ factors through $D(\Gamma)$, because the topological type of the bundle is constant on connected components. We denote this map by $\alpha : D(\Gamma) \to K(B\Gamma)$ which is also continuous for the augmentation ideal topology on $D(\Gamma)$, and thus there is a map $\hat{\alpha} : \widehat{D(\Gamma)} \to K(B\Gamma)$.

Consider the case that $\Gamma = \pi_1(M)$ for a surface M. Then $B\Gamma = M$ (which is also $K(\Gamma,1)$ since Γ is discrete). Now $K(M) = \mathbb{Z} \oplus \mathbb{Z}$ as a group and $\mathbb{Z}[T]/(T-1)^2$ as a ring where the bundle of rank n and degree (Chern class) d is $n + d(T-1)$. The image of α consists of the subring of flat bundles, i.e. those with $d = 0$, and is isomorphic to $\mathbb{Z}$. Since $D(\Gamma)$ contains a copy of $\mathbb{Z}$, knowing that α is injective would imply that $D(\Gamma) = \mathbb{Z}$. Merely knowing $\hat{\alpha}$ is injective would not give us that much without knowing the kernel of the completion $D(\Gamma) \to \widehat{D(\Gamma)}$.

However for a discrete group Γ the question is: to what extent is $\hat{\alpha} : \widehat{D(\Gamma)} \to K(B\Gamma)$ injective?

Another viewpoint for surface group representations in $GL_n(\mathbb{C})$ is through Chern classes. For a discrete group Γ and $\rho \in R_n(\Gamma)$ we can associate Chern classes $c_i(\rho) \in H^{2i}(\Gamma,\mathbb{Z}) = H^{2i}(B\Gamma,\mathbb{Z})$ which are simply the Chern classes of the associated vector bundle on $B\Gamma$ [At 1, Gr]. Obviously $c_i(\rho)$ only depends on the component containing ρ, and so $c_i(\rho)$ is a discrete invariant of $R_n(\Gamma)$. To what extent the Chern classes characterize the components is quite unknown, since even for Γ the fundamental group of a surface we do not know. (Then of course $c_1(\rho) = 0$ since it's a torsion class and $c_i(\rho) = 0$ for $i \geq 2$ since $B\Gamma = M$ has dimension 2.)

Example 10 This is another direction to generalize the conjecture that $\mathrm{Hom}(\pi_1(M),SL_n(\mathbb{C}))$ is connected for a surface M, replacing $SL_n(\mathbb{C})$ by an algebraic group G whose fundamental group is finite. As we have seen the obstruction to lifting $\rho : \pi_1(M) \to G$ to $\tilde{\rho} : \pi_1(M) \to \tilde{G}$ is given by $o_2 : \mathrm{Hom}(\pi_1(M),G) \to H^2(M,\pi_1(G)) = \pi_1(G)$. We may ask, as in [Go 2], to what extent are the fibers of o_2 the connected components? We have here two finite sets: the components of $\mathrm{Hom}(\pi_1(M),G)$ and $\pi_1(G)$ and also a map between them. The positive evidence that o_2 characterizes the components given in [Go 2] is for $G = SL_2(\mathbb{C})$ with $\pi_1(G) = 1$, as well as for $G = PSL_2(\mathbb{C})$ with $\pi_1(G) = \mathbb{Z}_2$, for $G = SU(2)$, $\pi_1(G) = 1$ and for $G = SO(3)$, $\pi_1(G) = \mathbb{Z}_2$. To this list we may add $G = SU(n)$, $\pi_1(G) = 1$, for all n, since the moduli space of special unitary representations $\mathrm{Hom}(\pi_1(M),SU(n))/SU(n)$ is connected by [Se]. It is the moduli space of semi-stable bundles of degree zero with holomorphically trivial determinant line bundle.

Example 11 For $SL_2(\mathbb{R})$ representations of a surface group we have seen that several connected components all give rise to the plane vector bundle with Euler number $g-1$ and $1-g$. Examples of this phenomenon for $GL_n(\mathbb{C})$ representations and complex vector bundles are given by spherical space forms - manifolds of the form S^k/Γ where Γ is a finite group. Thus they are manifolds M whose universal cover is a sphere and with finite fundamental group and they include the real projective spaces and lens spaces. Except for the antipodal actions of $\mathbb{Z}_2$ on even dimensional spheres giving the even dimensional real projective spaces, the spherical space forms all come from a finite subgroup Γ of $U(k)$ acting on $S^{2k-1} \subset \mathbb{C}^{2k}$ without fixed points. The classification of spherical space forms may be found in [Wf]. What we need is their K-theory found in [At 2, Gi]. The homomorphism from the representation ring of Γ to $K(M)$ is surjective and the kernel is the ideal generated by $\sum_{i=0}^{k} (-1)^i[\wedge^i \tau]$ where τ is the natural representation of Γ on $\mathbb{C}^k$ (remember Γ is a subgroup of $U(k)$) and $\wedge^i\tau$ is the exterior power.

Now Γ is finite so the components of $R_n(\Gamma)$ are just the isomorphism classes of representations. If two non-isomorphic representations ρ and σ differ by $\Sigma(-1)^i[\wedge^i\tau]$ and if the dimensions are large enough so that the associated vector bundles are in the stable range, then the bundles arising from ρ and σ are in different components.

For a concrete example consider $M = \mathbb{R}\mathbb{P}^3 = S^3/\mathbb{Z}_2$. Then $R[\mathbb{Z}_2] = \mathbb{Z}[x]/(x^2-1)$ where x is the class of the non-trivial character. The representation $\tau : \mathbb{Z}_2 \to GL_2(\mathbb{C})$ maps the generator to -I. Thus $[\tau] = 2x$ in $R[\mathbb{Z}_2]$, and $[\wedge^0\tau] = 1$, $[\wedge^1\tau] = [\tau] = 2x$, $[\wedge^2\tau] = 1$. So $K(\mathbb{R}\mathbb{P}^3) \cong \mathbb{Z}[x]/(x^2 - 1, 2x - 2)$. The rank 2 complex vector bundles arising from the trivial two-dimensional representation and from τ are the same in $K(\mathbb{R}\mathbb{P}^3)$, but that is in the stable range so they are isomorphic vector bundles and in fact the trivial bundle. The only two dimensional representations are the elements $2, 2x, x + 1$ in $R[\mathbb{Z}_2]$, but $x + 1$ defines a non-trivial bundle since $x + 1 \neq 2$ in $K(\mathbb{R}\mathbb{P}^3)$, so we see that the rank 2 trivial bundle has exactly two components of gauge-equivalence classes of flat connections. while the other rank 2 bundle has one component. This phenomenon cannot occur for line bundles; $H^2(\mathbb{R}\mathbb{P}^3,\mathbb{Z}) = \mathbb{Z}_2$ classifies line bundles by the first Chern class and there are just two one-dimensional representations which give rise to distinct elements of $K(\mathbb{R}\mathbb{P}^3)$ and must each, therefore, correspond to one of the line bundles.

In general for any manifold M and any line bundle the space of all flat connections is connected. (Of course it may be empty.) This is because the space of flat connections of a line bundle is an affine space modeled on the space of closed 1-forms which is a connected space. We can also see that there is a one-to-one correspondence between the components of $\mathrm{Hom}(\pi_1(M),\mathbb{C}^\times)$ and the torsion subgroup of $H^2(M,\mathbb{Z})$.

BIBLIOGRAPHY

[AM] S. Akbulut and J. McCarthy, "Casson's invariant for oriented homology 3-spheres," Lecture Notes from Michigan State Univ., 1985.

[Ar] M. Artin, "On Azumaya algebras and finite dimensional representations of rings," J. Algebra, 11(1969), 532-563.

[At 1] M. F. Atiyah, "Characters and cohomology of finite groups," Publ. Math. IHES, 9 (1961), 23-64.

[At 2] M. F. Atiyah, "K-theory," Benjamin, N.Y. 1967.

[AB] M. F. Atiyah and R. Bott, "The Yang-Mills equations over Riemann surfaces," Phil. Trans. R. Soc. London, A308, (1982), 523-615.

[AS] M. F. Atiyah and G. B. Segal, "Equivariant K-theory and completion," J. Diff. Geom., 3, (1969), 1-18.

[CS] M. Culler and P. B. Shalen, "Varieties of group representations and splittings of 3-manifolds," Ann. Math. 117 (1983), 109-146.

[Du] D. L. Dupont, "Simplicial de Rham cohomology and characteristic classes of flat bundles," Topology, 15(1976), 233-245.

[DT] "Differential topology, foliations and Gelfand-Fuks cohomology," ed. by Paul A. Schweizer, Springer Lecture Notes, vol. 652, 1978, 242-243.

[Ga] P. Gabriel, "Finite representation type is open," Proceedings of the International Conference on Representations of Algebras, vol. 488, Lecture Notes in Mathematics, Springer-Verlag, New York, 1976.

[Gi] P. B. Gilkey, "The eta invariant and the K-theory of odd dimensional spherical space forms," Invent. Math., 76, (1984), 421-453.

[Go 1] W. M. Goldman, "Discontinuous groups and the Euler class," Thesis, U. C. Berkeley, 1980.

[Go 2] W. M. Goldman, "Topological components of spaces of representations," preprint.

[Gr] A. Grothendieck, Classes de Chern et représentations linéaires des groupes discrets, "Dix exposées sur la cohomologie des schémas," North-Holland, Amsterdam, 1968, 215-305.

[Gu 1] R. C. Gunning, "Lectures on Riemann surfaces," Princeton Univ. Press, 1966.

[Gu 2] R. C. Gunning, "Lectures on vector bundles over Riemann surfaces," Princeton Univ. Press, 1967.

[He] J. L. Heitsch, "Secondary invariants of transversely homogeneous foliations," Michigan Math. J. 33(1986), 47-54.

[Hi] N. J. Hitchin, "The self-duality equations on a Riemann surface," Proc. London Math Soc. (3), 55(1987), 59-126.

[Ja] M. Jankins, "The space of homomorphisms from a Fuchsian group to $PSL(2,\mathbb{R})$, Thesis, University of Maryland, 1983.

[LM] A. Lubotzky and A. Magid, "Varieties of representations of finitely generated groups," Memoirs of Amer. Math. Soc. 336(1985).

[Mi] J. Milnor, "On the existence of a connection with curvature zero," Comm. Math. Helv. 32, (1958), 215-223.

[Mo 1] K. Morrison, "The scheme of finite dimensional representations of an algebra," Pac. J. Math, 91(1980), 199-218.

[Mo 2] K. Morrison, "Geometric equivalence relations on modules," J. Pure App. Algebra, 22(1981), 165-177.

[NS] M. S. Narasinhan and C. S. Seshadri, "Stable and unitary vector bundles on a compact Riemann surface," Ann. Math. 82, (1965), 540-567.

[Pr] C. Procesi, "Finite dimensional representations of algebras," Israel J. Math., 19(1974), 169-182.

[Ru] Z. Rudnik, "Representation varieties of solvable groups," J. Pure App. Algebra, 45(1987).

[Se] C. S. Seshadri, "Fibrés vectoriels sur les courbes algébriques," Astérisque no. 96, Soc. Math. de France, 1982.

[Su] D. Sullivan, "A generalization of Milnor's inequality concerning affine foliations and affine manifolds," Comm. Math. Helv., 51, (1976), 183-189.

[Wf] J. A. Wolf, "Spaces of constant curvature," Publish or Perish Inc., 1985.

[Wo] J. W. Wood, "Bundles with totally disconnected structure group," Comm. Math. Helv., 46, (1971), 257-273.

MATHEMATICS DEPARTMENT
CALIFORNIA POLYTECHNIC STATE UNIVERSITY
SAN LUIS OBISPO, CA 93407

Contemporary Mathematics
Volume **74**, 1988

A CHARACTERIZATION OF ORBIT CLOSURE

Joyce O'Halloran

Since algebraic group actions arise in many areas of mathematics and in applications of mathematics, it is important to develop techniques for describing the resulting orbit spaces. Such a description involves two problems: i) a description or parameterization of the set of orbits, and ii) a description of the closure of any orbit. When an algebraic group acts algebraically on a variety, the (Zariski) closure of an orbit is a union of orbits. (Over the fields $\mathbb{R}$ or $\mathbb{C}$, orbit closure in the Zariski topology is the same as orbit closure in the standard topology.) In most cases it is quite difficult to determine which orbits are in the closure of a given orbit. The purpose of this paper is to give an expository presentation of a characterization of orbit closure which has proven to be a useful approach to this problem. For particular algebraic group actions, this characterization appears in the literature (see, for instance, [8]). A general statement and its proof appear in [6].

1. SOME EXAMPLES OF ALGEBRAIC GROUP ACTIONS

A classical example of an algebraic group action is the conjugacy action of the general linear group Gl(n,k) on the variety M(n,k) of nxn matrices with entries in k.

$$g \cdot M = gMg^{-1}, \quad g \in Gl(n,k), \; M \in M(n,k)$$

If k is algebraically closed, then the orbits are parameterized by matrices in Jordan form; i.e. the set of orbits is in one–to–one correspondence with the set of matrices in Jordan form, modulo rearrangement of blocks. Orbit closure (established by Gerstenhaber in [4]) is given as follows. A matrix M in the closure of the orbit of a matrix N (where M and N are in Jordan form) has the same eigenvalues and eigenvalue multiplicities as N. If an eigenvalue α has multiplicity m, then the α–blocks of M and N are described by partitions of m. If M is in the closure of the orbit of N, then these two partitions of m must respect the following ordering on partitions of m:

$$(a_1, \ldots, a_m) \geq (b_1, \ldots, b_m), \text{ where } a_i \geq a_{i+1} \text{ and } b_i \geq b_{i+1}, \text{ if}$$
$$\Sigma_{i=1}^{k} a_i \geq \Sigma_{i=1}^{k} b_i \text{ for all } k.$$

1980 *Mathematics Subject Classification* (1985 *Revision*). 14L30.

Since a Lie algebra product (respectively an associative algebra product) on a vector space V is a bilinear transformation $V \times V \to V$, isomorphism classes of Lie algebras (respectively associative algebras) are orbits under the following action of Gl(n,k) on Hom(V⊗V,V), where V is a vector space over k of dimension n:

$$(g\cdot\mu)(x,y) = g(\mu(g^{-1}x,g^{-1}y)).$$

Identification of orbits and orbit closure for Lie algebras has been explored by Vergne, Carles, and Grunewald and myself, among others ([11], [1], [5]). Some results on orbit closure for associative algebras appear in, for instance, [10].

Cohomology classes can be realized as orbits under algebraic group actions. Consider the variety of 2–cocycles of an n–dimensional Lie algebra G with coefficients in k^r. In the cohomology class, a 2–cocycle z is equivalent to $z + \lambda([\ ,\ ])$, where $[\ ,\ ]$ is the Lie algebra product and λ is a linear transformation $G \to k^r$. We also want to consider equivalence under basis changes in G and in k^r. This equivalence on cohomology classes corresponds to the following action on $Z^2(G,k^r)$. Let G be the algebraic subgroup of $Gl(G \oplus k^r)$ consisting of elements of the form

$$g = \begin{bmatrix} \alpha & 0 \\ \lambda & \psi \end{bmatrix}, \text{ where } \alpha \in \text{Aut } G,\ \ \psi \in Gl(k^r).$$

The group G acts on $Z^2(G,k^r)$ by:

$$(g\cdot z)(x,y) = \psi(z(\alpha^{-1}x,\alpha^{-1}y)) + \lambda([\alpha^{-1}x,\alpha^{-1}y]).$$

Another example of an algebraic group action arises in the study of linear systems of differential equations. The standard control theory equations are of the form

$$\dot{x} = Ax + Bu,$$

where A is an nxn matrix, B is an nxm matrix, x is a function $\mathbb{R} \to \mathbb{R}^n$, and u is a function $\mathbb{R} \to \mathbb{R}^m$. The system is determined by the pair of matrices (A,B). If we change the basis for $\mathbb{R}^n$, i.e. if we replace x with Sx for some S in $Gl(n,\mathbb{R})$, we obtain an equivalent system via:

$$(S\dot{x}) = (SAS^{-1})(Sx) + SBu.$$

So the original system is equivalent to the system

$$\dot{x} = (SAS^{-1})x + SBu.$$

The resulting action of $Gl(n,\mathbb{R})$ on matrix pairs

$$g\cdot(A,B) = (gAg^{-1},gB)$$

has been studied by Hinrichsen and his colleagues [7].

Last, but not least, I should mention the algebraic group action which inspired this conference: the action of Gl(n,k) on the variety of n–dimensional representations of a finitely generated group, whose orbits are equivalence classes of representations. See [8] for a description of this action and results about orbit closure.

For other examples of algebraic group actions and for a very nice introduction to the subject, see Kraft's survey paper [9].

2. A CHARACTERIZATION OF ORBIT CLOSURE

In order to demonstrate this characterization of orbit closure, I will begin with a simple example in the case of the general linear group $Gl(2,\mathbb{C})$ acting by conjugation on the variety $M(2,\mathbb{C})$ of 2x2 matrices.

$$\begin{bmatrix}1&0\\0&1\end{bmatrix} \text{ is in the closure of the orbit of } \begin{bmatrix}1&1\\0&1\end{bmatrix}$$

via:

$$\begin{bmatrix}\epsilon&0\\0&1\end{bmatrix}\begin{bmatrix}1&1\\0&1\end{bmatrix}\begin{bmatrix}1/\epsilon&0\\0&1\end{bmatrix} = \begin{bmatrix}1&\epsilon\\0&1\end{bmatrix} \to \begin{bmatrix}1&0\\0&1\end{bmatrix} \text{ as } \epsilon \to 0.$$

Finding a curve whose intersection with an orbit is an open subset of the curve is a typical way of showing that an element is in the closure of an orbit. An alternate viewpoint in the example above is to consider the matrix $\begin{bmatrix}\epsilon&0\\0&1\end{bmatrix}$ as an element of the algebraic group $Gl(2,\mathbb{C}(\epsilon))$ which is acting on the variety $M(2,\mathbb{C}(\epsilon))$ and replace the limit $\epsilon\to 0$ with evaluation at 0:

$$\left[\begin{bmatrix}\epsilon&0\\0&1\end{bmatrix}\cdot\begin{bmatrix}1&1\\0&1\end{bmatrix}\right]\Bigg|_{\epsilon=0} = \left[\begin{bmatrix}1&\epsilon\\0&1\end{bmatrix}\right]\Bigg|_{\epsilon=0} = \begin{bmatrix}1&0\\0&1\end{bmatrix}.$$

The advantage of this viewpoint is that it replaces a family of group elements with a single entity. This example illustrates the basic idea behind the general theorem which follows. For the sake of uncluttered exposition, I have omitted the conditions on the functor (conditions which are satisfied in all applications I've considered); the reader will find the full statement and proof in [6].

THEOREM *Let* $\underline{S}$: k–algebras $\to$ sets *be a functor satisfying certain conditions, including:*

i) $\underline{S}(A)$ *has a* $\underline{G}(A)$ *action,*

ii) $\underline{S}(k)$ *is a variety with algebraic* $\underline{G}(k)$ *action.*

Let $x_0, x_1 \in \underline{S}(k)$. *Then* x_0 *is in the closure of the orbit of* x_1 *if and only if there is a discrete valuation* k–*algebra* A *with residue field* k, *whose quotient field* K *is finitely generated over* k *of transcendence degree one, and there is an element* $x_A \in \underline{S}(a)$ *such that*

$$j^*(x_A) = g\cdot((ji)^* x_1) \text{ for some } g \in \underline{G}(K), \text{ and}$$

$$\pi^*(x_A) = x_0,$$

where * *denotes the image of maps under the functor* $\underline{S}$, *the maps* i *and* j *are the inclusions* $k \to A$ *and* $A \to K$ *respectively, and* π *is the quotient map* $A \to A/M = k$.

In the example above, the functor $\underline{S}$ assigns an algebra B to the set of matrices over B, the algebra A is the local ring $k[\epsilon]_{<\epsilon>}$, and x_A is the matrix $\begin{bmatrix}1&\epsilon\\0&1\end{bmatrix}$.

The utility of this Theorem becomes clear when it is stated in terms of a particular functor. For instance, Proposition 1.24 of [8] is simply a statement of this theorem in the case of representation varieties. The proof of this theorem which appears in [6] is simply a generalization of the proof which appears in [8]. In [8], Lubotsky and Magid use this theorem to establish that the closed orbits are those of semisimple representations.

In determining relationships between orbit spaces given by different algebraic group actions, the ability to replace a family of objects with a single object can be very useful. It was this aspect which was exploited in [6]. There we considered central extensions of a Lie algebra G defined by elements of $H^2(G,k^r)$. Let C be the subset of $Z^2(G,k^r)$ consisting of those 2–cocycles which define central extensions with center k^r. There is an injection of the set of orbits in C (with the group action given in Section 1) into the set of orbits in the variety of Lie algebras (isomorphism classes of Lie algebras), and this correspondence respects orbit closure.

When the Theorem is stated in terms of variety of n–dimensional Lie algebras (where the orbits are the isomorphism classes), we have:

COROLLARY *Let* L_0, L_1 *be n–dimensional Lie algebras over* k. L_0 *is in the closure of the orbit of* L_1 *if and only if there is a discrete valuation* k*–algebra* A *with residue field* k, *whose quotient field* K *is finitely generated over* k *of transcendence degree one, and there is an n–dimensional Lie algebra* L_A *over* A *such that*

$$L_A \otimes K \cong L_1 \otimes K \quad \textit{and}$$
$$L_A \otimes k = L_0$$

This result has a striking resemblance to the definition of "formal deformation" of a Lie algebra (see [2]):

DEFINITION Let A be a local k–algebra which is either finite dimensional or complete. A *formal deformation parameterized by* A *of a Lie algebra* L_0 *over* k is a Lie algebra L_A over A such that $L_A \otimes k = L_0$.

This resemblance between orbit closure and formal deformations led Fialowski and myself to explore the similarities and differences of the two concepts (see [3]).

It is clear that this characterization of orbit closure is a powerful tool in the study of orbit spaces, and I would like to thank Andy Magid for pointing out the possibility of using this characterization in settings other than that of representation varieties.

BIBLIOGRAPHY

1. R. Carles, "Varietes d'algebres de Lie: point de vue global et rigidite," thesis, L'Universite de Poitiers (1984).

2. A. Fialowski, "Deformations of Lie algebras," Mat. Sbornyik USSR, 127 (169), N.4 (8) (1985), 476–482.

3. A. Fialowski, J. O'Halloran, "A comparison of deformations and orbit closure," preprint.

4. M. Gerstenhaber, "On nilalgebras and linear varieties of nilpotent matrices, III," Annals of Math. 70 (1959), 167–205.

5. F. Grunewald, J. O'Halloran, "Varieties of nilpotent Lie algebras of dimension less than six," J. of Algebra, to appear.

6. __________, "A characterization of orbit closure and applications," J. of Algebra, to appear.

7. D. Hinrichsen, "Canonical forms and parameterisation problems in linear systems theory," U. of Bremen preprint series, No. 123 (1984).

8. A Lubotsky, A. Magid, Varieties of representations of finitely generated groups, Memoirs of the AMS, No. 336, AMS, Providence, Rhode Island, 1985.

9. H. Kraft, "Geometric methods in representation theory" in: Representations of Algebras, Springer–Verlag, Berlin, Heidelberg, New York, 1982.

10. G. Mazzola, "The algebraic and geometric classification of associative algebras of dimension five," Manuscripta Math. 27 (1979), 81–101.

11. M. Vergne, "Varieties des algebres de Lie nilpotente," thesis, Fac. Sci. de Paris (1966).

DEPARTMENT OF MATHEMATICAL SCIENCES
PORTLAND STATE UNIVERSITY
PORTLAND, OREGON 97207

Contemporary Mathematics
Volume **74**, 1988

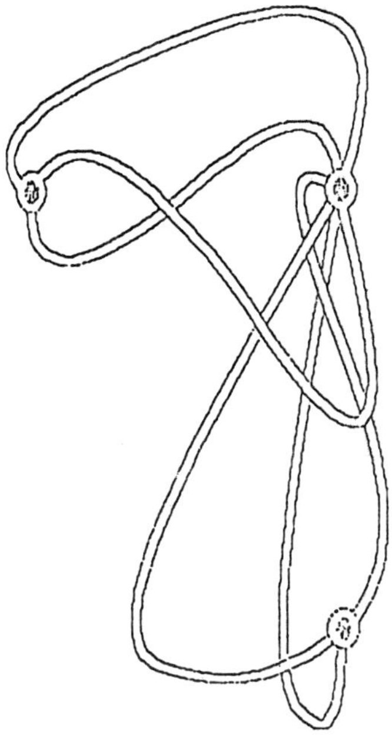

Calculus on Moduli Space

R. C. Penner

§ 1 INTRODUCTION.

It has become increasingly clear that two-dimensional conformal geometry and the (super-) string theory of particle physics are contiguous along numerous problems of mathematical and physical import. As a case in point, the problem of explicit differential calculus on the (super-) moduli space of Riemann surfaces is fundamental to explicit numerical computations in physics; analogously, there is much to be learned about the topology and geometry of the moduli space of Riemann surfaces from an explicit scheme of calculus. One goal of this note is to describe a numerically explicit scheme of integration of top-dimensional forms over various moduli spaces. (Actually, we give a substantial simplification of the integration scheme described in [12].)

Another point of contact between the mathematics and physics involves the construction of a "universal moduli space" (see [6] and [4]), which suitably generalizes the classical theory. We will also describe a model of a "(decorated) universal Teichmüller theory" with many nice properties and announce some recent results; this model is intended to give a proposed picture of some connections between the "combinatorial conformal geometry" described herein, the "Virasoro algebra" (i.e., the algebra of infinitessimal diffeomorphisms of the circle; see [7] for instance), and the solution to the Schottky problem provided by the phenomenal machinery of the KP heirarchy (see, for instance, [1], [9], [14], [15], and [16]).

§ 2 THE DECORATED TEICHMÜLLER SPACE

To begin, we define the *decorated Teichmüller space* $\tilde{\mathfrak{T}}_g^s$ as the total space of the canonical horocycle bundle

$$\begin{array}{c}\tilde{\mathfrak{T}}_g^s\\ \downarrow f\\ \mathfrak{T}_g^s\end{array}$$

over the classical Teichmüller space $\mathfrak{T}_g^s$ of the genus g, s-times punctured surface F_g^s, where $g \geq 0$, $s \geq 1$, and $2g - 2 + s > 0$; the fiber over a marked conformal type in $\mathfrak{T}_g^s$ is the space of *all* horocycles about *all* punctures of F_g^s. There is a natural action of the mapping class group MC_g^s of F_g^s on the bundle, and we are led to the "commutative square" of F_g^s

$$\begin{array}{ccc}\tilde{\mathfrak{T}}_g^s & \longrightarrow & \mathfrak{T}_g^s\\ (S_g^s)\downarrow & & \downarrow\\ \tilde{\mathfrak{M}}_g^s = \tilde{\mathfrak{T}}_g^s / MC_g^s & \rightarrow & \mathfrak{M}_g^s = \mathfrak{T}_g^s / MC_g^s\end{array}$$

To intuit about $\tilde{\Gamma}_m \in \tilde{\mathfrak{T}}_g^s$, let $\mathbb{M}^3$ denote Minkowski three-space with the pairing

$$< \underline{x}, \underline{y} > = -x_0y_0 + x_1y_1 + x_2y_2,$$

and define

$$L^+ = \{ v \in \mathbb{M}^3 : < v,v > = 0, \text{ and v has positive height}\}$$
$$\mathbb{H} = \{ v \in \mathbb{M}^3 : < v,v > = -1, \text{ and v has positive height}\}.$$

There is a canonical identification

$$L^+ \rightarrow \text{horocycle bundle over the circle } \mathbb{S}^1 \text{ at infinity}$$
$$w \rightarrow \{v \in \mathbb{H} : < v, w > = -1\} = h(w):$$

see Figure 1 and [10]. Identifying the Möbius group Möb with the identity component $SO^+(1,2)$ of isometries of $\mathbb{M}^3$ gives an identification

$$\tilde{\mathcal{T}}_g^s = \left\{ \begin{array}{l} (\Gamma_m, B_1,\ldots,B_s) : \Gamma_m \in \mathcal{T}_g^s \text{ and } B_i \subset L^+ \text{ is the } \Gamma_m \text{ -} \\ \text{orbit of points corresponding to the ith cusp} \end{array} \right\} / SO^+(1,2)$$

The "forgetful" map f is easily seen to be a principal $\mathbb{R}_+^s$-bundle.

This identification of $\tilde{\mathcal{T}}_g^s$ suggests the following coordinatization. Choose, one and for all, a decomposition Δ of F_g^s into ideal triangles with vertices at the punctures; see Figure 2 for examples. Fix $\Gamma_m \in \mathcal{T}_g^s$. After straightening to Γ_m-geodesics, Δ lifts to a tesselation $\tilde{\Delta}$ of the Poincaré disc $\mathbb{D}$; lift to $\mathbb{D}$ also the various distinguished horocycles arising from $\tilde{\Gamma}_m$; see Figure 3a. Given an arc $c \in \Delta$ connecting punctures, define the "λ-length coordinate"

$$\lambda(c; \tilde{\Gamma}_m) = \sqrt{2\, e^{\delta}} \ ,$$

where δ is the signed Poincare distance between horocycles along a lift $\tilde{c}$ of c to $\mathbb{D}$.

<u>REMARK</u>: The identification of $\tilde{\mathcal{T}}_g^s$ above motivates this definition by virtue of the identity $-\langle u,v \rangle = 2e^{\delta}$, where $u,v \in L^+$, and δ denotes the Poincaré distance between h(u) and h(v).

<u>THEOREM</u> [10]: *The map*

$$\tilde{\mathcal{T}}_g^s \quad \to \quad \mathbb{R}_+^{\Delta}$$

$$\tilde{\Gamma}_m \quad \to \quad \mathop{\times}_{c \in \Delta} \lambda(c; \tilde{\Gamma}_m)$$

is a surjective homeomorphism.

<u>REMARK</u>: There is a tractable algebraic description of the action of MC_g^s on these λ-length coordinates as a group of (integral) rational maps.

§ 3 h-LENGTHS

In preparation for the integration scheme, we next develop some other parameters on $\tilde{\mathfrak{T}}_g^s$; this is described more fully in [10]. An *end* of Δ is a Γ_m-orbit of ends (in the usual sense) of the triangles in the lift $\tilde{\Delta}$ of Δ to $\mathbb{D}$. Let $E(\Delta)$ denote the set of ends of Δ; notice that $\# E(\Delta) = 2 \# \Delta$. A specification of end E of Δ together with $\Gamma_m \in \mathfrak{T}_g^s$ determines a corresponding horocyclic segment (see Figure 3b), and we define the h-length (for "half horocyclic length") by $h(E; \tilde{\Gamma}_m) = 1/2 \times$ hyperbolic length of the corresponding horocyclic segment.

LEMMA: *Given linearly independent* $v_i \in L^+$, $i = 1,2,3$, *consider the corresponding h-lengths and λ-lengths indicated in Figure 4a. Then we have the equalities*

$$\alpha = a/bc$$
$$\beta = b/ac$$
$$\gamma = c/ab$$

Notice in particular that $\alpha\beta = 1/c^2$.

The proof, of course, is a computation. There is the immediate

COROLLARY: *The natural map*

$$\tilde{\mathfrak{T}}_g^s \to \mathbb{R}^{E(\Delta)}$$

$$\tilde{\Gamma}_m \to \times_{E\in E(\Delta)} h(E; \tilde{\Gamma}_m)$$

is an embedding onto the intersection $\mathcal{V}$ *(for "variety") of homogeneous quadrics determined by the "coupling equations"*

$$\alpha\beta = \gamma\delta$$

on coordinates $\alpha = h(A; \tilde{\Gamma}_m), \ldots, \varphi = h(F; \tilde{\Gamma}_m)$, *where* a,...,f *are the ends of* Δ *near an edge* $e \in \Delta$ *as in Figure 4b*; (*there is one coupling equation for each arc in* Δ).

REMARK: There are associated complex parameters (called "strand coordinates") arising from an embedding $\mathbb{R}_+^{E(\Delta)} \subset \mathbb{C}^{E(\Delta)}$ (which evolves from a simultaneous diagonalization of the quadrics) as follows. The orientation of F_g^s determines an orientation on each triangle of Δ and hence an induced orientation on each horocyclic segment (as in Figure 4a). An orientation $\tilde{c}$ on $c \in \Delta$ points towards a pair of ends E,E' (as in Figure 4c), and we define

$$\zeta(\tilde{c}; \tilde{\Gamma}_m) = h(E; \tilde{\Gamma}_m) + \sqrt{-1}\ h(E'; \tilde{\Gamma}_m).$$

This determines an IR-linear embedding $\mathbb{R}_+^{E(\Delta)} \subset \mathbb{C}^{E(\Delta)}$. An analytic change of coordinates identifies the variety $\mathcal{V}$ with the product $C^\Delta \subset (\mathbb{C}^2)^\Delta \approx \mathbb{C}^{E(\Delta)}$, where C is the "solid Clifford torus"

$$C = \{(z, w) \in \mathbb{C}^2 : |z| = |w|\}.$$

It is interesting that the action of MC_g^s is the restriction of a group of analytic motions of $\mathbb{C}^{E(\Delta)}$ preserving C^Δ at least in case s = 1.

§ 4 FATGRAPHS

We require one more combinatorial ingredient to describe the integration scheme; for more details, see [11]. A *fatgraph* G is a one-dimensional CW-complex so that each vertex has valence at least three, together with a cyclic order on the "half-edges" (i.e., edges of the first barycentric subdivision) about each vertex. A convenient pictorial representation is a planar projection (in the sense of classical knot theory) of a graph embedded in $\mathbb{R}^3$, where the cyclic order is that induced from a fixed orientation on the plane of projection. See Figure 5a for examples of fatgraphs. A fatgraph determines a pair (F(G), G) of the form (surface, spine of surface) in the nautral way as in Figure 5b; the obvious notion of isomorphism between fatgraphs corresponds to an isotopy class of orientation preserving homeomorphism of pairs. Let Aut(G) denote the fatgraph automorphism group and [G] the isomorphism class.

Our interest in fatgraphs stems from the existence of a MC_g^s-invariant cell-decomposition of the bundle $\tilde{\mathcal{T}}_g^s \to \mathcal{T}_g^s$, described in [10], [11]; see also [8], [3]. In fact, the collection of marked fatgraph classes forms a poset $\mathcal{G}_g^s$ under "Whitehead collapse" (or "contraction"); this combinatorial move is depicted in Figure 6.

<u>THEOREM</u>: *There is a geometric realization of the poset $\mathcal{G}_g^s$ which is naturally homeomorphic to the dual of a MC_g^s-invariant cell-decomposition of $\tilde{\mathcal{T}}_g^s$.* There is the

FUNDAMENTAL PROBLEM: *Fix appropriate g,s, and consider all isomorphism classes [G] of fatgraphs G with F(G) homotopy equivalent to* F_g^s. *Furthermore, fix the number e of edges of G. Develop a Dirichlet series of #{[G]; G has e edges and* $F(G) \approx F_g^s$} *(ideally including also the data #Aut(G)).*

REMARKS: 1) This problem is of direct relevance in the Physics; it is fair to say that these fatgraphs are some of Feynman diagrams of string theory, and the Fundamental Problem is related to growth of perturbative coefficients.

2) With Thurston's suggestion [18] of the relevance of Catalan numbers to the Fundmental Problem, there is some reasonable hope of an elegant solution (see, in particular, [5]); see also [11].

3) Using some old techniques, the Fundamental Problem above admits a reformulation in terms of the character theory of the symmetric group (see, for instance, [2; final appendix]). These techniques allow an easy algorithmic solution of the corresponding enumerative problem on the computer. Some sample numerical output in the on the form of isomorphism classes of fatgraphs for the surfaces F_1^2 and F_0^4 is in appendix.

§ 5 THE INTEGRATION SCHEME (an improvement over [12])

Before we pursue the primary applications of the techniques developed so far in this this paper, we must describe the connection between the fatgraph and h-length formalisms.

If Δ is an "ideal triangulation" of F_g^s as before, the formal dual of this cell-decomposition is, of course, a fatgraph $G \subset F_g^s$; see Figure 7. The edges of G are in bijective correspondence with the arcs in Δ, and in fact the "sectors" S(G) of G (see Figure 7) are in bijective correspondence with the ends $E(\Delta)$ of Δ. This correspondence induces an isomorphism

$$\mathbb{R}_+^{E(\Delta)} \approx \mathbb{R}_+^{S(G)}$$

and we may imagine "h-length coordinates" as defined on S(G).
A detail which might have been mentioned earlier is the existence of a canonical section σ

$$\begin{array}{c} \tilde{\mathcal{T}}_g^s \\ \sigma \uparrow \downarrow f \\ \mathcal{T} \end{array}$$

defined by the condition that all horocycles have unit hyperbolic length. Of course, a boundary component ∂ of $\partial F(G)$ corresponds to a horocycle on F_g^s, and the hyperbolic length of the corresponding horocycle is

$$h(\partial; \tilde{\Gamma}_m) = 2\Sigma \quad h(E; \tilde{\Gamma}_m),$$

where the sum is over all sectors of G "traversed" by ∂; see Figure 8. Clearly, the canonical section σ has image an affine locus $\mathcal{L} \subset \mathbb{R}_+^{E(\Delta)} \approx \mathbb{R}_+^{S(G)}$ determined by the level set $\times_{i=1}^{s} 1$ of the function

$$\tilde{\mathcal{T}}_g^s \to \mathbb{R}_+^s$$

$$\tilde{\Gamma}_m \to \times_{i=1}^{s} h(\partial_i; \tilde{\Gamma}_m)$$

where $\partial F(G) = \coprod \partial_i$.

<u>THEOREM</u>: *Suppose that* $\omega \in H^*(\mathcal{M}_g^s)$ *is a top-dimensional form. Then (provided* $F_g^s \neq F_1^1$*) we have*

$$\int_{\mathcal{M}_g^s} \omega = \sum_{\substack{[G] \text{ so that} \\ F_g^s \approx F(G) \text{ and} \\ G \text{ is trivalent}}} \frac{1}{\#\mathrm{Aut}(G)} \int_{\mathcal{D}(G)} \psi^* \omega ,$$

where

$$\tilde{\mathcal{T}}_g^s \xrightarrow{f} \mathcal{T}_g^s \to \mathcal{M}_g^s, \qquad \psi: \tilde{\mathcal{T}}_g^s \to \mathcal{M}_g^s$$

is the natural map, and $\mathcal{D}(G) \subset \mathbb{R}_+^{S(G)}$ *is determined by the following conditions on* $H \in \mathbb{R}_+^{S(G)}$

(i) H *satisfies the coupling equations.*

(ii) H *satisfies the "face constraint"*

$$\alpha + \beta + \gamma + \delta > \varepsilon + \varphi$$

for each edge e of G with adjacent horocyclic segments as in Figure 4b.

(iii) $H \in \mathfrak{L}$ *that is, for each component* ∂ *of* ∂ F(G), h(∂; H) = 1.

REMARKS: 1) The (substantial) improvement of this statement over the treatment in [12] is the linearity of the face constraint. In fact, there is a basis of $\mathbb{R}_+^{s(G)}$ so that (over each cell) the variety $\mathcal{V}$ is the graph over a cone on a coordinate simplex; thus, $\mathcal{D}(G)$ is simply an affine slice of the cone on a coordinate simplex. More precisely, there is (for each cell) a (non-orthogonal) linear decomposition

$$\mathbb{R}_+^{S(G)} = X \oplus Y$$

and a dissipative dynamical system (in fact a restricted gradient flow) in the Y-direction so that (the fixed cell in) $\mathcal{V}$ is the set of equilibrium points over the postitve orthant in X, and the projection $\mathcal{V} \to X$ is one-to-one; see [10].

2) The relevance of the Fundamental Problem of §4 to the integration scheme is apparent; see [11].

3) The computation of the Weil-Petersson Kähler masses of the various moduli spaces remains a promising application of the theorem (see [12]): the Kähler form is still pleasant in h-length coordinates.

4) In case $F_g^s = F_1^1$, there is only one appropriate fatgraph class [G] (namely, the fatgraph G_1 Figure 5a) and #Aut(G) = 6. The integration scheme takes the form

$$\int_{\mathcal{M}_1^1} \omega = \frac{1}{3} \int_{\mathcal{D}(G)} \psi^* \omega$$

in this case.

§ 6 UNIVERSAL OBJECTS AND SPECULATIONS

In this final section, we take the opportunity of announcing some recent work, to appear as "Universal Constructions in Teichmüller Theory", in preparation. Specifically, there is a proposed model for a universal Teichmüller space; precisely, there is a commutative square of topological spaces

$$\text{(US)}\qquad \begin{array}{ccc} U\tilde{\mathfrak{T}} & \longrightarrow & \text{Tess} \\ \downarrow & & \downarrow \\ U\tilde{M} & \longrightarrow & \text{Mod} \end{array}$$

which generalizes the "classical" squares (S_g^s) defined in §1. The spaces in (US) are defined presently; first, recall [10] that radial projection from (-1,0,0) to height zero in $\mathbb{M}^3$ induces a map $\bar{\cdot}: (\mathbb{H}, L^+) \to (\mathbb{D}, \mathbb{S}^1)$, where $\mathbb{D}$ denotes the Poincaré disc with boundary the circle $\mathbb{S}^1$ at infinity, and define

$$U\tilde{M}' = \{\text{countable } B \subset L^+ : \ \bar{\cdot}|_B \text{ is monic and } \bar{B} \subset \mathbb{S}^1 \text{ is dense}\},$$
$$\text{Mod}' = \{\text{countable dense subset of } \mathbb{S}^1\}.$$

The topologies on these sets are the subspace topologies induced from the infinite symmetric products. Tess' denotes the space of all tesselations (i.e., locally-finite decompositions of $\mathbb{D}$ into triangles by countable collections of disjoint ideal geodesics) with a distinguished triangle-edge pair and the obvious topology. A "universal tesselation" is given by the Farey tesslation $u\tau$ of $\mathbb{D}$ (see Figure 9); let $u\tau_\infty \subset \mathbb{S}^1$ denote the set of ideal points of $u\tau$. Define $U\tilde{\mathfrak{T}}' = \{\text{injections } \mu: u\tau_\infty \to L^+ : \mu \text{ is "order-preserving and } B = \text{Image}(\mu) \in U\mathfrak{M}'\}$ with the weak -star topology.

Finally, for each of the objects X' just defined, let $X = \text{Möb} \setminus X'$, where Möb denotes the Möbius group. The spaces in (US) are thus defined, and the maps in (US) are the obvious ones.

<u>META-THEOREM</u>: *(US) is a universal analogue of* (S_g^s) *for* $g \geq 0$, $s \geq 1$, *and* $2g - 2 + s > 0$.

Precisely, we assert

(1) There is a canonical map (defined generically) of (S_g^s) into (US) for each appropriate g,s. This map is *not* an embedding, and fibers correspond to commensurability classes of conformal surfaces.

(2) There is a universal analogue $\overline{Pt}$ (for "extended Ptolemy groupoid") of the mapping class groups so that the quotient of the top line of (US) by $\overline{Pt}$ is the bottom line. Furthermore, each MC_g^s embeds in $\overline{Pt}$ as a subgroup.

(3) There is a canonical non-degenerate closed two-form on (each space in) (US) which restricts to the classical Weil-Peterson Kähler two-form on (S_g^s), for each g,s.

To close, we repeat a parable mentioned to us by a colleague and indulge some speculation. There were three wise men from Baghdad, and one was given the trunk, another the tail, and the third the ears of an elephant. Attempting to reconstruct the animal from the observed anatomy, one wise man submitted that the elephant must be a snake, another that it must be a mouse, and the third was even convinced that the elephant was a butterfly. The anatomical trichotomy for us seems to involve the combinatorial conformal geometry and

The Virasoro algebra, which is of key interest in the string theory. This algebra acts naturally on each space in (US). Moverover, there is an infinitessimal Virasoro action on each $\tilde{\mathfrak{T}}s_g$ coming from the conformal field theory (described to me by Singer [17]), which *should be* tractable with our combinatorial conformal geometry.

The KP-heriarchy machine. This machine is relevant to both the string theory and to the classical problem of Schottky in conformal geometry.

FATGRAPH TABLES

How to read these tables:

(1) The permutation sig determines the valence tuple. (Permutations are enumerated by listing in order the image of 1,2,3,....) For instance,

sig = 2341
sig = 231564

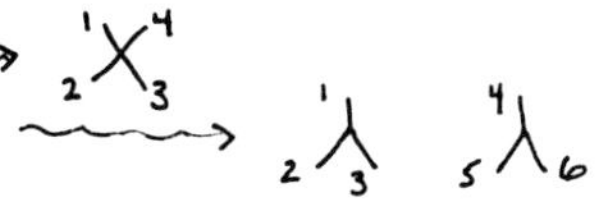

(2) The other permutation determines a "pairing on the stubs"; for instance:

sig = 2341
tau = 2143

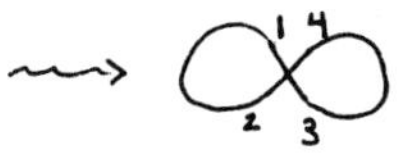

sig = 231564
tau = 214365

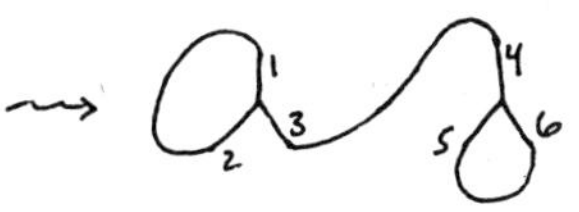

(3) The interpretation of the number (n) in parentheses following the permutation "tau" is as follows: from sig, compute v(k) = the number of k-valent vertices. Then the order #AUT of the fatgraph auto- morphism group is

$$\#AUT = \frac{1}{n} \prod_k k^{v(k)}\, v(k)!$$

For instance:

sig = 2315642
tau = 214365(9) $\rightsquigarrow$ $\#AUT = \frac{1}{9}(2! \cdot 3^2) = 2$

the genus is 0, and there are 4 punctures

with sig= 234561 ,there are 2 fatgraphs:
214365(2) 216543(3)

with sig= 23416785 ,there are 3 fatgraphs:
21563487(16) 21583764(16) 58761432(4)

with sig= 23451786 ,there are 3 fatgraphs:
21436587(15) 21563487(15) 21687354(15)

with sig= 23416759108 ,there are 7 fatgraphs:
21563487109(72) 21573846109(72) 21583764109(72) 21583109476(72)
35182764109(36) 57681324109(72) 57810192364(36)

with sig= 231564897111210 ,there are 6 fatgraphs:
214378561091211(972) 214379510681211(972) 214371059861211(648)
214371051211698(1944) 467110231211598(486) 471011282611395(162)

the genus is 1, and there are 2 punctures

with sig= 234561 ,there are 3 fatgraphs:
215634(6) 351624(3) 456123(1)

with sig= 23416785, there are 4 fatgraphs:
21573846(32) 35172846(8) 56781234(4) 56871243(16)

with sig= 23451786 ,there are 4 fatgraphs:
21678345(15) 34126587(15) 36178245(15) 36187254(15)

with sig= 23416759108 ,there are 8 fatgraphs:
21583910467(72) 35162487109(72) 35182910467(36) 35182109476(36)
56781234109(72) 56891210347(36) 56810129374(72) 58691310247(72)

with sig= 231564897111210 ,there are 5 fatgraphs:
214371051112689(1944) 457121031112689(486) 457121031211698(972)
471018112512369(486) 471018122511396(648)

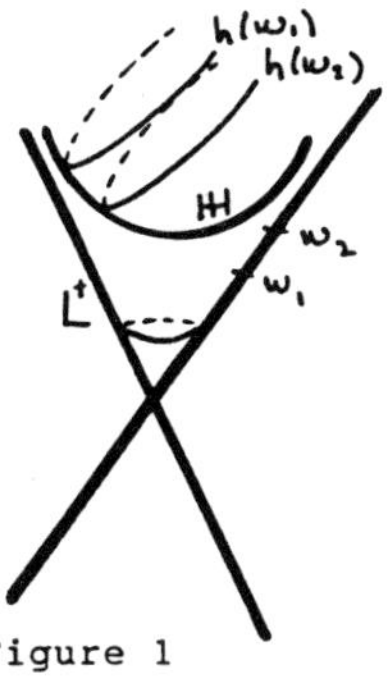

Figure 1

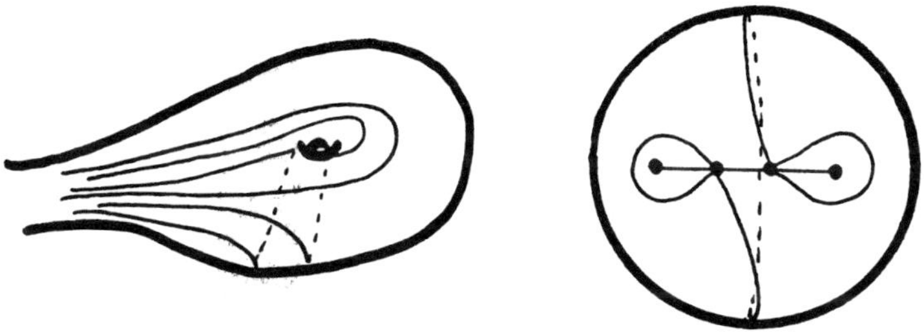

Figure 2

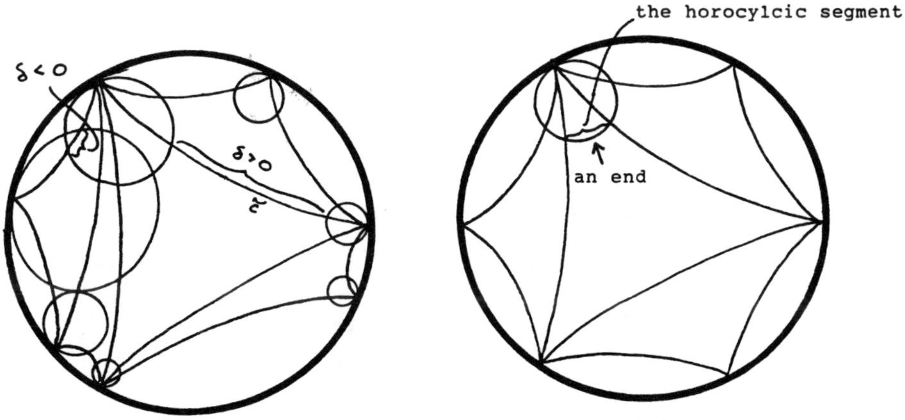

(a) (b)

Figure 3

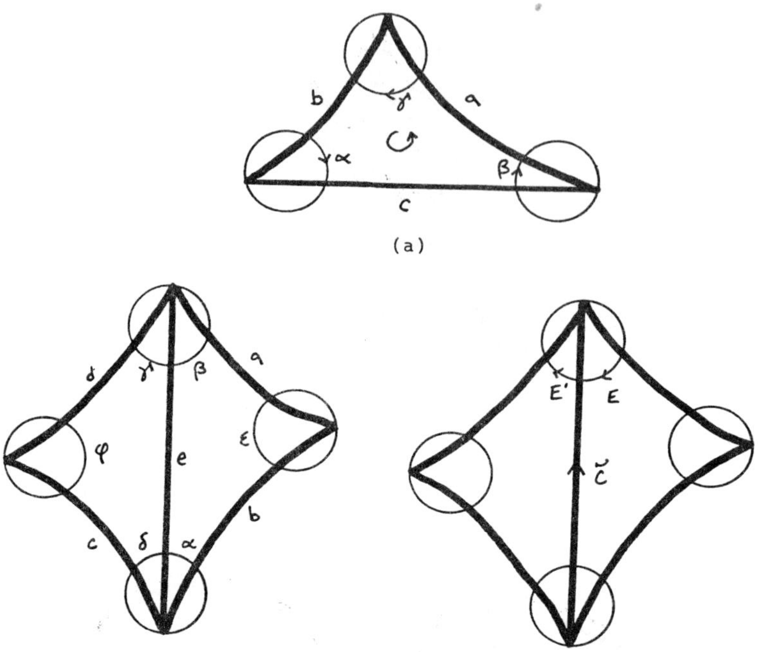

(a)

(b) (c)

Figure 4

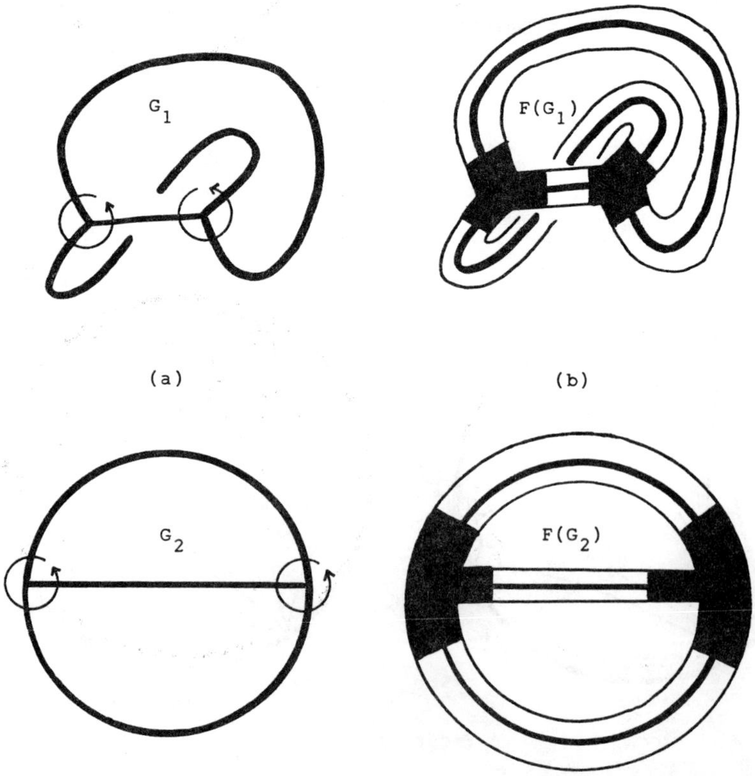

(a) (b)

Figure 6

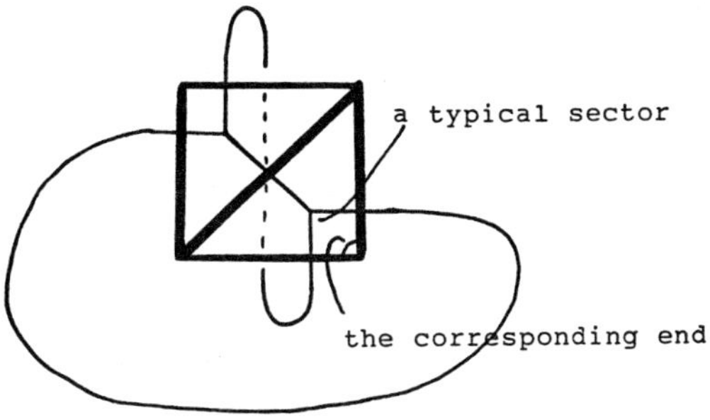

Figure 7

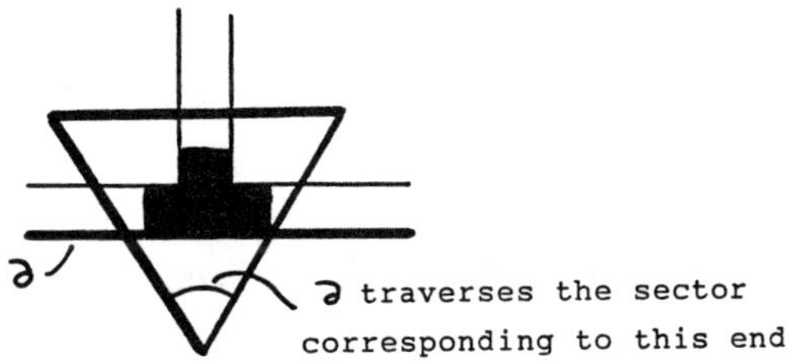

Figure 8

Figure 9

BIBLIOGRAPHY

1. E. Arbarello and C. De Concini, "On a set of equations characterizing Riemann matrices," *Annals of Math.* 120 (1984), 119-140.

2. D. Bessis, C. Itzykson, and J.B. Zuber, "Quantum field theory techniques in graphical enumeration," *Adv. Appl. Math* 1 (1980), 109-157.

3. B. Bowditch and D.B.A. Epstein, "Triangulations associated to punctured surfaces," *Topology* (1987).

4. M.J. Bowick and S.G. Rajeev, "The holomorphic geometry of bosonic string theory and Diff S^1/S^1, *Nuclear Physics B* , submitted.

5. A. Erdelyi and I.M.H. Etherington, "Some problem of non-associative combinations," *The Edinburgh Math. Notes* (1941).

6. D. Friedan and S. Shenker, "The analytical geometry of two dimensional conformal field theory," *Nuclear Physics B* (1987).

7. M.B. Green, J.H. Schwarz, and E. Witten, *Superstring Theory*, volume 1, Cambridge University Press (1987).

8. J. L. Harer, "The virtual dimension of the mapping class group of an oriented surface," *Invent. Math.* (1986).

9. M. Mulase, "The virtual dimension of the mapping class group of an oriented surface," *Invent. Math.* (1986).

10. R.C. Penner, "The decorated Teichmüller space of punctured surfaces," *Comm. Math. Phys.* (1987).

11. ———, "Perturbative series and the moduli space of punctured surfaces," *Jour. Diff. Geom* . (1987).

12. ———, "The moduli space of punctured surfaces," Proc. of Math. Aspects of String Theory, U.C.S.D. (1986), to appear.

13. ___________, "Universal constructions in Teichmüller theory," in preparation (1987).

14. M. Sato and Y. Sato, "Soliton equations as dynamical systems on infinite dimensional Grassmann manifold," preprint.

15. G. Segal and G. Wilson, "Loop groups and equations of KdV type," *Publ. I.H.E.S.*, number 61 (1985), 5-65.

16. T. Shiota, "Characterization of Jacobian varieties in terms of solition equations," Harvard Univ. preprint (1984).

17. I.M. Singer, private communication (1987).

18. W.P. Thurston, private communication (1987).

Department of Mathematics
University of Southern California
Los Angeles, California 90089

Contemporary Mathematics
Volume **74**, 1988

Affine Homogeneous Spaces

Dennis M. Snow†

ABSTRACT. A brief survey of the some characterizations of homogeneous affine varieties is given. A necessary condition for G/H to be affine is that the unipotent radical of H^o have only trivial intersections with reductive subgroups of G. The sufficiency of this condition is related to the existence of affine quotients for free triangular actions on C^n and holds when the dimension of the unipotent radical of G is ≤ 3.

A basic problem in the theory of homogeneous spaces is to find necessary and sufficient conditions on the algebraic groups G and H such that the quotient G/H is an affine variety. A closely related problem is to find conditions on connected complex groups G and H such that G/H is Stein, i.e. a closed *analytic* subspace of C^n. The first significant result in this direction is Matsushima's Theorem [8] which states that if G is a reductive complex Lie group, then G/H is Stein if and only if H is reductive. Since a reductive complex Lie group G is actually a linear algebraic group [6], and G/H is an affine variety if it is Stein [1], this result is essentially algebraic and has been generalized to reductive groups defined over algebraically closed fields of positive characteristic, see [11]. Surprisingly little progress has been made in finding such group theoretic conditions for arbitrary linear algebraic groups, although there are some cohomological characterizations. For example G/H is affine if and only if induction from H-modules to G-modules is an exact functor [5]. In this note we give a brief survey of some results in this area and show how a natural generalization of Matsushima's Theorem is related to conjectures about free triangular actions on affine space C^n. We include a simple proof of one of these conjectures when $n \leq 3$, and generalize Matsushima's Theorem to linear algebraic groups whose unipotent radical has dimension ≤ 3.

1. Matsushima's criterion

Perhaps the simplest criteria for G/H to be affine is that G be a unipotent linear algebraic group. Then, it is well-known that G/H is always isomorphic to affine space C^n for any algebraic subgroup H. More generally, if G is solvable, then $G/H \cong (C^*)^t \times C^s$ for any algebraic subgroup H. For complex Lie groups such statements are not automatically true, although a generalization of Matsushima's Theorem is possible for such groups, see [14] and the discussion in §4.

1980 *Mathematics Subject Classification* (1985 *Revision*). 14L30, 32M10

† Partially supported by NSF Grant DMS–8420315

Matsushima's Theorem, of course, handles the case of reductive groups. A typical example of how this theorem can be applied is the following: Let G be a reductive group acting algebraically on an affine variety X. Let $x \in X$ be a point whose orbit $G.x$ is closed. Then $G.x$ is an affine variety and the isotropy group G_x is reductive by Matsushima's Theorem. Therefore the isotropy representation of G_x on the tangent space T_x splits as a direct sum $V \oplus W$ where V is the tangent space of the orbit at x and W is a complementary subspace. It can be shown that there exists a locally closed G_x-invariant subspace S transversal to the orbit at x corresponding to the subspace W such that the twisted product $G \times_{G_x} S$ is isomorphic to a G-invariant neighborhood of the orbit $G.x$, [7], [13].

Any connected linear algebraic group G has a decomposition as a semi-direct product $G \cong M \times U$ where M is a maximal reductive subgroup and U is the unipotent radical of G. Even though strong results hold for these two "halves" of G, there does not yet appear to be a good group theoretic characterization of those algebraic subgroups H for which G/H is affine. A likely candidate, however, is the following.

Definition *We say that H satisfies "Matsushima's criterion" (in the group G) if the unipotent radical of the connected component of H has only trivial intersection with any reductive subgroup of G.*

Notice that if G is reductive then this criterion simply says that H is reductive, as in Matsushima's Theorem. If V denotes the unipotent radical of the connected component of H, then Matsushima's criterion is equivalent to $gVg^{-1} \cap M = 1$ for all $g \in G$. (The maximal reductive subgroups of G are all conjugate, so that any reductive subgroup of G is conjugate to a subgroup of M.) Geometrically, this means that M acts *freely* on G/V, i.e. all isotropy subgroups are trivial.

Proposition *If G/H is affine, then H satisfies Matsushima's criterion.*

Proof (see [14]): G/H is affine if and only if G/V is affine since $G/V \to G/H$ is a principal L-bundle where L is a maximal reductive subgroup of H. Now, the M-orbits of *minimal* dimension in G/V must be closed. Say $M.p$, $p = gV \in G/V$, is such a closed orbit. Then the isotropy subgroup $M_p = M \cap gVg^{-1}$ is a reductive subgroup of the unipotent group gVg^{-1} by Matsushima's Theorem. However, unipotent groups cannot have non-trivial reductive subgroups so $M \cap gVg^{-1} = 1$. This shows that the orbits of *minimal* dimension actually have *maximal* dimension, *dim* M, and therefore all M orbits in G/V have trivial isotropy so H satisfies Matsushima's criterion.

2. Free triangular actions

The observation that M acts freely on G/V when G/H is affine can also be expressed by saying that V acts freely (from the right) on $M \backslash G = U \cong \mathbf{C}^n$. Moreover, this action is *triangular* in the sense that there exist coordinates $x_1, \ldots, x_n$ for $\mathbf{C}^n$ such that for any $v \in V$, $x.v = (v_1(x), ,v_n(x))$ where v_i is a function of $x_i, \ldots, x_n$, $i=1,\ldots,n$. This is because each v has a unique decomposition $v = mu$ where $m \in M$ and $u \in U$ so that the action of v is the linear triangular action of m on U (isomorphic to the adjoint representation under the exponential map and preserving the central series of U) followed by the unipotent group multiplication by u (necessarily triangular), see [15]. Note that if a connected group V acts algebraically and freely on C^n then the group must be unipotent (otherwise there would be a copy of C^* which must have a fixed point in C^n, [3]). We are thus lead to asking questions about free group actions on C^n. Let us list some of the possibilities, which are all unsolved problems in general.

Wild conjecture: If V acts algebraically and freely on $U \cong C^n$ then the quotient variety U/V exists and is isomorphic to affine space C^k.

Generalized cancellation problem: If $V = C$ acts freely on $U \cong C^n$ then the quotient variety U/V exists and is isomorphic to C^{n-1}.

Cancellation problem: If $Y \times C \cong C^n$, then $Y \cong C^{n-1}$.

Tame conjecture: If V acts algebraically and freely on $U \cong C^n$ by triangular automorphisms then the quotient variety U/V exists and is affine.

Using a composition series for V and induction on dimension, it is not hard to prove that the 'wild conjecture' is equivalent to the 'generalized cancellation problem'. It is also easy to see that the 'tame conjecture' implies that Matsushima's criterion is sufficient for affine homogeneous spaces: If U/V is affine then $U \to U/V$ is a trivial principal bundle. In particular, there exists an V-equivariant map $U \to V$ which extends trivially to a V-equivariant map $G \to V$ so that $G \to G/V$ is trivial and G/V is affine. It is an interesting question whether a free triangular action on C^n can always be imbedded in a transitive action. If this were true, then the sufficiency of Matsushima's criterion would be *equivalent* to the 'tame conjecture'. A sobering thought is that *all* of these variations might be equivalent, and therefore equally hard to prove. For example, it is possible that free actions are necessarily triangular, and it is possible that the cancellation problem is always true. If that is the case then the 'tame conjecture' is equivalent to the 'wild conjecture'! The assumption that the action is *free* is important: In [2], Bass creates an example of a C action (with fixed points) which cannot be made triangular.

3. Special cases

The sufficiency of Matsushima's criterion is only known in the following special cases: 1) $V \subset U$, 2) $dim\ V \leq 1$, and 3) $dim\ U \leq 3$. The first case is the simplest and has been mentioned above: $U/V \cong C^k$ so that $G/V \cong M \times C^k$. We may assume that a maximal reductive subgroup L of H is contained in M. Then L acts on this C^k linearly via a projection of the adjoint representation of M on U, and $G/H \cong M \times_L C^k \to M/L$ is a homogeneous vector bundle over the affine variety M/L which is a minimal M orbit in G/H.

The second case, $dim\ V = 1$, was handled in [5], see also [14]. A rough idea of the proof is the following: If V is contained in the commutator U' of U, we are done, as in the previous paragraph. If it is not contained in U then we consider $G_1 = G/U'$ where $V_1 = VU'/U'$ still satisfies Matsushima's criterion. Therefore, we may assume that U is abelian. Now V acts on U as a one parameter subgroup of the affine transformation group (linear transformations and translations). In fact, we can realize V as the group $\{exp(tA) \mid t \in C\}$ where A is an some endomorphism of C^{n+1} ($n = dim\ U$). By inspecting the Jordan form of A it becomes clear that there is a coordinate, say x_1, such that the action of V transforms x_1 to $x_1 + t$. The hyperplane $S = \{x \in C^n \mid x_1 = 0\}$ then intersects each V orbit in exactly one point. Hence $S \cong C^{n-1}$ is isomorphic the quotient U/V.

If $dim\ U \leq 3$, the only case not discussed above is $dim\ V = 2$, $dim\ U = 3$ (Matsushima's criterion implies that $dim\ V \leq dim\ U$). In this case, V is isomorphic to two copies of the additive group C, $V = V_1 \times V_2$. Since V_2 satisfies Matsushima's criterion, we know that $U/V_2 = C^2 \times V_2$, as described above. It is known that free algebraic actions of C on C^2 must be conjugate to an affine actions, [10]; therefore the action of V_2 on C^2 has a quotient isomorphic to C. This implies that $U/V \cong C$ and hence that G/V (G/H) is affine.

In connection with this last case $dim\ U = 3$, there is the following related result:

Theorem *Let G be a connected complex Lie group acting algebraically on C^3 as a group of triangular automorphisms. If the action is free, then C^3 is equivariantly isomorphic to $G \times C^{3-dimG}$.*

This theorem also holds for quasi-algebraic actions (see [14]). Notice how this gives a positive answer to the question of imbedding free actions on C^3 into transitive ones. The proof of this theorem is easily reduced to the case $dim\ G = 1$ (see the previous paragraph). A sketch of the proof of this case is as follows: The action of G comes from a derivation D of the form $f_1D_1(x_2,x_3) + f_2(x_3)D_2 + f_3$ where D_i is the partial derivative with respect to the coordinate x_i, $i = 1,2,3$, f_1, f_2 are polynomials and f_3 is a constant. If $f_3 \neq 0$ then we can make obvious triangular changes of coordinates to transform D into the derivation f_3D_3 and we are done. Thus, we may assume that $f_3 = 0$, and for the same reason that neither f_1 nor f_2 is constant. We construct a quotient map $\pi: C^3 \to C^2$ explicitly by finding two appropriate G-invariant functions. The coordinate function x_3 is already invariant. A second invariant function h is the following: Let $F_1(x_2,x_3) = \int_0^{x_2} f_1(u,x_3)du$ and define $h(x) = f_1(x_2,x_3) - x_1f_2(x_3)$. Then $Dh = 0$, so h is invariant. The map $\pi: C^3 \to C^2$, $\pi(x) = (h(x),x_3)$ has the following properties: It is clearly surjective; it is smooth since its Jacobian is $\begin{pmatrix} -f_2 & f_1 & * \\ 0 & 0 & 1 \end{pmatrix}$ and f_1, f_2 cannot both be zero if there are no fixed points; and with some straightforward calculation it can be shown that the fibers are precisely the G orbits. Such a map is clearly locally trivial in the étale topology, so that it must be algebraically trivial, see [12].

4. Stein homogeneous spaces

The classification homogeneous Stein spaces is closely related to the classification of homogeneous affine varieties. If G and H are connected complex Lie groups and G/H is Stein with G acting effectively on G/H, then G is a Stein Lie group and H is a linear complex Lie group, [9]. Thus, H has a decomposition as a semi-direct product $L \times V$ where L is a maximal reductive subgroup of H and V is a simply-connected normal solvable of H. Of course, G/H is Stein if and only if G/V is Stein. As for algebraic groups, we say that H satisfies Matsushima's criterion (in G) if V has only trivial intersections with reductive subgroups of G.

Theorem [14] *If G/H is Stein then H satisfies Matsushima's criterion in G. In addition, there exist linear algebraic groups G_1 and H_1 with dim $R_u(G_1) \leq$ dim $R_u(G)$ and dim $H_1 \leq$ dim H such that H_1 satisfies Matsushima's criterion in G_1, and G/H is Stein if and only if G_1/H_1 is affine.*

In particular, if Matsushima's criterion is sufficient for affine algebraic groups, it is also sufficient for connected Stein Lie groups. A corollary of this reduction theorem is that if G is a solvable Stein Lie group, and H a connected subgroup, then G/H is Stein if and only if H satisfies Matsushima's criterion. With complex Lie groups one must be aware of examples such as $G = \mathbf{C}^*\times\mathbf{C}^* \supset H = \{(e^z, e^{iz}) \mid z \in \mathbf{C}\}$, for which the quotient G/H is a compact complex torus.

In the context of Stein homogeneous spaces, there is one way that Matsushima's criterion differs from the algebraic category. Recall that a subgroup H of an algebraic group G is called *observable* if every affine H-space can be equivariantly imbedded in an affine G-space. This property is equivalent to G/H being a quasi-affine variety, [4]. The corresponding property for Stein groups $G \supset H$ is that every Stein H-space can be equivariantly imbedded in a Stein G-space. It is somewhat surprising that observability for Stein groups $G \supset H$ implies that H satisfies Matsushima's criterion in G, see [14], since this places it very close to the condition of the quotient G/H being Stein.

References

[1] Barth, W., Otte, M.: Invariante holomorphe Funktionen auf reduktiven Liegruppen. Math. Ann. **201**, 97–112 (1973).

[2] Bass, H.: A non-triangular action of G_a on A^3. J. Pure Appl. Algebra **33**, 1–5 (1984).

[3] Bialyincki-Birula, A.: Remarks on the action of an algebraic torus on k^n, I and II. Bull. Acad. Pol. Sci **14** 177–181 (1966) and **15**, 123–125 (1967).

[4] Bialynicki-Birula, A., Hochschild, G., Mostow, G.D.: Extensions of representations of algebraic linear groups. Am. J. Math. **85**, 131-144 (1963).

[5] Cline, E.,Parschall, B.,Scott, L.: Induced modules and affine quotients. Math. Ann. **230**, 1–14 (1977)

[6] Hochschild, G., Mostow, G.D.: Representations and representative functions of Lie groups, III. Ann. Math. **70**, 85–100 (1959).

[7] Luna, : Slices étale, Bull. Soc. Math. France Mém. **33**, 81–105 (1973).

[8] Matsushima, Y.: Espaces homogenes de Stein des groupes de Lie complexes. Nagoya Math. J. **16**, 205–218 (1960).

[9] Matsushima, Y.: Espaces homogènes de Stein des groupes de Lie complexes, II. Nagoya Math. J. **18**, 153-164 (1961).

[10] Rentschler, Opérations du groupe additif sur le plan affine. C. R. Acad. Sci. Paris, Series A, **267**, 384–387 (1968).

[11] Richardson, R.: Affine coset spaces of reductive algebraic groups. Bull. London Math. Soc. **9**, 38–41 (1977).

[12] Serre, J. P.: Espaces fibrés algébriques. In: Anneaux de Chow et Applications. Séminaire Chevalley, Paris 1958.

[13] Snow, D.: Reductive group actions on Stein spaces. Math. Ann. **259**, 79–97 (1982).

[14] Snow, D. : Stein quotients of connected complex Lie groups. manuscripta math. **50**, 185–214 (1985).

[15] Snow, D.: Triangular actions on $\mathbf{C}^3$, to appear.

Contemporary Mathematics
Volume **74**, 1988

Deformations and discrete subgroups of loop groups

CHRISTOPHER W. STARK

Abstract. Discrete subgroups of loop groups on Lie groups and associated covering spaces are considered as methods of constructing infinite-dimensional Eilenberg-MacLane spaces $K(\Gamma, 1)$ with interesting differential geometry. Arithmetic examples of discrete subgroups of loops on complex linear groups are defined and studied and deformation/rigidity properties of finitely generated discrete subgroups of loop groups are examined.

The close kinship of Lie groups and loop groups on Lie groups extends to some aspects of the theory of discrete subgroups. This paper examines discrete subgroups of linear loop groups which are analogous to arithmetic subgroups and considers the rudiments of deformation theory for finitely generated subgroups of loop groups. Our point of view is topological and rather naive, and an underlying goal of this project was the construction of some explicit $K(\Gamma, 1)$'s as infinite-dimensional manifolds with attractive differentiable geometry (see sections 1 and 2). Deformations in loop groups and loops of deformations are closely related (section 3) and Mostow rigidity constrains some of the deformation spaces considered here (section 4).

Throughout this paper G denotes a (usually finite-dimensional) Lie group and LG denotes the space of free, continuous loops on G in the uniform metric with the operation of pointwise multiplication. Many of the statements below are also valid if we work in smooth or Sobolev loop spaces with appropriate topologies; these refinements are most interesting in the differential geometric context, to which we hope to return in another paper. The connection of loop groups with Kac-Moody Lie algebras is usually made through the space of polynomial loops on G, as in [4] or [6], and we do not consider this more algebraic formulation of loop groups here. Problems pointing in these and other directions are posed in the body of the paper.

Most of the work reported here was done or sketched during a visit to the University of Edinburgh. I am most grateful to Andrew Ranicki, the Department of Mathematics in Edinburgh, and the N.R.C. for their hospitality.

1980 *Mathematics subject classifications*. Primary 57S30, 22E40. Secondary 22E65.

1. Some examples.

Let G be a finite dimensional Lie group and let LG be the group of continuous, free loops on G (i.e. the space of all continuous maps from S^1 to G, in the uniform metric and with the operation of pointwise multiplication). The first family of examples of discrete subgroups of LG is provided by the discrete subgroups Γ of G, thought of as groups of constant loops. The second family of examples derives from the first by looking at loops of deformations of an imbedding i of Γ in G as a discrete subgroup, i.e. by taking a loop in the representation variety $\mathrm{Hom}(\Gamma, G)$. If Γ is not rigid in G under the imbedding i then we should not be surprised if non-trivial loops of deformations of i were to exist.

This section exhibits groups which are in a reasonable sense arithmetic subgroups of $L\mathrm{SL}(n, \mathbf{C})$. We begin with the observation that if $\mathbf{F} = \mathbf{R}$ or $\mathbf{C}$ then $L\mathrm{SL}(n, \mathbf{F}) \cong \mathrm{SL}(n, L\mathbf{F})$ by considering matrix coefficients, and the analogous statement holds for all the linear groups. Let $\mathbf{Z}[x]$ denote the ring of integer polynomials and let $\mathbf{Z}[x, x^{-1}]$ denote the ring of Laurent polynomials over the integers. Let t denote a complex variable taking values on the unit circle, and observe that $\mathbf{Z}[x] \cong \mathbf{Z}[t]$ and $\mathbf{Z}[x, x^{-1}] \cong \mathbf{Z}[t, t^{-1}]$. Consider $V = L(\mathbf{C}^n) \cong (L\mathbf{C})^n$ in either the uniform or the L_2 topology; $A = (\mathbf{Z}[x, x^{-1}])^n$ is a $\mathbf{Z}$-submodule of this vector space V such that $\mathbf{C} \otimes A$ is dense in V, so A may be viewed as the analog of an integral lattice in a finite-dimensional vector space and the automorphism group of A, $GL(n, \mathbf{Z}[x, x^{-1}])$ is to be regarded as an arithmetic subgroup of $LGL(n, \mathbf{C})$. Our attention will in fact be confined to the subgroups $\Gamma_1 = \mathrm{SL}(n, \mathbf{Z}[x, x^{-1}])$ and $\Gamma_0 = \mathrm{SL}(n, \mathbf{Z}[x])$; the first is of index two in $GL(n, \mathbf{Z}[x, x^{-1}])$ so it stands in for the full group of automorphisms and the second will have particular virtues below.

PROPOSITION 1. *Γ_0 and Γ_1 are discrete subgroups of $LSL(n, \mathbf{C})$ in either the uniform metric or the L_2 norm.*

PROOF: Consider the L_2-inner products of components of elements of these groups with the (unnormalized) basis $\{t^k : k \in \mathbf{Z}\}$ for $L_2(S^1)$.

The evaluation maps $ev_w : L\mathrm{SL}(n, \mathbf{C}) \to \mathrm{SL}(n, \mathbf{C})$ $(w \in S^1)$ are continuous (with respect to the uniform metric on the loop group) and, via evaluation, every subgroup Γ of LG determines a loop in $\mathrm{Hom}(\Gamma, G)$. If Γ is a discrete group of constant loops or a loop of deformations of a discrete subgroup of $\mathrm{SL}(n, \mathbf{C})$ (through discrete subgroups) then clearly each image $ev_w(\Gamma)$ is a discrete subgroup of $\mathrm{SL}(n, \mathbf{C})$ isomorphic to Γ. Our arithmetic subgroups diverge from the relatively trivial examples at this point.

PROPOSITION 2. *Suppose that $n \geq 2$.*

(1) *If $w \in S^1$ is an algebraic integer then $ev_w(\Gamma_j) \not\cong \Gamma_j$ for $j = 0, 1$.*
(2) *If $w \in S^1$ is transcendental then $ev_w(\Gamma_j) \cong \Gamma_j$ for $j = 0, 1$ but $ev_w(\Gamma_j)$ is not a discrete subgroup of $SL(n, \mathbf{C})$.*

PROOF: If w is an algebraic integer and $p(t)$ is a monic polynomial with integral coefficients such that $p(w) = 0$ then ev_w carries the matrix

$$\begin{bmatrix} 1 & p(t) & 0 \\ 0 & 1 & 0 \\ 0 & 0 & J \end{bmatrix}$$

(J = (n–2) $\times$ (n–2) identity) in Γ_0 to the identity in $\mathrm{SL}(n, \mathbf{C})$, so ev_w is not monic.

If w is a transcendental number then the set of matrices of the form

$$\begin{bmatrix} 1 & w^k & 0 \\ 0 & 1 & 0 \\ 0 & 0 & J \end{bmatrix}$$

($k \in \mathbf{Z}(\geq 0)$, J = (n–2) $\times$ (n–2) identity) in Γ_0 is mapped by ev_w to an indiscrete subset of $\mathrm{SL}(n, \mathbf{C})$. Clearly the kernel of ev_w on $\mathrm{SL}(n, \mathbf{Z}[t])$ is trivial for transcendental w.

2. Actions and $K(\Gamma, 1)$'s. Suppose G is a connected, finite-dimensional Lie group with maximal compact subgroup K. LG/LK is a contractible space on which any discrete subgroup Γ of LG will act, and quotient spaces for these actions may provide geometrically interesting candidates for $K(\Gamma, 1)$'s, if all goes well. It is a commonplace that a discrete subgroup Γ of a Lie group G with maximal compact subgroup K will act on the coset space G/K properly discontinuously, so that the quotient map $G/K \to \Gamma \setminus G/K$ is a branched covering projection. We need to consider the details of this familiar statement.

Let X be a manifold with a locally smooth action of a discrete group Γ. We will say that the quotient map $q : X \to \Gamma \setminus X$ is a "branched covering" (and that Γ acts by "branched covering transformations") if for each $x \in X$ there is an open neighborhood U of x such that $\{\gamma \in \Gamma : U \cap \gamma U \neq \emptyset\}$ is a finite set. (For locally smooth group actions this only differs from the translation of Fox's definition [3] in allowing reflections. In particular each point $[x] \in \Gamma \setminus X$ has a neighborhood $W \cong \Gamma_x \setminus V$ for a neighborhood V of x in X, where $q^{-1}(W) = \Gamma V$ and $V \cap \gamma V \neq \emptyset$ if and only if $V = \gamma V$.) We do not assume that the underlying manifold is finite-dimensional in this formulation, and we remark that any translation action by a subgroup of the Lie group G on G or on a coset space G/H (H a closed subgroup) is smooth even if G is infinite-dimensional.

Most of us tend to think of these matters in terms of "properly discontinuous actions" and some important distinctions must be drawn in the present setting. We refer to Bourbaki [1] and Kulkarni [7] for most of the following definitions and results. Here H is a topological group and (H, X) is a continuous action of H on the topological space X.

(1) The action (H, X) is said to be "proper" if and only if the mapping

$$\theta : H \times X \to X \times X$$
$$\theta : (h, x) \mapsto (x, hx)$$

is proper, i.e. if and only if $\theta^{-1}(C)$ is compact in $H \times X$ for each compact subset C of $X \times X$.

(2) A discrete group H acts "properly discontinuously" on X if and only if it acts properly in the sense above.

(3) If X is a Hausdorff, locally compact space then a Hausdorff group H acts properly on X if and only if for each pair of compact subsets $C, D \subseteq X$ the set $\{h \in H : C \cap hD \neq \emptyset\}$ is compact in H. (So if H is discrete, we

ask that this subset of H be finite – this is the familiar condition from discussions of discrete subgroups and covering spaces.)

(4) If K is a subgroup of H then the left translation action $(H, H/K)$ is proper if and only if K is compact.

Observe that "proper discontinuity" is neither automatic nor the right notion to use in the setting of $(\Gamma, LG/LK)$, since (a) $(LG, LG/LK)$ is not a proper action and (b) LG/LK is not locally compact; the version of "branched covering" formulated above is what we want and agrees with the usual definition of a regular covering projection posed in terms of evenly covered open neighborhoods should all the isotropy subgroups be trivial.

LEMMA 3. *If $X \to X/H$ is a principal fiber bundle with H connected then $LX/LH \cong L(X/H)$.*

Two observations prove Lemma 3: since $X \to X/H$ is a fiber bundle with connected fiber, loops in the quotient lift to loops in X, and the difference of two such lifts is both continuous and H-valued. We apply the Lemma to $G \to G/K$. The next Lemma is familiar in the usual context and easily proved here.

LEMMA 4. *Suppose H is a discrete group acting locally smoothly on X. If H' is a normal subgroup of finite index in H such that the restricted action (H', X) is an action by branched covering transformations then (H, X) is also an action by branched covering transformations.*

We return to the groups considered in section 1: let $G = \mathrm{SL}(n, \mathbf{C})$ and let $\Gamma_0 = \mathrm{SL}(n, \mathbf{Z}[t])$, $\Gamma_1 = \mathrm{SL}(n, \mathbf{Z}[t, t^{-1}])$ be the discrete subgroups of LG defined above, and let $K = \mathrm{SU}(n)$ be a maximal compact subgroup of G.

THEOREM 5. *Suppose $n \geq 2$.*

(1) *Γ_1 does not act on LG/LK by branched covering transformations.*

(2) *Γ_0 acts on LG/LK by branched covering transformations.*

PROOF: The first claim is established easily by considering the (2×2)-matrix

$$\begin{bmatrix} t & 0 \\ 0 & t^{-1} \end{bmatrix}$$

which is an element of LK, so it stabilizes the base coset in LG/LK, and which has infinite order.

Prove the second claim by considering first any elements $\gamma = \gamma(t) \in \Gamma_0$ which stabilize a loop $l = l(t)$ in $LG/LK \cong L(G/K)$. Write $l(t) = \lambda(t) \cdot K$ for a suitable lift $\lambda : S^1 \to G$ of l; the stabilization statement then translates into the conjugacy statement:

$$\lambda(t)^{-1} \cdot \gamma(t) \cdot \lambda(t) \in K$$

(for all $t \in S^1$). Consider the invariant polynomials (trace, determinant, etc.) $P_1(\gamma)(t), P_2(\gamma)(t), \ldots, P_n(\gamma)(t)$ of the polynomial matrix γ. These invariants are elements of $\mathbf{Z}[t]$ but their values are those of unitary matrices; in particular, $P_j(\gamma^{-1})(t) = \overline{P_j(\gamma)(t)} = P_j(\gamma)(\bar{t}) = P_j(\gamma)(t^{-1})$ is an element of $\mathbf{Z}[t] \cap \mathbf{Z}[t^{-1}]$, so all these invariants are integral constants. Thus for any $w \in S^1$, $P_j(ev_w(\gamma)) =$

$P_j(\gamma)$, so each element of Γ_0 stabilizing a loop maps by any of the evaluation maps to an element of the special linear group stabilizing a point in G/K. $ev_1(\Gamma_0) = \mathrm{SL}(n, \mathbf{Z})$ has a finite-index subgroup H acting freely on G/K, so we conclude that $\Gamma_0' = ev_1^{-1}(H)$ acts freely on LG/LK.

We claim that this finite-index subgroup Γ_0' of Γ_0 acts on LG/LK as a group of covering transformations, i.e. that each loop l has a neighborhood U (in the uniform metric) such that $\gamma \cdot l \notin U$ if $\gamma \in \Gamma_0'$ and $\gamma \neq 1$. If $\gamma \cdot l$ is uniformly close to l, then γ is uniformly close to a "moving conjugate" $\lambda \cdot K \cdot \lambda^{-1}$ of K because the action $(G, G/K)$ is proper. Therefore the values of the polynomial invariants $P_1(\gamma)(t), P_2(\gamma)(t), \ldots, P_n(\gamma)(t)$ are uniformly close to the invariants of the projection $\delta(t)$ of $\gamma(t)$ on its $\lambda(t) \cdot K \cdot \lambda^{-1}(t)$ factor under the polar decomposition of G with respect to this maximal compact. Similarly the inverse of γ has $P_1(\gamma^{-1})(t), P_2(\gamma^{-1})(t), \ldots, P_n(\gamma^{-1})(t)$ uniformly close to the invariants of $\delta^{-1}(t)$ – but this implies that $P_j(\gamma^{-1})(t)$ and $\overline{P_j(\gamma)(t)} = P_j(\gamma)(\bar{t}) = P_j(\gamma)(t^{-1})$ are uniformly close, implying that these invariants are integral constants, and in fact are the invariants of the identity element if our uniform neighborhood is sufficiently small, since $\gamma \in \Gamma_0'$. The invariants of the projection $\delta(t)$ of $\gamma(t)$ are thus uniformly near the invariants of the identity. The only unitary matrix with all eigenvalues equal to 1 is the identity, and unitary matrices with every eigenvalue close to 1 must be near the identity, so if our uniform neighborhood is sufficiently small then $\gamma \cdot l$ uniformly close to l implies γ is uniformly close to I. Since Γ_0' is discrete, we therefore cannot have $\gamma \cdot l$ arbitrarily close to l, and so we have shown that $(\Gamma_0', LG/LK)$ is an action by covering transformations. Lemma 4 finishes the proof of (2).

Observe that the ring $\mathbf{Z}$ of integers could be replaced in the discussion above by a ring of algebraic integers and that we could carry on the discussion within a subgroup of $\mathrm{SL}(n, \mathbf{C})$ (e.g. within an orthogonal group such as $\mathrm{SU}(p, q)$) so we have a number of examples produced by this construction. The Theorem suggests some problems.

Problem 1. Find an explicit CW-decomposition of $\Gamma_0' \setminus LG/LK$, preferably by exhibiting a Morse function.

Problem 2. Is Γ_0 rigid in LG?

Problem 3. Γ_1, not Γ_0, seems the more truly "arithmetic" group. A good criterion testing subgroups of LG for the property of acting on LG/LK by branched covering transformations is wanted, and it should account for the phenomena of the Theorem. A good discreteness criterion for subgroups of LG is also desirable; here arithmeticity does not seem to be misleading.

Problem 4. The C^k version of the Theorem should be true, for essentially the same reasons as the C^0 version considered here. More interestingly, suitable care with the Sobolev Multiplication Theorem should produce a version of the Theorem for L_2 and Sobolev spaces of loops on G.

3. Representation and character varieties. Most of the deformation theory captured in the representation and character varieties for finitely generated groups and finite-dimensional Lie groups works out in the loop group setting, although it is perhaps abusive to refer to these objects as "varieties" in the naive treatment given here (see the problems posed below). Suppose Π is a finitely generated group and G is a finite-dimensional linear Lie group. Let $\{g_1, \ldots, g_r\}$ be a set of generators for Π and, following the usual recipe, topologize the representation space $\text{Hom}(\Pi, LG)$ as the subspace of the product $\prod_1^r LG$ consisting of all (λ_i) such that $\lambda_1, \ldots, \lambda_r$ satisfy all the relations satisfied by $g_1, \ldots, g_r$. Let $R(\Pi, G) = \text{Hom}(\Pi, G)$ denote the space of representations and let $R(\Pi, G)/G = \text{Hom}(\Pi, G)/G$ be the "character space" of equivalence classes of representations $\Pi \to G$ under the adjoint action of G. We reserve the $X(\Pi, G)$ notation for the categorical or algebro-geometric quotient $R(\Pi, G)//G$, when that quotient is defined. We will consider the relationships between $R(\Pi, LG)$, $R(\Pi, G)$, $R(\Pi, LG)/LG$, and $R(\Pi, G)/G$.

For each $t \in S^1$ we restrict or evaluate an element h of $R(\Pi, LG)$ to get a homomorphism $ev_t(h) = h_t : \Pi \to G$, and so we have a map

$$\begin{aligned} \phi : R(\Pi, LG) &\to LR(\Pi, G) \\ h &\mapsto (t \mapsto ev_t(h)) \end{aligned}$$

from the representation variety for LG to the space of free loops on the representation variety of G. ϕ is a homeomorphism and induces a map

$$\psi : R(\Pi, LG)/LG \to L(R(\Pi, G)/G)$$

which is the first object to study. LG acts by pointwise conjugation on $LR(\Pi, G)$, and this action is identified via ϕ with the action on $R(\Pi, LG)$. Testing ψ for surjectivity can then be rephrased into the loop-lifting question for the quotient projection $R(\Pi, G) \to R(\Pi, G)/G$. The next Lemma is slightly modified from [2], Theorem II.6.2:

LEMMA 6. *Let G be a compact Lie group and let X be a G-space.*

(1) *Let $f : I \to X/G$ be any path. Then there exists a lifting $f' : I \to X$ of f.*
(2) *If G is connected and f is a loop in X/G then there exists a lift f' to X of f which is also a loop.*

Such path-lifting results are ordinarily derived from slice theorems, which are familiar for compact group actions [2] and which are available for proper actions of noncompact Lie groups [10]. If G is not compact then $(G, R(\Pi, G))$ is not proper (consider the trivial representation) and path lifting problems seem to be little studied for such actions. The context considered by Johnson and Millson [5] is both important and convenient from this point of view. Recall that they define a representation $\rho \in R(\Pi, G)$ to be "stable" if the orbit $G \cdot \rho$ is closed in $R(\Pi, G)$ and if the isotropy subgroup $Z(\rho)$ of ρ in G is finite and they declare a stable representation ρ to be "good" if $Z(\rho) = Z_G$, the center of G. (If G is a real linear Lie group then G is to be replaced in these definitions by its complex form.) Let $S(\Pi, G)$ be the set of stable representations in $R(\Pi, G)$; Johnson and Millson show that the conjugation action of G on $S(\Pi, G)$ is proper and admits slices, so Lemma 6 and these slices yield path-lifting for $S(\Pi, G) \to S(\Pi, G)/G$ and loop-lifting if G is connected. We have shown that $\eta : LS(\Pi, G)/LG \to L(S(\Pi, G)/G)$, the restriction of ψ, is surjective if G is a complex, connected linear group. A more complete analysis of this map seems to be possible in terms of tangent spaces.

Recall the description of tangent spaces going back to Weil (Johnson and Millson offer a review). Let ρ_t be a differentiable curve in $R(\Pi, G)$ with $\rho_0 = \rho$ (a curve deforming ρ) and define a function from Π to the Lie algebra $\mathcal{G}$ of G

$$\begin{aligned} c : \Pi &\to \mathcal{G} \\ \gamma &\to \dot{\rho}(\gamma)\rho(\gamma)^{-1} \end{aligned}$$

where the indicated group operation on the Lie algebra is the adjoint representation of G. This c is a $\mathcal{G}$-valued 1-cocycle. One continues by identifying "trivial" deformations as those paths ρ_t in $R(\Pi, G)$ which are of the form $\rho_t = \mathrm{Ad}g_t \cdot \rho$ for a differentiable path $t \mapsto g_t$ in G with $g_0 = 1 \in G$. If $\dot{g}$ is the tangent vector to g_t at time $t = 0$ then $\dot{g} \in \mathcal{G}$ and differentiating ρ_t gives

$$c(\gamma) = \dot{g} - \mathrm{Ad}\rho(\gamma)\dot{g}$$

so if c is tangent to a trivial deformation then c is a 1-coboundary. Observe: up to this point the group G here could just as well be an infinite-dimensional Lie group, since differentiation of curves was all that we required, so the Weil imbedding of the tangent cone TC_ρ at ρ of the representation variety in the space $Z^1(\Pi; \mathcal{G})$ of 1-cocycles extends to arbitrary infinite-dimensional Lie groups G. The next step in the Weil program is a converse statement, obtained by exponentiating: every 1-coboundary is the tangent vector to a trivial deformation. Here we must require that our infinite-dimensional Lie group has a suitably good exponential map, and this is a substantial restriction [8], but for loop groups we can recapitulate all of this infinitesimal analysis of deformations. Recall that the Lie algebra of LG is identified with $L\mathcal{G}$, the space of loops on the Lie algebra of G [11].

PROPOSITION 7.

(1) *The tangent cone TC_ρ at $\rho \in LS(\Pi, G)$ is identified with a subset of $Z^1(\Pi; \mathcal{G})$ (with the $Ad\rho$ action).*
(2) *The quotient map $\pi : LS(\Pi, G) \to LS(\Pi, G)/LG$ annihilates $B^1(\Pi; \mathcal{G})$.*

If ρ is a loop of good representations then Lemma 2.3 of [5] implies that the differential of π maps the (imbedded) tangent cone TC_ρ onto the tangent cone of $LS(\Pi, G)/LG$ at the image of ρ, and this second tangent cone is identified with a subset of $H^1(\Pi; \mathcal{G})$. $\eta : LS(\Pi, G)/LG \to L(S(\Pi, G)/G)$ now appears tractable, since the tangent cone to the left-hand space has been (almost) characterized above and the tangent cone for the right-hand space is the space of loops in the tangent cone to $S(\Pi, G)/G$, so this is a subspace of the space of continuous sections of a sheaf with cohomology groups $H^1(\Pi; \mathcal{G})$ as stalks.

Problem 5. Complete this analysis and describe the kernel of the differential of η. This kernel seems likely to be related to the kernel of the change-of-coefficients homomorphism

$$H^1(\Pi; \mathcal{G}) \to H^1(\Pi; \mathcal{G}^{S^1})$$

where $\mathcal{G}^{S^1}$ denotes the set of all (not necessarily continuous) functions $S^1 \to \mathcal{G}$.

Problem 6. Garland and Kac discuss algebraic formulations of loop groups and extensions of loop groups in the context of Kac-Moody Lie algebras [4], [6]. Do the representation and character spaces formulated in that setup become infinite-dimensional varieties in the sense of Shafarevich? (See [6] for definitions.)

4. Consequences of Mostow rigidity. Suppose that Π is a lattice in a semi-simple linear analytic group G having no compact factors, no three-dimensional factors, and no compact normal subgroup of positive dimension, and let $i_0 : \Pi \to G$ be the inclusion. Let $R_0(\Pi, G)$ be the subset of $R(\Pi, G)$ consisting of imbeddings ρ of Π onto lattices in G. Mostow's rigidity theorem [9] shows that $R_0(\Pi, G)/Aut(G)$ is a single point; what can we say about $LR_0(\Pi, G)/LG$ and $LR_0(\Pi, G)/LAutG$? For each $\rho \in R_0(\Pi, G)$ the Borel density theorem ([9], §8) shows that only one automorphism of G carries $i_0 \mapsto \rho$ and yields the next result.

PROPOSITION 8. *Let Π, G, and $R_0(\Pi, G)$ be as above. We have these homeomorphisms:*

(1) $R_0(\Pi, G) \cong Aut(G)$,
(2) $LR_0(\Pi, G) \cong LAut(G)$,
(3) $LR_0(\Pi, G)/LG \cong LAut(G)/LG \cong LOut(G)$, *and*
(4) $LR_0(\Pi, G)/LAut(G) \cong \{pt.\}$.

It is known that $Out(G)$ is finite in these cases, so the space $LR_0(\Pi, G)/LG$ in the proposition is finite. These identifications drain most of the potential interest from the loops on the subspace of lattice representations of Π, but it remains to be seen whether faithful non-lattice representations of Π typically give rise to interesting loops of deformations. The examples of section 1 of this paper show that faithful but indiscrete representations might lie on interesting loops; the deformation theory of such representations considered in [5] suggests that worthwhile monodromy problems may be attached to these loops.

REFERENCES

1. N. Bourbaki, "General Topology, Part 1," Hermann–Addison-Wesley, Reading, Mass., 1966.
2. G. E. Bredon, "Introduction to Compact Transformation Groups," Academic Press, New York, 1972.
3. R. H. Fox, *Covering spaces with singularities*, in "Algebraic Geometry and Topology," R. H. Fox, D. C Spencer, and A. W. Tucker, eds., Princeton University Press, Princeton, pp. 243–257.
4. H. Garland, *Arithmetic theory of loop groups*, Publ. Math. IHES **52** (1980), 5–136.
5. D. Johnson and J. Millson, *Deformation spaces associated to compact hyperbolic manifolds*, in "Discrete Groups in Geometry and Analysis," R. Howe, ed., Birkhäuser, Boston, 1987, pp. 48–106.
6. V. G. Kac, *Constructing groups associated to infinite-dimensional Lie algebras*, in "Infinite Dimensional Groups with Applications," V. G. Kac, ed., Springer–Verlag, New York, 1985, pp. 167–216.
7. R. S. Kulkarni, *Proper actions and pseudo-Riemannian space forms*, Advances in Math. **40** (1981), 10–51.

8. J. Milnor, *Remarks on infinite-dimensional Lie groups*, in "Relativity, Groups and Topology II," B. S. DeWitt and R. Stora, eds., North-Holland, Amsterdam, 1984, pp. 1007–1057.
9. G. D. Mostow, "Strong Rigidity of Locally Symmetric Spaces," Princeton University Press, Princeton, 1973.
10. R. S. Palais, *On the existence of slices for actions of non-compact Lie groups*, Annals of Math. **73** (1961), 295–323.
11. A. Pressley and G. Segal, "Loop Groups," Oxford University Press, Oxford, 1986.

University of Florida, Gainesville, FL 32611

CLOSING REMARKS AT THE BOULDER CONFERENCE ON GEOMETRY OF GROUP REPRESENTATIONS JULY 9, 1987

Murray Gerstenhaber

Some of the participants in this conference have asked me to thank the organizers for their diligent and extraordinarily successful efforts. These instigators have, I think, chosen the person with the thinnest hair and fewest neurons for this task, on the principle that no further intellectual effort should be required after a week of intense excitement. Before offering our official thanks, I would like to evoke some vignettes of the week past:

– Bill Abikoff's creative use of both screens, totally occluding the blackboard and making necessary his effective disappearance from the lecture hall as he scrambled behind them to retrieve a piece of chalk.

– A small group of us crossing the peak of Green mountain three times looking for the way down.

– Joyce O'Halloran's using her rubber featherless chicken to illustrate how one wrings the neck of a Riemann surface, and finally

– Charlie Frohman's ecstatic comment last Wednesday at the dining room, We're over the hump—there are fewer meals to face than we've already downed.

Enough—let us now express my personal warmest thanks and the gratitude of all the participants for the unstinting service of Hy Bass, Bill Goldman, and Andy Magid, and our collective appreciation to Carole Kohanski, to her AMS staff, and to our hosts here at the University of Colorado for what has been one of the most pleasant and most inspiring conferences that it has been my pleasure to attend! Thank you all.

(Editors' note: Professor Gerstenhaber contributed this transcript of his closing remarks, which we include here as an afterword.)